GALACTIC AND EXTRAGALACTIC INFRARED SPECTROSCOPY

ASTROPHYSICS AND SPACE SCIENCE LIBRARY

A SERIES OF BOOKS ON THE RECENT DEVELOPMENTS
OF SPACE SCIENCE AND OF GENERAL GEOPHYSICS AND ASTROPHYSICS
PUBLISHED IN CONNECTION WITH THE JOURNAL
SPACE SCIENCE REVIEWS

Editorial Board

J. E. BLAMONT, *Laboratoire d'Aeronomie, Verrières, France*

R. L. F. BOYD, *University College, London, England*

L. GOLDBERG, *Kitt Peak National Observatory, Tucson, Ariz., U.S.A.*

C. DE JAGER, *University of Utrecht, The Netherlands*

Z. KOPAL, *University of Manchester, England*

G. H. LUDWIG, *NASA Headquarters, Washington, DC, U.S.A.*

R. LÜST, *President Max-Planck-Gesellschaft zur Förderung der Wissenschaften, München, F.R.G.*

B. M. McCORMAC, *Lockheed Palo Alto Research Laboratory, Palo Alto, Calif., U.S.A.*

L. I. SEDOV, *Academy of Sciences of the U.S.S.R., Moscow, U.S.S.R.*

Z. ŠVESTKA, *University of Utrecht, The Netherlands*

VOLUME 108
PROCEEDINGS

GALACTIC AND EXTRAGALACTIC INFRARED SPECTROSCOPY

PROCEEDINGS OF THE XVIth ESLAB SYMPOSIUM,
HELD IN TOLEDO, SPAIN, DECEMBER 6–8, 1982

Edited by

M. F. KESSLER

and

J. P. PHILLIPS

*Astronomy Division, Space Science Department of ESA,
Noordwijk, The Netherlands*

D. REIDEL PUBLISHING COMPANY

A MEMBER OF THE KLUWER ACADEMIC PUBLISHERS GROUP

DORDRECHT / BOSTON / LANCASTER

Library of Congress Cataloging in Publication Data

ESLAB Symposium (16th : 1982 : Toledo, Spain).
    Galactic and extragalactic infrared spectroscopy.
    (Astrophysics and space science library : v. 108)
    Includes bibliographical references and index
1. Infra-red astronomy—Congresses. 2. Galaxies—Congresses.
3. Interstellar matter—Optical properties—Congresses. 4. Astronomical
spectroscopy—Congresses. I. Kessler, M. F. (Martin Francis), 1954–
II. Phillips, J. P. (John Peter), 1949–   . III. Series
QB470.A1E85   1982   522'.68   83–24542
ISBN 90-277-1704-4

Published by D. Reidel Publishing Company,
P.O. Box 17, 3300 AA Dordrecht, Holland.

Sold and distributed in the U.S.A. and Canada
by Kluwer Academic Publishers,
190 Old Derby Street, Hingham, MA 02043, U.S.A.

In all other countries, sold and distributed
by Kluwer Academic Publishers Group,
P.O. Box 322, 3300 AH Dordrecht, Holland.

All Rights Reserved
© 1984 by D. Reidel Publishing Company, Dordrecht, Holland
No part of the material protected by this copyright notice may be reproduced or
utilized in any form or by any means, electronic or mechanical
including photocopying, recording or by any information storage and
retrieval system, without written permission from the copyright owner

Printed in The Netherlands

TABLE OF CONTENTS

List of Participants		vii
Foreword		xi
Welcome Address	D.E. Page	1

SECTION I: INTERSTELLAR DUST AND CHEMISTRY

Absorption and Emission Characteristics of Interstellar Dust	L.J. Allamandola	5
Observed Spectral Features of Dust	S.P. Willner	37
Formation of Molecules on Dust	D.A. Williams	59
Gas Phase Chemistry in the Interstellar Medium	W.D. Watson	69
Observational Constraints on Interstellar Chemistry	G. Winnewisser	83

SECTION II: EMISSION PROCESSES AND THEIR INTERPRETATION

Atomic and Ionic Emission Processes	D.R. Flower	89
Infrared Spectroscopy of Interstellar Shocks	C.F. McKee, D.F. Chernoff and D.J. Hollenbach	103
Infrared Hydrogen Emission Lines from HII Regions and "Protostars"	C.G. Wynn-Williams	133
The Significance of Far-Infrared Spectra of the Interstellar Medium	M. Harwit	145
Astrophysical Interpretation of Molecular Spectra	N.Z. Scoville	167
Excitation Conditions in Molecular Clouds	G. Winnewisser and H. Ungerechts	177

SECTION III: GALACTIC SOURCES

Far-Infrared Spectroscopy of Neutral Interstellar Clouds	D.M. Watson	195
Molecular Astronomy at Submillimeter Wavelengths	D. Buhl	221
Submillimeter Observations of Molecules and the Structure of Giant Molecular Clouds	P.F. Goldsmith	233
Energetic Outflows, Winds and Jets Around Young Stars	C.J. Lada	251
Near-Infrared Spectroscopy of H_2 and CO in Molecular Clouds	D.N.B. Hall	269
Spectroscopy of HII Regions in the 1 - 15 μm Region	J.H. Lacy	281
Far-Infrared Spectroscopy of HII Regions	R.J. Emery and M.F. Kessler	289
Infrared Spectroscopy of Late Type Stars	S.T. Ridgway	309
Infrared Spectroscopy of Evolved Objects	D.K. Aitken and P.F. Roche	331

SECTION IV: GALACTIC CENTRE AND EXTRAGALACTIC SYSTEMS

The Galactic Center	I. Gatley	351
Ground-Based Extragalactic Infrared Spectroscopy and Related Studies	M.G. Smith	369
A Comparison of the FIR Line and Continuum Emission from External Galaxies with the Emission from our Galaxy	P.G. Mezger	423
Molecular Gas Distribution in Spiral Galaxies	N.Z. Scoville	443
Titles of Contributed Papers		457
Subject Index		459
Object Index		469

LIST OF PARTICIPANTS

AITKEN D.K.	Physics Dept., RAAF Academy, Melbourne University, Parkville, Victoria 3052, Australia
ALLAMANDOLA L.J.	Sterrewacht Leiden, Huygens Laboratorium, Postbus 9513, 2300 RA Leiden, The Netherlands
BALUTEAU J-P.	Laboratoire d'Astronomie Infrarouge, Observatoire de Meudon, 92190 Meudon, France
BARLOW M.J.	Dept. of Physics and Astronomy, University College, Gower Street, London WC1E 6BT, England
BECKMAN J.	Physics Dept. Queen Mary College, Mile End Road, London E1 4NS, England
BIANCHI L.	ESA Villafranca Satellite Tracking Station, P.O. Box/Apartado 54065, Madrid, Spain
BUHL D.	Goddard Space Flight Center, NASA, Greenbelt, Maryland 20771, USA
CACCIARI C.	ESA Villafranca Satellite Tracking Station, P.O. Box/Apartado 54065, Madrid, Spain
CAUX E.	C.E.S.R., 9-11 Avenue du Colonel Roche, 31029 Toulouse Cedex, France
CESARSKY C.	Dph/EP/AP CEN Saclay, 91191 Gif sur Yvette Cedex, France
COLINO L.	ESA Villafranca Satellite Tracking Station, P.O. Box/Apartado 54065, Madrid, Spain
COSMOVICI C.B.	DFVLR-AWF, D-8031 Wessling, W. Germany
EIROA C.	Centro Astronomico Hispano-Aleman, Apartado de Correos 511, Almeria, Spain
ELSASSER H.	Max-Planck-Institut für Astronomie, D-6900 Heidelberg 1, Konigstuhl, W. Germany
EATON N.	Astronomy Dept., Leicester University, Leicester, England
EMERY R.J.	Space Science Dept., ESTEC, European Space Agency, Noordwijk, The Netherlands
FELLI M.	Osservatorio Astrofisico di Arcetri, Largo Enrico Fermi 5, 50125 Firenze, Italy
FERRARI-TONIOLO M.	Instituto Astrofisica Spaziale-CNR, CP 67, 00044 Frascati, Italy
FITTON B.	Space Science Dept., ESTEC, European Space Agency, Noordwijk, The Netherlands

FLOWER D.R.	Dept. of Physics, The University, Durham DH1 3LE, England
GATLEY I.	U.K. Infrared Telescope Unit, 900 Leilani Street, Hilo, Hawaii 96720, USA
GILLESPIE A.	Max-Planck-Institut für Radioastronomie, Auf dem Hügel 69, D-5300 Bonn 1, W. Germany
GREWING M.	Astronomisches Institut, Waldhauser strasse 64, D-7400 Tübingen, W. Germany
GOLDSMITH P.F.	Dept. of Physics and Astronomy, University of Massachusetts, GRC Tower B, Amherst, MA 01002, USA
GRAAUW Th. DE	Space Science Dept., ESTEC, European Space Agency, Noordwijk, The Netherlands
HALL D.	Milton S. Eisenhower Library, The Space Telescope Science Institute, John Hopkins University, Rowland Hall, Baltimore, MD 21218, USA
HARTQUIST T.W.	Dept. of Physics and Astronomy, University College, Gower St., London WC1E 6BT, England
HARWIT M.	Dept. of Astronomy, Space Sciences Building, Cornell University, Ithaca, New York 14853, USA
HAWARDEN T.G.	UKIRT, Royal Observatory, Blackford Hill, Edinburgh EH9 3HJ, Scotland
ISRAEL F.P.	Space Science Dept., ESTEC, European Space Agency, Noordwijk, The Netherlands
JOSEPH R.D.	Astronomy Group, Physics Dept., Imperial College, London SW7 2BZ, England
KESSLER M.F.	Space Science Dept., ESTEC, European Space Agency, Noordwijk, The Netherlands
KOORNNEEF J.	ESO, Casilla 16317, Santiago 9, Chile
KOLLATSCHNY W.	Universitats-Sternewarte Gottingen, Geismarlandstr. 11, D-3400 Gottingen, W. Germany
LACY J.H.	Space Sciences Laboratory, University of California, Berkeley, CA 94720, USA
LADA C.	Steward Observatory, University of Arizona, Tucson, AZ 95721, USA
LEGER A.	Université de Paris VII, Group de Physique des Solides de l'E.N.S. Tour 23, 2, Place Jussieu, F-75221 Paris Cedex 05, France
LEMKE D.	Max-Planck-Institüt für Astronomie, Konigstuhl, D-6900 Heidelberg, W. Germany
LENA P.J.	Observatoire de Meudon, 92190 Meudon, France

LIST OF PARTICIPANTS

LENZEN R.	Max-Planck-Institüt für Astronomie, Konigstuhl, D-6900 Heidelberg, W. Germany
McGREGOR P.J.	Mt. Wilson and las Campanas Observatories, 813 Santa Barbara St., Pasadena, CA 91101, USA
McKEE C.	Physics Dept., University of California, Berkeley, CA 94720, USA
MAMPASO, A.	Instituto de Astrofisica de Canarias, Universidad de la Laguna, La Laguna, Tenerife, Spain,
MEZGER P.G.	Max-Planck-Institut für Radio-Astronomie, Auf dem Hügel 69, 5300 Bonn 1, W. Germany
MUIZON M. DE	Observatoire de Paris, 92190 Meudon, France
MILLAR T.J.	UMIST, P.O. Box 88, Manchester M60 1QD, England
MOORWOOD, A.F.M.	ESO, Karl-Schwarzschild-strasse 2, D-8046 Garching bei München, W. Germany
MOUNTAIN M.	Astronomy Group, Physics Dept., Imperial College, London SW7 2BZ, England
NATTA A.	Osservatorio di Arcetri, Largo Enrico Fermi 5, 50125 Firenze, Italy
NORDH H.L.	Stockholm Observatory, S-13300 Saltsjobaden, Sweden
NUSSBAUMER H.	Institute for Astronomy, ETH Zentrum, CH-8092 Zürich, Switzerland
OLIVA E.	E.S.O., Karl-Schwarzschild-strasse 2, D-8046 Garching bei München, W. Germany
OLOFSSON G.	Stockholm Observatory, S-13300 Saltsjobaden, Sweden
PAGE D.E.	Space Science Dept., ESTEC, European Space Agency, Noordwijk, The Netherlands
PANAGIA N.	Istituto di Radioastronomia, via Irnerio 45, I-40126, Bologna, Italy
PAPOULAR R.	DPH/EP/SAP, CEN Saclay, 91191 Gif sur Yvette Cedex, France
PATRIARCHI R.	ESA Villafranca Satellite Tracking Station, P.O. Box/Apartado 54065, Madrid, Spain
PERINOTTO M.	Osservatorio Astrofisico, Arcetri, Largo Enrico Fermi 5, Firenze, Italy
PERSI P.	Istituto Astrofisica Spaziale, CNR, C.P. 67, 00044 Frascati, Italy
PHILLIPS J.P.	Space Science Dept., ESTEC, European Space Agency, Noordwijk, The Netherlands
PHILLIPS T.G.	Caltech 320-47, Pasadena, CA 91125, USA
POULTER G.	University College, Dept. of Physics and Astronomy, Gower St., London WC 1E 6BT, England

REAY N.K.	Astronomy Group, Physics Dept., Imperial College, London SW7 2BZ, England
RIDGWAY S.T.	Kitt Peak National Observatory, P.O. Box 26732, Tucson, Arizona 85726, USA
ROCHE P.F.	University College, Dept. of Physics and Astronomy, Gower St., London WC1E 6BT, England
ROUEFF E.	Dept. d'Astrophysique Fundamentale, Observatoire de Meudon, 92190 Meudon, France
SANCHEZ MAGRO C.	Institut de Astrofisica de Canarias, Universidad de la Laguna, Tenerife, Spain
SANZ FERNANDEZ DE CORDOBA L.	ESA Villafranca Satellite Tracking Station, P.O. Box/Apartado 54065, Madrid Spain
SCOVILLE N.Z.	Dept. of Physics and Astronomy, Grad. Res. Center, University of Massachusetts, Amherst, MA 01003, USA
SMITH M.G.	Royal Observatory, Blackford Hill, Edinburgh EH9 3HJ, Scotland
STASINSKA G.	DAF, Observatoire de Meudon, 92190 Meudon, France
VRIES J. DE	Kapteyn Laboratory, Hoogbouw WSN, P.O. Box 800, 9700 AN Groningen, The Netherlands
WADE R.	Royal Observatory, Blackford Hill, Edinburgh EH9 3HJ, Scotland
WAMSTEKER W.	ESA Villafranca Satellite Tracking Station, P.O. Box/Apartado 54065, Madrid, Spain
WATSON D.M.	Caltech, Pasadena, CA 91125, USA
WATSON W.D.	Dept. of Physics, University of Illinois, Urbana, Illinois 61801, USA
WHITTET D.C.B.	Division of Physics and Astronomy, Preston Polytechnic, Preston PR1 2TQ, England
WILLIAMS D.A.	Mathematics Department, UMIST, Manchester M60 1QD, England
WILLNER S.	60 Garden St., Cambridge, MA 02138, USA
WINNEWISSER G.F.	I. Physikalisches Institut, Universitat zü Koln, Zulpcher Strasse 77, 5000 Cologne 41, W. Germany
WYNN-WILLIAMS G.	Institute for Astronomy, 2680 Woodlawn Drive, Honolulu HI 96822, USA
ZADRONZNY A.	Astronomy Group, Physics Dept., Imperial College, London SW7 2BZ, England

FOREWORD

The last major conference on infrared astronomy was the IAU Symposium No. 96 in June 1980. Since then, the discipline has continued to mature and to contribute to all branches of astrophysics. One particular area of growth has been in spectroscopic capabilities at all infrared wavelengths. The purpose of the Symposium in Toledo was to review the scientific questions to be addressed via infrared spectroscopy and to provide, in the proceedings, a useful summary of the field.

The sensitivity of infrared spectroscopic observations is still generally limited by detector characteristics or by thermal background radiation. However in recent years improvements in detector technology together with developments in spectroscopic instrumentation have made possible both quite detailed spectroscopy of the brighter members of many classes of galactic sources and also begun to open up some infrared spectroscopy of extragalactic sources.

The potential of the field in the next decade or two is clear. The IRAS mission has completed one of the pre-requisites, namely an all-sky photometric survey. Major space missions utilising cryogenic infrared telescopes have been approved in Europe (ISO) and seem likely in the USA (SIRTF); plans for space submillimeter telescopes are firming up. On the ground large telescopes optimized for infrared observations are now in operation at high altitude sites and specialized submillimeter facilities are under construction. The particular advantages of planned, very large telescopes for infrared observations are widely accepted.

Within this context, the selection of "Galactic and Extragalactic Infrared Spectroscopy" as the topic for the XVIth ESLAB Symposium was particularly timely. The sessions during the three day symposium were made up entirely of invited papers by active researchers in their fields.

The time was divided between papers dealing with the physical processes which lead to spectroscopic features in the infrared and those reviewing the current status of observations of a specific class of objects and their interpretation. These Proceedings consist of manuscripts submitted by all of the invited speakers and, in many cases, they include additional material intended to make them suitable for a broader audience than the symposium attendees.

Contributed papers were presented as posters, but the full manuscripts of these were published in advance of the meeting and circulated to participants. The titles of these papers are listed at the end of this volume.

Very many people helped to make this Symposium a success and the organisers (B. Fitton, M.F. Kessler and J.P. Phillips) wish to express their gratitude to all involved, particularly to Dr. Willem Wamsteker for the local arrangements. Thanks are also due to the conference secretaries, Lucy de Boer, Anne van den Eijkel and Carmen Ramirez-Palacios; to Anne Cijsouw for much of the post-meeting typing; and to the management of the Parador Hotel in Toledo for providing such a pleasant atmosphere in which to work.

D.N.B. Hall
Space Telescope
   Science Institute

M.F. Kessler
Astronomy Division,
Space Science
   Department of ESA

WELCOME ADDRESS

ESA's Space Science Department, which used to be called ESLAB, provides the study and project scientists for ESA's scientific satellite programme. Each year this Department organises an "ESLAB" symposium and it was decided that the 1982 symposium, held in Toledo, should be devoted to Galactic and Extragalactic Infrared Astronomy.

ESLAB symposia programmes are frequently based on the results arriving from ESA spacecraft. Sometimes they are designed to help guide ESA in obtaining and disseminating data from a spacecraft due for launch in the near future. On other occasions the programme is used to give ESA a better chance to understand what a particular scientific community would like to do in space and at the same time to give the community an opportunity to put its ideas together in public. What we hear also helps us in steering the internal research carried out in Space Science Department so that research can be as relevant as possible for future satellite projects.

The ESLAB symposium of 1972 was called: "Infrared Detection Techniques for Space Research" and with that event an attempt was made to get together the interested scientific community. About the same time an infrared research effort was set in motion in Space Science Department. It was felt that in 1982 it was time again to bring our ideas up to date, in at least part of the infrared astronomy field, and that to get the community together again was appropriate in view of the imminent decision required on the selection of ISO as the next ESA scientific project.

The decision to come to Toledo was governed not merely by the attractions of that historic city. We would gratefully acknowledge the help received with IUE from the Spanish authorities at Villafranca.

D. Edgar Page,
Head Space Science Department

SECTION I: INTERSTELLAR DUST AND CHEMISTRY

ABSORPTION AND EMISSION CHARACTERISTICS OF INTERSTELLAR DUST

Louis J. Allamandola

Laboratory Astrophysics, Leiden University, The Netherlands

1. INTRODUCTION

Molecular transitions which occur in the middle infrared region of the spectrum correspond with the characteristic frequencies of molecular vibrations. Thus, moderate resolution spectroscopy of the interstellar medium offers unique evidence about the molecules in the condensed and gaseous phases and their distribution. It is the purpose of this paper to discuss the spectral properties of the condensed phase. However, in the astrophysical literature, it is difficult to find a qualitative description of the effects the solid state has on molecular vibrations, and since it is these which largely determine the spectroscopic properties of the interstellar dust, this discussion will begin with a general description of these effects and then be directed toward describing the optical characteristics of the molecular ice component of the dust. The properties of this component of the dust will be stressed, rather than those expected from more homogeneous components such as silicates, graphite, or amorphous carbon since these have been discussed in considerable detail elsewhere (e.g. Krätschmer and Huffman, 1979, Knacke and Krätschmer, 1980, Day, 1979, Borghesi et al., 1983 and references therein).

On the average, the dust we shall be concerned with here is comprised of a ~ 0.05 µm radius core covered by a ~ 0.05 µm thick mantle (e.g. Greenberg, 1978). The composition of the core has little or no effect on the spectroscopic properties of mantles more than a few monolayers thick (a 0.05 µm mantle is equivalent to about 1000 monolayers) and scattering effects on spectral band shapes are only important when the grain radius is on the order of or greater then 0.05 times the wavelength. Since we are dealing with the 2-15 µm wavelength region and particle sizes on the order of 0.1 µm, scattering effects, when significant, will only influence absorption and emission bands in the short wavelength end of the spectrum (see section 3.1.1) These dimensions, as well as mantle composition are expected to vary from region to region and depend on local conditions as well as history. In a dense cloud, most gas phase molecules which strike the dust particles should stick and, since the dust temperature ($\sim$ 10K) is so much lower than the melting point of most interstellar molecules, these accretion mantles should have an amorphous rather than crystalline structure. In addition to simple accretion, photochemical processes which create new molecules and alter the structure of the solid will take place in all but the densest of regions and, even here, it is possible that UV produced by cosmic-rays (Greenberg, 1971) or X-rays (Montmerle et al., 1983) will process the mantle. Thus, in U.V. rich regions, highly processed mantles should be present, with the diffuse medium representing the extreme case with completely processed mantles, while in denser regions, only slightly modified, amorphous accretion mantles should make up the outer portion of the grain. Of course, boundaries are rarely well defined and it is to be anticipated that interstellar ice will be a mixture of the above. These types of molecular mantles are sufficiently different from each other, both in solid state structure and composition, that each possesses its own peculiar set of infrared spectroscopic characteristics and, despite the problems associated with unraveling the frequency shifts and intensity changes in the spectra of such complicated mixtures, the spectral study of the interstellar medium provides important information about the type of molecular ices present.

2. INFRARED SPECTROSCOPY OF MOLECULAR ICES.

2.1. Laboratory Production of Molecular Ices.

Since a detailed description of the experimental procedure appears elsewhere (Hagen et al., 1979), only the key points, schematically represented in figure 1, will be summarized here. The 11±1 K substrate upon which the molecular ice is formed, processed, and studied and which substitutes for the core of an interstellar grain, is either a polished aluminum block (for reflection-transmission spectra) or a cesium iodide window (for transmission studies) mounted on a closed cycle helium refrigerator and suspended in a high vacuum sample chamber. Various gases or mixtures of gases which have been previously prepared (allowing ample time for thorough mixing) are then deposited onto the cooled substrate at a rate of about 5×10^{-3} moles/hour which corresponds to a thickness growth rate of 6 µm/hour. When optical constants are to be determined, sample thickness is measured by monitoring the temporal development of the thin film interference pattern of a He-Ne laser beam reflected off the sample during deposition. The infrared spectrum is measured with a Fourier transform infrared spectrometer. Laboratory spectra are plotted as $\log[I(\lambda)/I_o(\lambda)]$ versus wavelength (µm) or frequency (cm^{-1}), where $I_o(\lambda)$ is the single beam spectrum measured from the substrate before sample deposition and $I(\lambda)$, the spectrum measured after. Vacuum ultraviolet photoprocessing is achieved by means of a microwave excited hydrogen flow lamp which has a sharp emission peak at 121.6 nm (Lyman α) and an approximately 50 nm broad component centered at about 160.0 nm, a spectrum which reproduces quite well that of the diffuse interstellar medium (Grewing, 1983). The integrated ultraviolet photon flux ($\lambda \lesssim 200$ nm) is a few times 10^{15} s^{-1} cm^{-2}.

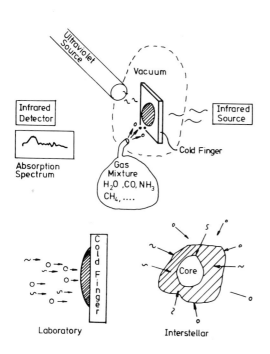

Fig. 1. Upper half: schematic representation of the experimental method used to study mixed molecular ices. Circles represent accreting atoms and molecules, the ~ sign photons. Lower half: Comparison of the configuration of the ice prepared in the laboratory with that in space.

2.2. Summary of Vibration-Rotation Spectroscopy

The frequencies of molecular vibrations are determined by the masses of the vibrating atoms, the molecular geometry and the forces holding the atoms in their equilibrium position. For example, to a first approximation, the vibrations of a diatomic molecule can be represented as a quantum mechanical harmonic oscillator with the fundamental vibrational frequency given by

$$\nu = \frac{h}{2\pi} \sqrt{k/\mu} \qquad (1)$$

where h is Planck's constant, k the chemical bond force constant and μ the reduced mass $(m_1 m_2/(m_1+m_2))$, where m_1 and m_2 are the masses of atoms 1 and 2. This formalism can also be used to approximate the stretching frequencies of polyatomic molecules. The combination of the masses of the most cosmically abundant atoms, H, O, C and N with the kinds of chemical bonds they form restricts most of the fundamental vibrational frequencies for virtually all molecules made up of these atoms to the 4000 to 500 cm^{-1} region. However, all molecular vibrations do not give rise to the absorption or emission of infrared photons, but only those for which a dipole moment oscillates during the vibration. Thus, the vibrations of all homonuclear diatomic molecules (e.g. O_2, N_2, ..) as well as the centrosymmetric vibrations of symmetric molecules are inaccessible to the infrared spectroscopist. In general, since certain groups of atoms consistently absorb at or near the same frequency regardless of the rest of the molecule, virtually all molecules containing these groups possess a characteristic vibrational absorption spectrum (e.g. Silverstein and Bassler, 1967; Bellamy, 1958). The reason for this, in rather oversimplified language, is that chemical bonds generally fall into three distinct types which are classified as single, double and triple bonds, depending upon the number of electrons participating in the bond. The term single bond is used to represent a bond in which a total of two electrons occupy the bonding molecular orbital, the double bond has four and the triple bond six. The more bonding electrons, the stronger the force constant and, as implicit in equation (1) if one keeps the mass of the vibrating atoms constant, the higher the vibrational frequency. Conversely, keeping the type of bond constant but increasing the masses of the vibrating atoms results in a reduction of the fundamental vibrational frequency. This is illustrated in figure 2 where the characteristic frequencies of various groups are schematically represented. The stretching vibration of the single bonds -C-H, -O-H, and N-H all occur near 3000 cm^{-1} whereas the single bonded carbon-oxygen, carbon-nitrogen or carbon-carbon stretch lie in a lower frequency range centered about 1000 cm^{-1}. Inspection of figure 2 also shows that increasing the bond strength from a singly bonded carbon through doubly bonded to triply bonded carbon implies an increase of the stretching frequency from $\sim$ 1000 cm^{-1} to somewhat above 2000 cm^{-1}. Figure 2 also shows that absorption in each of these specific regions can be used to deduce the type of molecules which are present.

In interstellar clouds, molecules are in both the gaseous and solid phases. In addition to undergoing internal vibrations, molecules in the gas phase can also rotate. Since the energies involved in rotational transitions correspond with frequencies of only a few wavenumbers and selection rules simultaneously allow rotational transitions to occur with vibrational transitions, rotational structure is superimposed giving rise to rotation-vibration bands. Since a more detailed description of these bands and the information they contain form the subject of the two chapters by D. Hall and N. Scoville in this volume, the discussion here will be restricted only to those aspects which will be of use in understanding the solid state features.

In the case of symmetric, polyatomic molecules the lines making up a rotation-vibration band produce a recognizable pattern which resembles the rotational level population distribution (see fig. 3). The transitions involving an increase in the rotational quantum number J give rise to the so-called R branch, those corresponding to a decrease in J the P branch and those with no change in J, the central Q branch. For all heteronuclear diatomics and certain vibrations of linear polyatomics, the Q branch is absent. For a thorough description of infrared molecular spectroscopy, the reader is referred to Herzberg (1968), or, on a more introductory level, Barrow (1962).

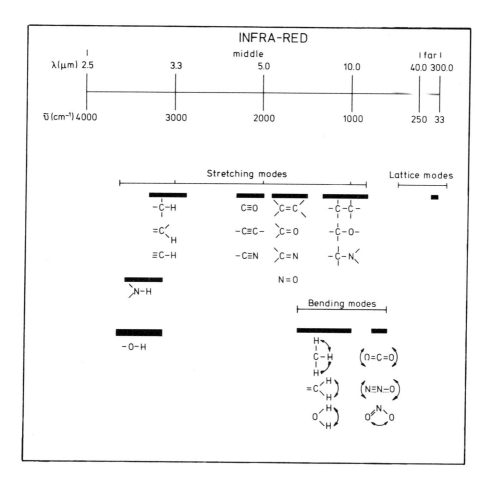

Fig. 2. *Characteristic frequency ranges of various groups.*

2.3. Solid State Effects on Rotation-Vibration Bands

Three dramatic changes occur in the infrared spectrum of a molecule in going from the gas to the solid phase:
1) Rotational structure suppression
2) Line Broadening
3) Line Shifting

1) Rotational Structure Suppression: Molecules are not free to rotate in most ices at 10 K. The familiar P and R branches are replaced by one broad line due to vibration alone. This behavior is illustrated in figure 3 where the region of the vibration-rotation band of CO is shown.

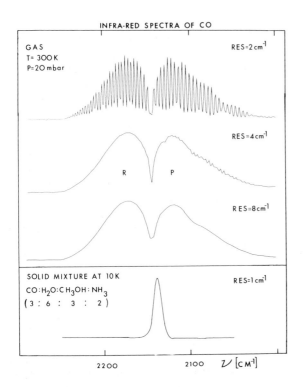

In the upper frame the spectrum of gas phase CO is shown with different resolution. At 2 cm^{-1} resolution, the rotational lines making up the P and R branches are resolved. At lower resolution, while the individual lines are no longer resolvable, the rotational profile due to the P and R branch is still recognizable. The spectrum of CO frozen in a molecular ice is shown in the lower frame. Note that, in this case, the vibration occurs at nearly the same frequency as that of gas phase CO i.e. in the null gap or missing Q branch region between 2135 and 2151 cm^{-1}.

Fig. 3. The spectrum of CO in the gas phase and in a molecular ice between 2000 and 2300 cm^{-1}. RES signifies resolution.

2) Line Broadening: Figure 4 shows the effects of concentration on the spectrum of CO in CO/NH$_3$ ices. The changes indicate the influence of both the uniformity of and degree of interaction with the environment on the transition. While the peak absorption frequency undergoes little change in the series, the full width at half height (FWHH) does, providing a crude measure of the concentration. For CO/NH$_3$ = 100, it is 1.4 cm^{-1}, slightly larger than that of pure solid CO, where the width is due largely to disorder in the CO crystal (see Legay-Sommaire and Legay, 1982 and references therein). As one increases the amount of impurity (in this case NH$_3$), the solid CO structure is altered and becomes much less regular than it was originally, increasing the types of site that CO can occupy in the solid and the band becomes correspondingly broader. This effect increases with increased concentration of impurity as shown in the figure where, for CO/NH$_3$ = 10, the FWHH is 4 cm^{-1}. In mixed molecular ices where the ratio of CO to the sum of all the other molecules (CO/m) is less then 10, broader lines are measured. For example, in the mixture CO/CH$_3$OH/NH$_3$/H$_2$O (5/3/1/3) where CO/m is 0.7 the FWHH is about 10 cm^{-1}.

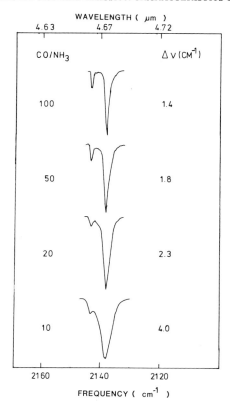

3) Line Shifting: This is the result of interactions between the particular molecule under consideration with other species in the solid. These interactions give rise to shifts because they influence the bond force constant; the stronger the interaction, the greater the effect. The magnitude of this effect can vary over a large range. In solid CO, for example, the CO stretch occurs at 2138 cm^{-1}, about 5 cm^{-1} lower then for gaseous CO while in the dirty ice $CO/CH_3OH/NH_3/H_2O$ (5/3/1/3), it is shifted to $\sim 2137 \pm 2$ cm^{-1}. This rather small shift indicates that there is only a weak interaction between CO and even such strongly polar molecules such as H_2O and NH_3. For molecules which can form weak chemical complexes, or even stronger hydrogen bonds (H-bonds), the shift can be much larger (Pimentel and McClellan, 1960).

Fig. 4. *The spectrum of the fundamental vibration of CO in CO/NH_3 at 10 K, showing the effect of concentration on the FWHH, $\Delta \nu$. Resolution = 1 cm^{-1}.*

In view of its importance in astrophysics, the behavior of the OH stretching mode of H_2O, with particular emphasis on the large frequency shift, will now be discussed in some detail. The different ways in which water can be present in the solid phase are shown schematically in figure 5. The diagram on the left represents H_2O mixed with CO at 10 K, but in sufficiently low concentration that each water molecule is completely surrounded by CO, preventing $H_2O...H_2O$ interactions. In this case ν_3 is 3707 cm^{-1}, ~ 50 cm^{-1} lower than in the gas phase. In the second frame, CO has been removed and H-bond formation between the H-atoms of one water molecule with the oxygen atom of an adjacent molecule occurs. This attraction of the H-atom by another water molecule causes a substantial reduction of the bond force constant and a correspondingly large shift (nearly 500 cm^{-1}) to lower frequencies. Of course, an amorphous H_2O structure consists of a wide variety of interaction geometries and strengths which give rise to an extremely broad band (FWHH $\simeq$ 300 cm^{-1}). The influence of this behavior on the spectrum in the OH stretching region is shown in figure 6. The upper trace in figure 6 shows the spectrum for the mixture CO/H_2O (100/1) in which nearly all of the water molecules are isolated. The absorptions at 3707 and 3673 cm^{-1} are due to monomers and dimers respectively (Hagen and Tielens, 1981). Upon warm up to 20 K, diffusion occurs permitting

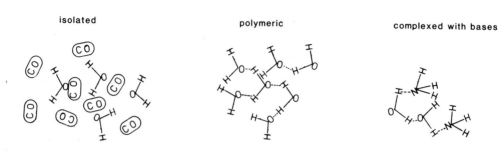

Fig. 5 Schematic diagram of the various possible arrangements of the water molecule in low temperature ices.

some of the water molecules to form H-bonds as shown by the increase in absorption due to dimers and higher polymers in the second spectrum. This process continues as the sample is warmed up to higher temperatures. Note also that the absorption coefficient per water molecule undergoes a substantial, anomalous increase as water goes from the monomeric to ice form (van Thiel et al., 1957; Allamandola et al., 1979).

Returning to figure 5, the third frame shows the situation in which water is now mixed with ammonia, a chemical base. In acid-base terminology, water is neutral and in the presence of a base, at least one of the hydrogen atoms will experience an attractive force greater than that involved in H-bond formation, giving rise to an even larger red shift of the OH stretching frequency. In the case of water-ammonia mixtures, the shift is on the order of 800 cm^{-1} and, the band is again a few hundred wavenumbers wide. A mixture consisting of comparable amounts of water and ammonia will be made up of some water molecules H-bonded to other water molecules and some forming acid-base complexes with the ammonia. Since absorption due to both forms of water are so broad, the absorption band will be a blend of the two. Because the base-water complex absorption is shifted more then the H-bonded water band, the overall band will resemble that due to amorphous water distorted by a wing on the long wavelength side (e.g. see figure 10). The absorption strength of the wing with respect to that of the 3 μm component will depend upon how much of each species is present. Such a long wavelength wing only implies the presence of a base, not necessarily ammonia.

In concluding this section, it must be stressed that while the spectroscopic study of pure substances (e.g. solid CO, CO_2, NH_3 and H_2O, Fink and Sill, 1982, Wood and Roux, 1982) or 1:1 complexes suspended in inert matrices are indicative of the type of effects the solid state environment has on a particular transition, in order to understand interstellar ice spectra, one must study the spectroscopic properties of molecular mixtures, and not simply add absorption spectra of pure substances or binary mixtures. For example, in the H_2O-NH_3 (1:1) complex, the OH stretching frequency is shifted to ~3400 cm^{-1} (Nelander and Nord, 1982), whereas in the H_2O/NH_3 (3/1) mixture, the hydrogen stretching frequency is shifted to 2900 cm^{-1}.

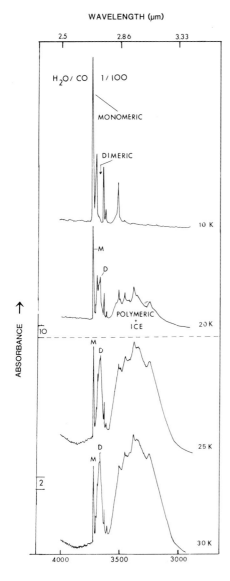

Fig. 6 The effects of thermal cycling on the spectrum in the OH stretching region of H_2O/CO. The first spectrum is that of $H_2O/CO = 1/100$ measured just after deposition. The rest of the spectra were measured after the sample was momentarily warmed up to the temperature indicated and recooled to 10 K. M and D refer to H_2O monomer and dimer respectively.

The column density of individual components of the ice can only be determined for species in which the integrated absorbance A, has been measured in various mixtures and shown to be only slightly effected by the local environment. For a discussion of the problems associated with solid state vibrational band intensity determination, see Person (1981). The column density of any solid state species can be estimated by:

$$N_{sol} = \frac{\tau_{max} \Delta \nu}{A} \quad (2)$$

where τ is the optical depth of the band at maximum absorbance, $\Delta \nu$ the FWHH in cm^{-1}, and A the integrated absorbance in $cm\ molecule^{-1}$.

3. ABSORPTION SPECTROSCOPY OF INTERSTELLAR ICE.

The 2-15 μm spectra of many interstellar objects associated with dust show, in addition to the broad 10 μm silicate band, absorptions due to the molecular ice dust component generally labeled as the 3.1, 4.67, 6.0 and 6.8 μm bands (e.g. Aitken, 1981, Willner, this volume). Spectra from several different types of object showing

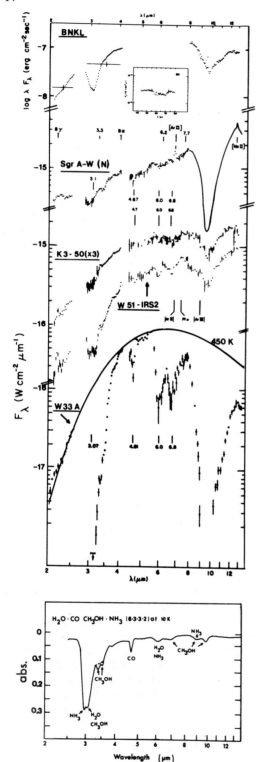

some of these bands are compared with that of the ice $CO:CH_3OH:NH_3:H_2O$ (3/3/2/6) in figure 7 (reprinted from Hagen et al., 1980a, hereafter referred to as paper I). Taken together with the impressive collection of protostellar spectra published by Willner et al., (1982) these spectra illustrate two points. First, except for the 10 μm silicate band, all the major features of the interstellar spectra agree with those of the laboratory ice, which of course reflect the solid state effects described above, showing that the characteristic groups present in the molecules comprising the lab made ice are also present in the interstellar ice. Second, the large variation in the interstellar spectra shows the unique potential of IR spectroscopy as a diagnostic tool in understanding grain composition and cloud evolution. The observational windows showing these bands will now be discussed in turn.

Fig. 7. Spectra of various interstellar objects from 2-14 μm compared with that of the ice $CO:CH_3OH:NH_3:H_2O$ (3:3:2:6) at 10 K. Note the absence of the 10 μm silicate band in the laboratory spectrum. This figure is reproduced from paper I. The interstellar spectra were originally reported in Gillett et al. (1975; BN), Russell et al. (1977a; BN insert), Willner et al. (1979; Sgr A-W(N), Puetter et al. (1979); K3-50, W51-IRS2) and Soifer et al. (1979; W33A).

3.1. The 2.8 - 4.0 μm Region.

3.1.1 Scattering effects.

Greenberg et al. (1983) have carried out a detailed study of the importance of parameters such as index of refraction and particle size and shape on the extinction cross section across the 3.1 μm band. A few of the salient features will be discussed here. Figure 8 shows the behavior of the extinction efficiency (Q_{ext}) and absorption efficiency (Q_{abs}) calculated for a homogeneous silicate type sphere as a function of $2\pi a/\lambda$, where a is the particle radius and λ is the wavelength. The difference between Q_{ext} and Q_{abs} is the scattering efficiency (Q_{sca}). Figure 8 shows that for all values of $2\pi a/\lambda$ less than 0.3, extinction and absorption efficiencies are equal and that only for greater values is scattering important. Therefore scattering will influence the extinction profile of the 3.1 μm band only for particles with radii greater than 0.15 μm. This is demonstrated in figure 9 where the extinction efficiency across this band is plotted for three different sizes of spherically shaped silicate core-amorphous water ice mantle particles. Note that even for the 0.5 μm radius case the extinction curve differs only slightly from the profile of the absorption band (0.14 μm curve) while for the 1 μm radius particles, there is sufficient scattering to provide a substantial long wavelength wing.

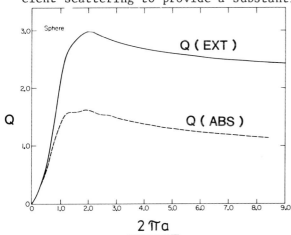

Fig. 8. *Extinction and absorption efficiencies for a silicate type spherical grain (Greenberg, 1978)*

To illustrate the effect of particle shape on the extinction profile, in figure 10 the absorption band measured from a mixture of H_2O/NH_3 (3/1) is compared with the extinction efficiency calculated using spherical and cylindrical silicate core-ice mantle particles. Since polarization is measured across this band, implying the presence of non-spherical grains in the interstellar medium, calculations using cylindrical particles are more relevant than those using spherical particles. Note that the profile calculated using the cylindrically shaped particles matches the absorption profile of a film more closely then that for spheres (Greenberg, 1972).

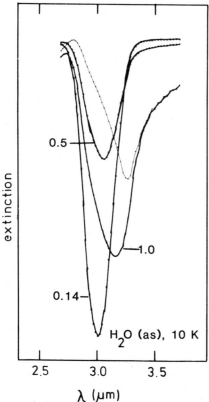

Fig. 9. Extinction efficiencies calculated for silicate core-amorphous water ice mantle spheres of various sizes in the 2.5 to 3.8 μm region. Core radius is 0.12 μm, core plus mantle radii are 0.14, 0.5 and 1.0 μm. The vertical scale for the 0.14 μm radius particles is exaggerated to facilitate the comparison. The index of refraction of the core is 1.5. The light curve is the scattering efficiency of the 1.0 μm particle.

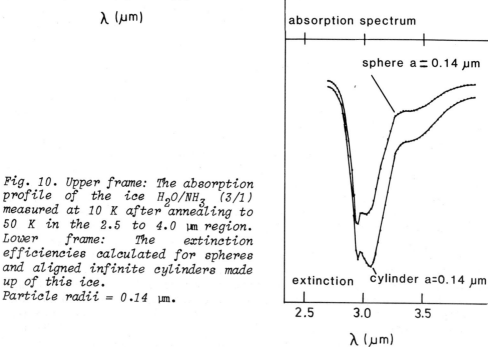

Fig. 10. Upper frame: The absorption profile of the ice H_2O/NH_3 (3/1) measured at 10 K after annealing to 50 K in the 2.5 to 4.0 μm region. Lower frame: The extinction efficiencies calculated for spheres and aligned infinite cylinders made up of this ice.
Particle radii = 0.14 μm.

For spheres, the absorption at 3.1 μm due to the OH-stretch is reduced with respect to the 2.9 μm feature as compared with the absorption profile. Thus, if one uses a spherical model for comparison with observations, there would tend to be an underestimate of the amount of NH_3 relative to H_2O.

3.1.2 <u>Absorption Bands</u>. Figure 2 shows that absorption in this region corresponds to the stretching vibration of hydrogen bonded to a heavier atom. In paper I it was pointed out that in many of the interstellar spectra, shown in figure 7, the 3.1 μm band appears to have two minor peaks centered at 2.98 and 3.1 μm. In the laboratory spectrum, the 2.98 μm absorption is due to the NH stretch in NH_3 and is superimposed on the much broader 3.1 μm band due to the OH stretch in H-bonded H_2O and CH_3OH. Recent, airborne observations of the B.N. object from 2.4 to 3.3 μm by Knacke et al. (1982) have confirmed the presence of a separate absorption centered at 2.9 μm. This absorption probably arises from the NH stretching vibration and indicates the presence of ammonia or amines ($-NH_2$, $>NH$). Another important aspect of this band in both the laboratory and interstellar spectra is the long wavelength wing. In section 2.3 it was shown that in ices containing H_2O and a base (here NH_3), the 3.1 μm band is a blend of two very broad bands, one due to H-bonded H_2O and the other to the base-H_2O complex. In the ice spectrum, only the "fine structure" at about 3.4 to 3.5 μm are due to the C-H stretch in the $-CH_3$ groups of methanol. These bands are superimposed on the much stronger absorbing long wavelength wing of the ice band. Thus, in measuring the equivalent width of the absorption due to the C-H stretch in hydrocarbons, the baseline must be estimated (see dotted line in figure 7) and not taken to be the same as the continuum used for the 3.1 μm band.

The source of the long wavelength wing on the interstellar 3.1 μm band is controversial. Publications from the Leiden Group have consistently argued that the wing is a spectroscopic feature inherent in complex molecular ices. Long wavelength wings have been measured in photolyzed samples not containing NH_3 (presumably due to the formation of strong bases such as -COOH, Hagen et al., 1979) as well as in ices containing NH_3 (Hagen et al., 1980b; and Paper I). Recently Hagen et al. (1983) report a thorough study of the optical properties of this band in mixtures with various other species and conclude that <u>both</u> the extinction and polarization profile of the interstellar 3 μm band are due to H_2O mixed with other molecules. These authors also consider the circumstellar spectra of OH 0739-14 (Soifer et al., 1981) which is unique since it is that of essentially pure amorphous water: there is no wing on the 3.1 μm band, the 6.0 μm band is somewhat peaked and the 12 μm band, perturbated and unrecognizable under certain conditions in mixed molecular ices (Paper I), is strong. Additional evidence that the long wavelength wing originates from interactions between water and other mantle constituents is that absorption at 4.67, 6.0 and 6.8 μm show that interstellar ices are molecular mixtures, with the 2.98 μm band implying that a base is present as well. Furthermore a general feature of the interstellar long wavelength wing is the inflection at

3.3 μm (Willner et al., 1982; Whittet et al., 1982, 1983) which is also evident in the laboratory spectra (e.g. see figures 7 and 10) and finally, the strong correlation between the optical depth at 3.08, 2.92, 3.35 and 3.47 μm measured with respect to the underlying continuum in protostellar objects by Willner et al. (1982) show a dependence on the water ice abundance which dominates this entire region. Alternatively, Léger et al. (1983) propose that the long wavelength wing is due to scattering from anomalously large spherical grains. These authors match the extinction profile of the 3.1 μm band in B.N. using a grain size distribution cutoff at a_{max} = 1.2 μm.

Perusal of the 3.4 μm region of interstellar spectra (e.g. Willner et al. 1982; Whittet et al., 1982, 1983) shows that determining the equivalent width of hydrocarbon absorption in the 3.4 to 3.5 μm region is presently rather uncertain. In this respect, the spectrum of IRS 7 towards the galactic center (Soifer et al. 1976, Allen and Wickramasinghe 1981 and Jones et al. 1983), which lacks a strong ice band but shows structure in the 3.4 μm region, is particularly important. Since this observation is associated with the diffuse interstellar medium, these features, showing the presence of hydrocarbons, probably arise from a non-volatile component of the dust. While all of the components of freshly prepared laboratory ices are volatile, upon photolysis new, larger, less volatile molecules are formed. As one warms irradiated ices up to room temperature under vacuum, the infrared spectrum continuously changes since individual components evaporate at somewhat different temperatures. Ultimately, a non-volatile residue remains on the substrate even at room temperature. Van de Bult (1984) who has been studying the photochemistry and infrared spectroscopy of molecular ices has shown that saturated hydrocarbons are present in essentially all residues formed from ices which initially contain a hydrocarbon such as CH_4. In Figure 11 is shown a comparison between the 2.9-3.9 μm spectrum of GC-IRS 7 with two different residue spectra. This comparison demonstrates that, as higher resolution spectra of interstellar dust become available, one can begin to unravel the actual types of molecules which are present. For example, these spectra suggest that the dust in the diffuse medium contains saturated hydrocarbons in which the $-CH_2-$ groups are more abundant than the $-CH_3$ groups. A detailed discussion of these residue spectra as well as the spectroscopic properties of photolyzed ices can be found in Van de Bult (1984). A preliminary analysis of the strength of the 3.4 μm absorption for one residue was applied to the galactic center spectrum by Greenberg (1982). Proton radiolysis of similar ices also produce residues which absorb in the 3.4 μm region (Moore and Donn, 1982), however this is not a dominant process in the general interstellar medium.

3.2. The 4-5 μm Region.

Figure 2 shows that absorption in this region corresponds to stretching vibrations between triply bonded heavier atoms. Although unresolved, CO was identified with absorption at 4.5 μm in this region as it is such an abundant interstellar molecule. Quite recently, observations with 2 cm^{-1} resolution have been made of 7 compact infrared sources (McGregor et al.

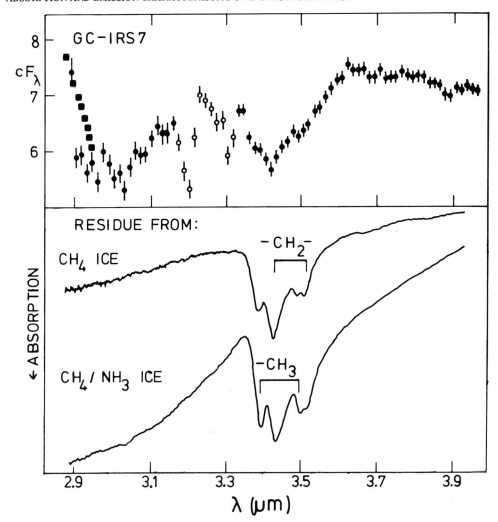

Fig. 11. Comparison of the 2.9-3.9 µm spectrum of IRS-7 towards the galactic center with spectra of the non-volatile residue produced during extended photolysis of pure CH_4 ice and a CH_4/NH_3 (3/1) ice. The spectrum of IRS-7 is reproduced from Allen and Wickramasinghe (1981). Except for the sharp features (open circles), the spectrum from 3.0 to 3.6 µm has been confirmed (Jones et al. 1983). Residue absorptions are weak compared to the parent ice and only become dominant after most of the volatile components have evaporated. The structure between 3.35 and 3.55 µm is characteristic of saturated hydrocarbons, where absorption at about 3.4 and 3.5 µm are due to the $-CH_3$ group, and those at about 3.45 and 3.52 µm to the $-CH_2$-group. The relative strength of these bands can be used to estimate the relative amounts of these groups (Bellamy, 1958, and Van de Bult, 1984).

1982, Lacy et al., 1983) showing that this band consists of the three features shown in figure 12; a strong broad absorption at 2165 cm^{-1}, a weaker broad band estimated at 2135 cm^{-1} and a sharper feature at 2140 cm^{-1}. The match in peak absorption (2140 ± 2 cm^{-1}) and FWHH ($\simeq$ 5 cm^{-1}) of the sharpest band with that of solid mixtures of CO. (section 2.3) indicates that the interstellar feature is due to ices in which CO/m is less than 10, whereas the frequency ($\sim$ 2135 cm^{-1}) and FWHH ($\sim$ 10 cm^{-1}) of the broader band implies ices in which CO/m is less than 1.0. The broad absorption centered at 2165 cm^{-1} is produced only when ices containing carbon and nitrogen are photolyzed, strongly pointing to the cyano group (-C$\equiv$N), incorporated in a larger molecule (X-C$\equiv$N), as the carrier (Van de Bult, 1984). Upon warm-up, the 2165 cm^{-1} band remains until the temperature of the substrate is above 200K. A comparison between the 2100-2200 cm^{-1} spectrum of W33A with the laboratory spectrum measured after the sample has been warmed up to 150 K to remove most of the CO and recooled to 10 K is shown in figure 13. The absorption at about 2140 cm^{-1} in the lab spectrum is presumably due to CO trapped in the material remaining on the substrate.

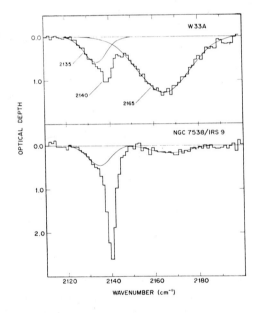

Fig. 12. Spectra of W33A and NGC 7538/IRS9 between 2110 and 2200 cm^{-1} with resolution $\simeq$ 2.6 cm^{-1}. This figure has been reproduced from Lacy et al (1983). The smooth curves are Gaussian fits to the absorptions centered at 2135 and 2165 cm^{-1}.

The amount of CO present in the dust can be determined since the integrated absorbance (A) of CO is only slightly effected by the environment. Jiang et al (1975) have found that A for CO varies from $\sim$ 53 to 68 km/mole depending upon whether CO is in solid CO, as polymers, as the nearly pure CO-H$_2$O complex, or isolated in argon matrices. Legay-Sommaire and Legay (1982) measure A for solid CO to be somewhat lower. Using a value of A equal to 64 km/mole in equation 2, Lacy et al. (1983) show that, for the sharp 2140 cm^{-1} component N(CO)$_{sol}$ is 10^{18}cm^{-2} in W33A and 1.8×10^{18} cm^{-2} toward NGC 7538 IRS 9, while for the broader 2135 cm^{-1} component, it is $>3.6 \times 10^{18}$ cm^{-2} and $>0.4 \times 10^{18}$ cm^{-2} respectively.

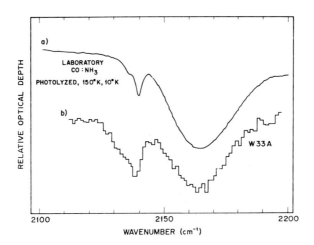

Fig. 13. Comparison between the spectrum of W33A and the laboratory spectrum of a CO/NH_3 (3/1) ice which had been photolyzed at 10 K, warmed up to 150 K (to remove the more volatile components of the ice) and then recooled to 10 K under vacuum. This figure has been reproduced from Lacy et al. (1983). Prior to photolysis, the laboratory spectrum in this region consists only of the CO absorption (figures 3 and 4), upon photolysis a moderately strong absorption grows in at 2165 cm^{-1}, adjacent to the strong CO band. During warmup, the species responsible for the 2165 cm^{-1} band remains while most of the CO evaporates

The presence of a considerable amount of CO in grains implies rich mantle chemistry since it is well known that this molecule is quite reactive in solids at cryogenic temperatures (e.g. Milligan and Jacox, 1971). Since the radial HCO is readily produced upon photolysis of mixed molecular ices and Van IJzendoorn et al. (1983) have shown that the reaction H+CO → HCO proceeds smoothly in the solid state between 10-15 K, the large cosmic abundance of hydrogen implies that HCO is quite likely to be present in grain mantles. This species absorbs at 2490, 1860 and 1090 cm^{-1} in solid CO (4.00, 5.38 and 9.17 μm respectively). The 2490 cm^{-1} band is broad (FWHH ~ 20 cm^{-1}) and weak, that at 1860 cm^{-1} sharper (FWHH ~ 3-5 cm^{-1}) and about five times stronger then the 2490 cm^{-1} band, while that at 1090 cm^{-1} has a FWHH ~ 10 cm^{-1} and is weaker then the 1860 cm^{-1} band (Milligan and Jacox, 1964; for the spectrum in a dirty ice see Hagen et al, 1979). Another species which is also formed in significant quantities upon photolysis of ices containing CO is CO_2 which absorbs strongly at about 660, 2340, 3600 and 3710 cm^{-1}. Unfortunately these are all obscured by atmospheric CO_2.

3.3. The 5 to 8 μm Region.

Figure 2 shows that absorptions in this region are less specific since different types of molecular vibrations absorb here. The spectrum in the 4-8 μm region of W33A compared with that of a laboratory ice, before and after photolysis, are shown in figure 14.

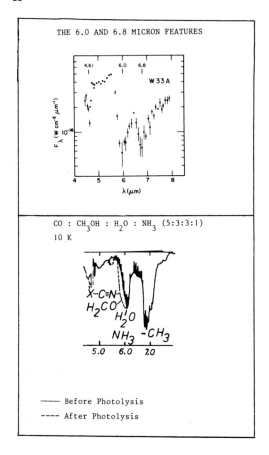

Fig. 14. Comparison of the 4-8 μm spectrum of W33A with that of the ice $CO:CH_3OH:H_2O:NH_3$ (5:3:3:1) before and after photolysis at 10 K. The spectrum of W33A is reproduced from Soifer et al. (1979).

The interstellar bands at 3.1 and 2.98 μm show that water and ammonia (and/or amines) are present, all of which have hydrogen bending modes at approximately 6μm. However, they cannot account for the entire band alone, since the maximum interstellar absorption lies shortward of 6 μm, while in unphotolyzed ices it is slightly above 6.0 μm. Upon photolysis of virtually all ices which contain carbon, oxygen and hydrogen, H_2CO, which absorbs strongly just below 6.0 μm, is readily made. This absorption blends with the initial 6.0 μm band and shifts the peak below 6.0 μm. In addition to H_2CO, another strong absorption also grows in at about 6.0 μm upon photolysis which seems to be correlated with the X-C≡N band at 2165 cm^{-1} (Van de Bult, 1984). Thus it appears that at least four species contribute to the 6.0 μm band: H_2O, NH_3 (-NH_2, $\geq$N-H), H_2CO and X-CN. In different laboratory spectra, structure is evident which depends on how much of the various constituents are present, thus better profiles of the 6.0 μm interstellar band are of critical importance.

Figure 14 also shows that the 6.8 μm absorption in the laboratory spectrum undergoes little or no change upon photolysis and matches the interstellar band rather well, supporting the suggestion that the interstellar band is due to the scissoring and bending vibrations of $-CH_2-$ and $-CH_3$ groups, respectively. Note also the shoulder in the laboratory spectrum at 7.0 μm and the suggestion of similar structure in the spectrum of W33A, which, if confirmed would provide insight into the type and amount of hydrocarbons present in the dust. Of course, a confident assignment of the interstellar band with saturated hydrocarbons is only possible once the corresponding C-H stretching vibrations are positively detected in the 3.4 μm region. Unfortunately these are generally weak (laboratory spectra show that in ices, the integrated absorbance of the 3.4 μm bands are comparable to those measured at 6.8 μm) and, in most interstellar spectra, are superimposed on the much stronger ice band but account for only a small part of the absorption. In this sense, the identification of the 3.4 μm bands towards the galactic center is significant.

Finally, as pointed out in the previous section, the strongest absorption of the radical HCO falls in this region. In a mixed molecular ice it occurs at 1850 cm^{-1} (5.41 μm) close to that for NO which absorbs at 1870 cm^{-1} (5.35 μm) in a N_2 matrix.

3.4. Probing Grain Mantles and Cloud Conditions

The wide variation in UV radiation field in going from the diffuse to dense interstellar medium should be reflected in an equally broad range of grain mantles since it is the UV which photolyzes the grains and determines the relative abundance of the different types of species in the gas (i.e. molecules, radicals, ions, electrons). Figure 15, reproduced from Greenberg et al. (1983), shows three types of core-mantle grains. The drawing on the left represents a diffuse cloud grain which is coated with a highly processed non-volatile material. That in the center corresponds to molecular cloud grains in which the ratio of the number of UV photons striking the grain to the number of atoms accretting per unit time is on the order of 10^3, and finally the last drawing represents grains in denser regions, where photoprocessing has essentially stopped, and the photoprocessed mantle is capped by a pure accretion mantle.

While the development of computor models which correctly include processes which take place both in the gas phase as well as in the mantle is at an early stage, initial results are available. Tielens and Hagen (1982), including grain surface reactions, have calculated the equilibrium composition of accretion mantles and have shown that the relative concentration of species in the mantle depends strongly on the physical conditions of the gas. d'Hendecourt (1984), using a more complete time dependent analysis including processing within the mantle itself has shown that mantle composition varies with time and depends on both cloud conditions and previous history.

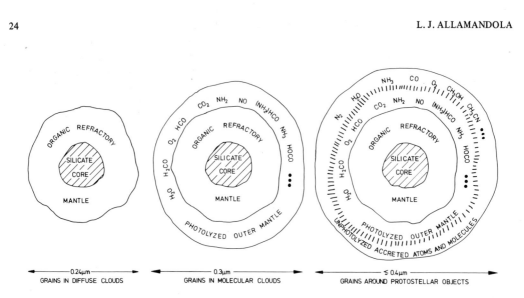

Fig. 15. Schematic representation of the types of core-mantle grains expected in different regions of the interstellar medium. This figure is reproduced from Greenberg et al. (1983).

The large variation of absorption features attributed to the molecular ice component of the interstellar medium shows that the composition of interstellar grains does indeed vary considerably from one region to the other. These spectra, when properly interpreted, provide the key to the determination of grain mantle composition and conversly, can be used to probe the conditions in interstellar clouds.

4. EMISSION SPECTROSCOPY OF INTERSTELLAR DUST

The discovery of broad emission features at 8.6 and 11.3 μm in the planetary nebula NGC 7027 by Gillett et al. in 1973 started an interesting chapter in astronomical infrared spectroscopy. Since that time similar emission bands at 3.28, 3.4, 3.5, 6.2 and 7.7 μm have been discovered in a number of stellar objects, planetary and reflection nebulae, H-II regions and extragalactic sources. Examples of such spectra are shown in figures 16 and 17. The interrelationship among the bands as observed and the theoretical explanations which have been proposed to describe the phenomenon form the subject of some controversy and equally strong statements supporting and condemning the various viewpoints can be found in the literature. Aitken (1981) has summarized the salient features of both facets of the problem, while Willner (this volume) including more recent results, has concentrated mainly on observational questions. In this review I shall attempt to present a summary of the development of our "understanding" of this phenomenon.

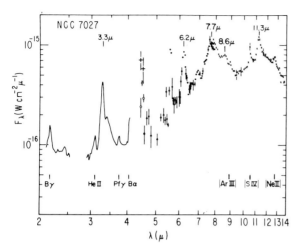

Fig. 16 2-14 μm spectrum of NGC 7027 showing the emission features at 3.3, 3.4, 6.2, 7.7, 8.6 and 11.3 μm (Russell et al., 1977c).

Fig. 17. 2-13 μm spectrum of HD 44179 showing the emission features at 3.3, 6.2, 7.7, 8.6 and 11.3 μm (Russell et al., 1978).

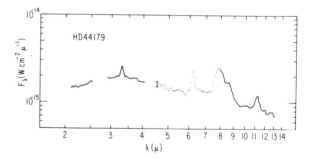

From the outset the peaks at 6.2, 7.7, 8.6, and 11.3 μm were considered to be characteristic of the dust responsible for the infrared continuum, however as there was no plausible candidate due to the lack of agreement between these emission bands and absorption or emission features of known materials, these bands became classified as unidentified (Gillett et al., 1973, Aitken and Jones, 1973, Russell et al., 1977c). The origin of the 3.28 and 3.4 μm emission was less firmly connected with the dust until higher spectroscopic resolution observations failed to show structure in the band (Merrill et al., 1975, Grasdalen and Joyce, 1976, Russell et al., 1977b, Tokunaga and Young, 1980).

4.1 Band Origin and Excitation Mechanism.

The match in frequency of the emission bands to the characteristic vibrational frequencies of simple molecules condensed in low temperature solids, the lack of P and R-like rotational structure expected from vibrationally excited gas phase molecules and the breadth of the bands were taken as evidence that these lines arise from vibrationally excited molecules frozen on grains (Allamandola and Norman, 1978a,b, Allamandola et al. 1979, hereafter AGN). Initially it was proposed that, once excited, radiative relaxation (infrared fluorescence) from specific vibrations of some of the molecules trapped in the mantles of interstellar grains would be possible provided that non-radiative relaxation was inefficient. Three, non-thermal excitation mechanisms were considered: 1) ultraviolet-visible (UV-VIS) photon absorption, 2) radical reactions, including reactions with hydrogen atoms, and 3) collisions between interstellar grains and molecules. Mechanisms 1 and 2 were shown to be the most efficient. However since the interstellar features arise in regions close to UV sources, it was concluded that mechanism 1 (hereafter referred to as the UV-VIS fluorescence model) was most likely and that at a distance of 100 pc, detectible levels of infrared emission could be produced by a variety of stellar sources (AGN).

In 1980, Dwek et al. pointed out that, for NGC 7027, which is at a distance of ~ 1 kpc, the efficiency (α^{ph}) implied for conversion of 10 ev photons into infrared photons emitted in all of the bands varies from 1.6 to 2.24, values they felt unacceptably high. Furthermore they stressed that, although the 3.28 μm emission could be attributed to UV pumped IR fluorescence since α^{ph}(3.28 μm) ranges from 0.03 to 0.08, all the other bands require efficiencies comparable to unity and since these bands generally seem to occur together (see Willner, this volume, for a discussion of this point) they argue that the 3.28 μm band must be excited in the same way. In view of the unreasonably high efficiencies required, they proposed that the features are due to thermal emission from very small grains (a $\simeq$ 0.01 μm) which are heated to about 300 K by UV radiation (hereafter referred to as the UV pumped thermal emission model). The small variation of the observed flux ratios for the different bands implies that they arise in materials which emit over a very narrow range of temperatures, suggesting a volatile mantle on the small grains. They further speculate that the seed grain upon which this mantle forms is graphite and, if the initial layers of the mantle are formed by chemical reactions between incident atoms and the surface, these layers would form carbon rich ices independent of the C/O ratio of the region. For example, Bar-Nun et al. (1980) report the formation of CH_4, C_2H_6, C_2H_4 and C_2H_2 by the reaction of cold (T $\geq$ 7K) H atoms with graphite particles.

Recently, Olofsson (1983) has criticized this mechanism on the grounds that the ratio of the 10 μm flux to that in the 3.28 μm band must be at least one order of magnitude more than that which is observed in the Orion Bar and S 235 A and concludes that the 3.28 μm emission cannot be explained as due to thermal emission from dust at 300 K. He also points out that the intensity of the 3.28 μm band can only be accounted for in a thermal excitation model by invoking an unreasonably high value for the cross section (or oscillator strength) per C-H bond. The abundance of the species giving rise to the 3.28 μm band is inversely proportional to the oscillator strength, f, which can never exceed unity for any particular transition (e.g. Barrow, 1962, pg 81). In their calculations Dwek et al. (1980) and Sellgren (1981) assumed an oscillator strength comparable to unity (Baas, 1981), which implies an allowed electronic transition which is roughly equivalent to a geometric cross section ($\sim 10^{-16}$ cm^2). Infrared cross sections for vibrational transitions are typically several orders of magnitude less ($\sim 10^{-19}$ cm^2), implying f values of $\sim 10^{-3} - 10^{-4}$. Thus the abundance of 10^{-7} relative to hydrogen calculated for the species giving rise to the emission at 3.28 μm within the framework of the thermal emission model should be increased by about 10^3.

Duley and Williams (1981), assuming that small (a $\tilde{=}$ 0.01 μm) amorphous carbon particles in diffuse clouds are not coated by mantles, suggested that it is the surface groups formed by the reaction of atoms with these amorphous carbon particles that give rise to the features which are pumped by the thermal emission model as proposed by Dwek et al. (1980). While focusing on the possible reactions between hydrogen atoms and amorphous carbon, they point out that other surface groups are also likely, providing characteristic frequencies spanning the middle infrared. Barlow (1983) has speculated that reactions between atomic hydrogen and carbon grains both produce and excite the molecules which emit from the grain surface.

4.2 Recent Observations.

4.2.1 The 2 - 5 μm region.
Emission at 3.28 and 3.4 μm superimposed on a smooth continuum extending from 2 to 5 μm has recently been discovered in the visible reflection nebulae NGC 2023, 2068 and 7023 by Sellgren et al. (1983). In these nebulae α^{ph} (3.28 μm), calculated using 10 ev-photons, ranges from 0.05 to 0.16. These observations expose a new facet of the problem as well. While the color temperature (1000 K) of the 2.2-3.8 μm continuum is similar at all positions in the nebula, it was found that the flux in the continuum has the same surface brightness distribution as the visual reflected light. This observation firmly establishes the connection between UV-VIS photons and the infrared continuum and poses some difficulties for the thermal emission model pumped by the continuous heating of grains by UV which implies a different spatial dependence (Sellgren, private communication). Both the implied high temperature and the apparent lack of color temperature dependence on position in the nebula also appear difficult to reconcile with thermal emission. The luminosity emitted in the continuum is 10^{-2} of the total luminosity of the illuminating stars and the authors suggest that such a continuum might be a widespread phenomenon which is connected to the 3.28 μm band but whose small contribution to the total luminosity in other more complex regions may have been masked by other near IR continuum emission. Several possible sources of continuum radiation (free-free emission, reflected light and imbedded faint stars) were considered and rejected as accounting for that observed in the reflection nebulae. The authors speculate that either thermal fluctuations in very small grains induced by absorption of individual UV photons or UV-VIS induced broad band fluorescence might account for both the constancy of color temperature and the 3.28 μm band.

Broad emission at 3.5 μm has only been detected in two objects. It was discovered by Blades and Whittet (1980) in HD 97048, a pre-main sequence star associated with a reflection nebula in the Chamaeleon association.

Subsequently Allen et al. (1982) in the course of carrying out a survey of 3 μm emission features found the 3.5 μm band in Elias 1 which is believed to be a Herbig Ae star which illuminates the nebula IC 359 in the Taurus dark cloud complex. Both objects also show weaker emission at 3.28 and 3.4 μm (Whittet et al., 1982). Aitken and Roche (1981) have observed the 8-14 μm spectrum of HD 97048 and show that the intensity ratio of the 3.28 to 11.3 μm features is similar to that found in other objects. Elias 1 is located behind a cloud with $A_v = 16$, and the emission features are superimposed on the 3.1 μm ice band associated with the cloud (Whittet et al., 1982, Whittet et al., 1983). Based on the model of UV-VIS pumped IR fluorescence from molecules in grain mantles (Allamandola and Norman, 1978a,b), Blades and Whittet (1980) assigned the 3.5 and 3.4 μm bands to formaldehyde (H_2CO). This assignment was later questioned by Aitken and Roche (1981) due to the non-detection of emission near 9 μm which would be expected if H_2CO was in its polymerized, less volatile form, the form they felt would most likely be found on grains. Higher resolution spectra of the 3.4 and 3.5 μm bands in HD 97048 by Baas et al. (1983) shows that there is a very strong resemblance between these emission bands and laboratory absorption spectra of H_2CO suspended in solids at 10 K. Since the UV radiation from HD 97048 peaks at about 300 nm and solid H_2CO absorbs strongly at wavelengths shortward of 360 nm, Baas et al. (1983) conclude that the emission at 3.5 μm arises from UV pumped IR fluorescence from H_2CO frozen on grains located in a circumstellar shell of one magnitude extinction.

4.2.2. The 8-14 μm region. Aitken and Roche (1983) have carried out a very thorough high spatial resolution study of NGC 7027 in the 10 μm region. The 8.6 and 11.3 μm bands originate from a more extended region than the shell of ionized gas, suggesting that they arise entirely from the neutral region. A similar conclusion has been drawn concerning the region of origin of these two features (Aitken et al., 1979) and the 3.28 μm band (Sellgren, 1981) in the Orion nebula. Aitken and Roche also point out that if one includes visible as well as UV photons in calculating α^{ph}, the overall efficiency required is reduced but still remains somewhat greater than unity (1.1). They also contend that in the planetary nebula environment the grains have not had a history conducive to either mantle formation or the growth of processed carbon compounds. They argue that these two points place severe constraints on UV-VIS pumped IR fluorescence of molecules frozen on grains, and suggest that hydrogen atom sputtering reactions with the surface of amorphous carbon grains (Barlow, 1983) is particularly attractive in this situation.

4.3. Laboratory Studies of Candidate Materials

Common to all of the models currently considered as likely to account for the infrared emission features is that they arise from vibrationally excited molecules which are located within the mantle or on the surface of dust grains. To the best of the author's knowledge laboratory measurements of the infrared emission spectrum from likely grain materials has not yet been reported. Consequently it is worthwhile to compare the interstellar emission bands with absorption spectra of potential candidate materials.

4.3.1. <u>Molecular Ices</u>. The assignment of the 3.5 μm (2833 cm^{-1}) feature to formaldehyde implies a blue shift of the C-H stretching frequency from its gas phase value which is slightly larger than has been reported for dimeric and complexed forms of formaldehyde isolated in inert gas matrices (Baas et al. 1983). In order to investigate this shift under conditions likely to be prevelant in interstellar dust, Van der Zwet has begun a study of the absorption spectrum of H_2CO suspended in various molecular mixtures at 10 K as a function of composition and irradiation. Preliminary results indicate that both the peak position and profile of the 3.5 μm emission band can be satisfactorily matched by absorption of H_2CO suspended in processed molecular ices which contain some CO_2 and O_2. A discussion of the effects these mixtures have on the rest of the spectrum and related results will be presented elsewhere (Van der Zwet et al. 1983).

Tokunaga and Young (1980) have shown that the 3.28 μm band has a FWHH of 50±5 cm^{-1} and is centered at 3040±5 cm^{-1}. The proposed assignment of this feature to methane (Allamandola and Norman, 1978a) implies a blue shift of the C-H stretching frequency of approximately 20 cm^{-1} from its gas phase value and substantial band broadening. As pointed out by Tokunaga and Young (1980), the narrower width (18 cm^{-1}) quoted by AGN is in error. Only a limited study of the spectral behavior of this molecule in processed molecular ices has been carried out. To date, the most extreme case of band broadening and shifting of the methane spectrum was found when a $CH_4:NH_3:O_2$ (20:1:5) ice deposited with simultaneous vacuum ultraviolet radiation was warmed up to 30 K. At this point, prominent, broad features at 3025±15 and 1310±10 cm^{-1} (3.3 and 7.6 μm respectively) appeared in the spectrum. The FWHH of the 3025 cm^{-1} band is 100 cm^{-1} and that at 1310 cm^{-1} is about 40 cm^{-1}. Upon further warming of this sample to 200 K, these two features disappeared and the spectrum of gaseous CH_4 appeared. The comparison of the laboratory measured absorption band at 7.6 μm with the even broader emission band in HD 44179 can be found in Russell et al. (1982).

4.3.2. <u>Refractory Materials</u> Some minerals and mineral carbonates have been considered as carriers of the bands (Gillett et al., 1973, Aitken and Jones, 1973, Russell et al., 1977c) however, the absence, in the interstellar spectra, of the strongest features measured in the laboratory which occur near the maximum of the emissivity curve has led to their rejection.
The absorption efficiency of small, amorphous carbon particles (a ≃ 10^{-2} μm) has recently been reported (Koike et al., 1980, Borghesi et al., 1983). While there are some differences from sample to sample, the wavelength dependence of Q_{ext}/a from 2.5 to 40 μm is smooth and proportional to λ^{-1}, showing small structure in the 6-15 μm range. These features are shown in figure 18 where they have been enhanced by plotting the ratio R of $\lambda(Q_{abs}/a)$ normalized to the maximum value near 8 μm. The amorphous carbon particles are produced in two ways, 1) a car-

bon arc discharge in an atmosphere of argon and 2) burning the aromatic hydrocarbons benzene and xylene in air. The arc-produced amorphous carbon spectrum reported by Borghesi et al. shows a broad peak centered at about 8 μm, a weaker feature at 11.3 μm and possibly one at 12 μm. Comparing the profiles and wavelengths of maximum absorption of the features from sample to sample indicates that the spectrum is somewhat dependent upon the way in which the sample is prepared. In this sense, it is interesting that the amorphous carbon samples studied by Koike et al. all showed an extremely weak feature at about 6.2 μm which is absent in the data of Borghesi et al. Conversely, the amorphous carbon produced by the later group all showed a band at 11.35 μm which is completely absent in the Koike et al. spectra. This probably reflects the presence of different molecular subgroups in the different samples, with a particular characteristic group favored by the synthetic techniques used by the one set of researchers versus those employed by the other.

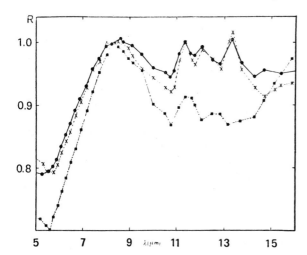

Fig. 18. Structure in the amorphous carbon absorption efficiency between 5 and 15 μm. R is the ratio of $\lambda(Q_{abs}/a)$ divided by the value of $\lambda(Q_{abs}/a)$ at the maximum which occurs near 8 μm. Squares represent the spectrum for arc-produced amorphous carbon, dots and crosses amorphous carbon made by burning respectively, benzene and xylene in air (Borghesi et al., 1983).

4.4. Speculations and Conclusions

While it appears that neither UV-VIS pumped IR nor UV pumped thermal emission alone can account for the observed intensity in all of the IR emission features, combining elements from both mechanisms may remove many of the difficulties.

Dwek et al. (1980) and Aitken and Roche (1983) have questioned the UV-VIS pumped IR mechanism on the grounds that α^{ph} is too large. By far the largest contribution to α^{ph} comes from the flux in the 7.7 and 11.3 μm bands, rather close to the wavelengths of the structure in the spectrum of arc-produced amorphous carbon (figure 18). In the laboratory spectrum, the occurance of the moderate feature at 11.3 μm, the weaker one at 6.3 μm and the peak position of the feature near 8 μm are all apparently dependent upon the techniques used to synthesize the material. Thus, as has been suggested by Duley and Williams (1981) in general terms, it is quite conceivable that under interstellar conditions amorphous carbon particles could be produced which exhibit features at the appropriate wavelengths. However these features are very weak in the laboratory absorption spectrum and, in view of the much greater abundance requirement implied when realistic values of the oscillator strength are assumed in the thermal emission model, it is perhaps more likely that reactions between gaseous species and carbon particles produce the molecules which ultimately give rise to the emission features. To test whether or not the time required for the accretion of such molecules is compatible with the time available it is worthwhile to calculate how long it would take to form a 100 monolayer thick (~ 0.01 μm) accretion mantle in clouds of density, $n_H = 10^3$ and 10^4 cm^{-3}. Greenberg (1978, equation 6.2.5) has shown that the mantle thickness growth rate is given by

$$\frac{da}{dt} = 3.4 \times 10^{-22} \, n_H \, \text{cm s}^{-1}. \qquad (3)$$

for grains with a >0.01 μm, and assuming that $T_{gas} = 100K$, grain density = 1g cm^{-3}, sticking probability = 1, and the abundance of the accreting molecules all taken together is 1.2×10^{-3} the hydrogen abundance. This shows that a 0.01 μm thick mantle can grow in about 10^5 years at a density of 10^3 cm^{-3} and 10^4 years at 10^4 cm^{-3}. Thus there is ample time for the accretion and photoprocessing of a substantial mantle in a region such as that surrounding NGC 7027 provided the molecules are available. If the molecules originate in the same region, the molecule production rate by the reaction of H atoms with the carbon grains can be as much as a few hundred times faster than the accretion rate since the collision rate of H atoms with the same grains is about 1000 times greater. Thus even if the molecules are local rather than interstellar in origan is should be possible to produce them in quantities sufficient to satisfy the 100-1000 times greater abundances required by the thermal emission model to account for the strength of the most intense longer wavelength bands.

This scenario however does not relieve the difficulties pointed out by Olofsson (1983) which are associated with trying to account for the intensity of the 3.28 μm emission by the UV pumped thermal emission model. This, in addition to the result that $\alpha^{ph}(3.28)$ is nearly two orders of magnitude lower than that for the 7.7 and 11.3 μm features suggests that the UV-VIS pumped IR fluorescence mechanism can account for the 3.28 μm as well as the other short wavelength bands, all which possess even lower α^{ph} values.

UV-VIS pumped IR fluorescence should give rise to features throughout the middle infrared, however the intensity in the 3 μm region may well be greater than that at longer wavelengths for several reasons. Allamandola and Norman (1978b) have pointed out that radiative relaxation of vibrationally excited molecules (IR fluorescence) may compete with non-radiative relaxation if acceptors are not readily available in the solid which can take this energy and couple it into the lattice. In general, the smaller the frequency mismatch between acceptor and donor, the more efficient the coupling and the faster the decay. As can be inferred from figure 2, on purely statistical grounds the density of vibrational states which are available in a mixed molecular ice to act as efficient acceptors increases as one goes to lower frequencies. If, for example, the ice is made up of molecules which contain OH, NH and CH groups, it is conceivable that an OH stretching motion can relax non-radiatively by exciting a lower lying OH or NH stretch which, in turn, can excite a lower lying CH stretch with the small difference in energy exciting a lattice mode (phonon). Once the lower lying CH stretching vibrations are excited, overlap with other fundamental modes is negligible and the non-radiative energy transfer process can be inefficient enough to permit competition by radiative relaxation. This is not all of the story however as the precise nature of the vibration must be taken into account, including aspects such as the symmetry of the particular vibrational mode involved and how it is coupled with its neighbors. Thus, since there is a much higher density of states as one goes to longer and longer wavelengths (up to the break between the lowest energy vibrations of simple molecules and the lattice modes), the fluorescence efficiency should be lower here, giving rise to mode selective vibrational relaxation even within the same molecule. Consequently it is in the longer wavelength region where emission from 300 K grains can dominate fluorescence. This picture of stepwise non-radiative relaxation competing with radiative relaxation implies that absorption into bands at all frequencies higher than those observed in emission could contribute to the fluorescence pumping process, reducing α^{ph} even further.

Although there are some differences in relative intensities among the IR emission bands, the fact that they generally occur together has been taken to imply a common origin and pumping process (Aitken, 1981; Willner, this volume). This observation can be reconciled with a dual pumping mechanism provided that the conditions which exist in the molecule formation region also satisfy the requirements imposed by both mechanisms. Thus, the presence of one set of lines implies the other and the slight variation between intensity of the features can be accounted for.

ACKNOWLEDGEMENT

It is a pleasure to acknowledge Mayo Greenberg whose insight and understanding of the interstellar grain problem, conveyed in countless stimulating discussions, has been invaluable in providing the basis for many of the experiments described here, and Fred Baas whose tireless efforts to bridge the gap between experimentalist and observer have met with success to the benefit of both. I express my sincere appreciation to C.E.P.M. van de Bult for permitting me to quote some of his results on the photochemistry of ices, and L. d'Hendecourt for permitting me to present some results of his time dependent mantle evolution calculation prior to publication. I also wish to thank Leo van IJzendoorn and G.P. van der Zwet who have carried out numerous experiments to check many of the results presented here and who have critically read an earlier version of this manuscript. Finally, I would like to thank T. Geballe, W. Krätschmer and K. Sellgren for sending useful comments on the manuscript and the conference organizers for the opportunity of presenting this work.

REFERENCES

Aitken, D.K. and Jones, B., 1973, Mon. Not. R. astro. Soc. 165, 363.
Aitken, D.K., Roche, P.F., Spenser, P.M. and Jones, B., 1979, Astron. Astrophys. 76, 60.
Aitken, D.K., 1981, in Infrared Astronomy, IAU Symposium, No. 96, eds. Wynn-Williams, C.G. and Cruickshank, D.P., (Reidel, Dordrecht), 207.
Aitken, D.K. and Roche, P.F., 1981, Mon. Not. R. astr. Soc., 196 39p.
Aitken, D.K. and Roche, P.F., 1983, Mon. Not. R. astr. Soc. 202, 1233.
Allamandola, L.J. and Norman, C.A., 1978a, Astron. Astrophys. 63, L23.
Allamandola, L.J. and Norman, C.A., 1978b, Astron. Astrophys. 66, 129.
Allamandola, L.J., Greenberg, J.M. and Norman, C.A., 1979, Astron. Astrophys. 77, 66, AGN
Allen, D.A. and Wickramasinghe, D.T., 1981, Nature, 294, 239.
Allen, D.A., Baines, D.W.T., Blades, J.C. and Whittet, D.C.B., 1982, Mon. Not. R. astr. Soc. 199, 1017.
Baas, F., 1981 The author wishes to thank Dr. F. Baas for drawing the high value of the assumed oscillator strength to his attention.
Baas, F., Allamandola, L.J., Geballe, T.R., Persson, S.E. and Lacy, J.H., 1983, Astrophys. J. 265, 290.
Barlow, M.J., 1983, in Planetary Nebulae, (ed.) Flower, D.R., (Reidel, Dordrecht)
Bar-Nun, A., Litman, M. and Rappaport, M.L., 1980, Astron. Astrophys. 85, 197.
Barrow, G.M., 1962, Introduction to Molecular Spectroscopy (McGraw-Hill, New York).
Bellamy, L.J., 1958, The Infra-red Spectra of Complex Organic Molecules, (John Wiley, 2nd ed. New York).
Blades, J.C. and Whittet, D.C.B., 1980, Mon. Not. R. astr. Soc. 191, 701.
Borghesi, A., Bussoletti, E., Colangeli, L., Minafra, A. and Rubini, F., 1983, Infrared Physics, in press.
Day, K.L., 1979, Astrophys. J. 234, 158.
Duley, W.W. and Williams, D.A., 1981, Mon. Not. R. astr. Soc. 196, 269.
Dwek, E., Sellgren, K., Soifer, B.T. and Werner, M.W., 1980, Astrophys. J. 238, 140.
Fink, U. and Sill, G.T., 1982, in Comets, (ed.) Wilkening, L.L. (University of Arizona Press) 164.
Gillett, F.C., Forrest, W.J. and Merrill, K.M., 1973, Astrophys. J. 183, 87.
Gillett, F.C., Jones, T.W., Merrill, K.M. and Stein, W.A., 1975, Astron. Astrophys. 45, 77.
Grasdalen, G.L. and Joyce, R.R., 1976, Astrophys. J., 205, L11.
Grewing, M, 1983, Diffuse Matter in Galaxies, NATO Advanced Study Institute, Cargese, France (preprint).

Greenberg, J.M., 1971, Astron. Astrophys. 12, 240.
Greenberg, J.M., 1972, J. Coll. and Interface Science, 39, 513.
Greenberg, J.M., 1978, in Cosmic Dust, (ed.) McDonnell, J. (Wiley, New York), 187.
Greenberg, J.M. 1982, in Comets, (ed.) Wilkening, L.L. (University of Arizona Press) 131.
Greenberg, J.M., van de Bult, C.E.P.M. and Allamandola, L.J., 1983, J. Phys. Chem., in press.
Hagen, W., Allamandola, L.J. and Greenberg, J.M., 1979, Astrophys. Space Sci. 65, 215.
Hagen, W., Allamandola, L.J. and Greenberg, J.M., 1980a, Astron. Astrophys. 86, L3, paper I.
Hagen, W, Allamandola, L.J., Greenberg, J.M. and Tielens, A.G.G.M., 1980b, J. Mol. Struct. 60, 281.
Hagen, W. and Tielens A.G.G.M., 1981, J. Chem. Phys. 75, 4198.
Hagen, W., Tielens, A.G.G.M. and Greenberg, J.M., 1983, Astron. Astrophys. 117, 132.
d'Hendecourt, L.B. (1984), Ph.D. Thesis, Leiden University, Leiden, The Netherlands.
Herzberg, G.H., 1968, Infrared and Raman Spectra of Polyatomic Molecules, (D. van Nostrand Company, Princeton).
Jiang, G.J., Person, W.B. and Brown, K.J. 1975, J. Chem. Phys. 62, 1201.
Jones, T.J., Hyland, A.R. and Allen, D.A., 1983, Mon. Not. R. astr. Soc. (preprint).
Knacke, R.F. and Krätschmer, W., 1980, Astron. Astrophys. 92, 281.
Knacke, R.F., McCorkle, S., Puetter, R.C., Erickson, E.F. and Krätschmer, W., 1982, Astrophys. J. 260, 141.
Koike, C., Hasegawa, H. and Manabe, A., 1980, Astrophys. Space Sci. 67, 495.
Krätschmer, W. and Huffman, D.R., 1979, Astrophys. Space Sci. 61, 195.
Lacy, J.H., Baas, F., Allamandola, L.J., van de Bult, C.E.P.M., Persson, S.E., McGregor, P.J., Lonsdale, C.J. and Geballe, T.R., 1983, Astrophys. J. (preprint).
Legay-Sommaire, N. and Legay, F., 1982, Chem. Phys. 66, 315.
Léger, A., Gauthier, S., Défourneau, D. and Rouan, D., 1983, Astron. Astrophys. 117, 164.
McGregor, P.J., Persson, S.E., Lacy, J.H., Baas, F., Allamandola, L.J., Lonsdale, C.J. and Geballe, T.R., 1982, 16th ESLAB Symposium: Galactic and Extragalactic Infrared Spectroscopy. Eds. Kessler, M.F., Phillips, J.P., and Guyenne, T.D., ESA SP-192, p. 51.
Merrill, K.M., Soifer, B.T. and Russell, R.W., 1975, Astrophys. J. 200, L37.
Milligan, D.E. and Jacox, M.E., 1964, J. Chem. Phys. 41, 3032.
Milligan, D.E. and Jacox, M.E., 1971, J. Chem. Phys. 54, 927.
Montmerle, T., Koch-Miramond, L., Falgarone, E. and Grindlay, J., 1983, Astrophys. J. in press.
Moore, M.H. and Donn, B. 1982, Astrophys. J. 257, L47.
Nelander, B. and Nord, L., 1982, J. Phys. Chem. 86, 4375.
Olofsson, G. 1983, Astron. Astrophys. (preprint).
Person, W.B., 1981, in Matrix Isolation Spectroscopy, (eds.) Barnes, A.J. et al. (Reidel, Dordrecht), p 415.

Pimentel, G.C. and McClellan, A.L., 1960, The Hydrogon Bond, (W.H. Freeman and Co., San Francisco).
Puetter, R.C., Russell, R.W., Soifer, B.T. and Willner, S.P., 1979, Astrophys. J. 228, 118.
Russell, R.W., Soifer, B.T. and Puetter, R.C., 1977a, Astron. Astrophys. 54, 959.
Russell, R.W., Soifer, B.T. and Merrill, K.M., 1977b, Astrophys. J. 213, 66.
Russell, R.W., Soifer, B.T. and Willner, S.P., 1977c, Astrophys. J. 217, L149.
Russell, R.W., Soifer, B.T. and Willner, S.P., 1978, Astrophys. J. 220, 568.
Russell, R.W., Gull, G., Beckwith, S. and Evans II, N.J., 1982, Publ. Astron. Soc. Pacific 94, 97.
Sellgren, K., 1981, Astrophys. J. 245, 138.
Sellgren, K., Werner, M.W. and Dinerstein, H., 1983, Astrophys. J. preprint.
Silverstein, R.M. and Bassler, G.C., 1967, Spectrometric Identification of Organic Compounds, (John Wiley and Sons, New York) Chapter 3, Infrared Spectrometry.
Soifer, B.T., Russell, R.W. and Merrill, K.M., 1976, Astrophys. J. 207, L83.
Soifer, B.T., Puetter, R.C., Russell, R.W., Willner, S.P., Harvey, P.M. and Gillett, F.C., 1979, Astrophys. J. 232, L53.
Soifer, B.T., Willner, S.P., Capps, R.W. and Rudy, R.J., 1981, Astrophys. J., 250, 631.
Tielens, A.G.G.M. and Hagen, W., 1982, Astron. Astrophys. 114, 245.
Tokunaga, A.T. and Young, E.T., 1980, Astrophys. J. 237, L93.
Van de Bult, C.E.P.M. (1984), Ph.D. Thesis, Leiden University, Leiden, The Netherlands.
Van der Zwet et al. (1983) in preparation.
Van IJzendoorn, L.J., Allamandola, L.J. Baas, F. and Greenberg, J.M., 1983, J. Chem. Phys., in press.
Van Thiel, M., Becker, E.D. and Pimentel, G.C., 1957, J. Chem. Phys. 27, 486.
Willner, S.P., Russell, R.W., Puetter, R.C., Soifer, B.T. and Harvey, P.M., 1979, Astrophys. J. 229, L65.
Willner, S.P., Gillett, F.C., Herter, T.L., Jones, B., Krassner, J., Merrill, K.M., Pipher, J.L., Puetter, R.C., Rudy, R.J., Russell, R.W. and Soifer, B.T., 1982, Astrophys. J. 253, 174.
Whittet, D.C.B., Davies, J.K., Bode, M.F., Evans, A. and Longmore, A.J., 1982, 16th ESLAB Symposium: Galactic and Extragalactic Infrared Spectroscopy. Eds. Kessler, M.F., Phillips, J.P., and Guyenne, T.D., ESA SP-192, p. 59.
Whittet, D.C.B., Bode, M.F., Longmore, A.J., Baines, D.W.T. and Evans, A., 1983, Nature, (preprint).
Wood, B.E. and Roux, J.A., 1982, J. Opt. Soc. Am. 72, 720.

OBSERVED SPECTRAL FEATURES OF DUST

S.P. Willner

Harvard-Smithsonian Center for Astrophysics, Cambridge, Massachusetts, U.S.A.

I. INTRODUCTION

This review will concentrate on the observed properties of dust spectral features. Related laboratory results are reviewed by Allamandola (1983) and will not be discussed here, but identifications, which are based on laboratory data, will be given whenever plausible ones exist. There are a very large number of papers in the literature of even such a young field as infrared spectroscopy, and it is impossible to refer to all of the original sources in a short review such as this one. I have therefore referred only to the most recent paper on a topic or to another review, and the reader will have to go back through successive reference lists to find the original work.

Readers should keep in mind that infrared spectroscopy is a new field, and in many cases the number of objects observed to show particular spectral features is small. With small samples, it is often unclear whether a pattern that has been noticed is universal or not. It would be redundant to express this uncertainty for every topic discussed in this review, and it must suffice to say here that in what follows, there is hardly any statement that cannot be challenged. Statements involving "appears" or "seems" should be regarded especially warily. I have tried to draw conclusions when the data warrant, but ultimately the reader must consult the original sources to find out the size and variety of the sample and the quality of the data on individual objects.

II. "NORMAL" DUST ABSORPTION

Determining the "normal" extinction curve is much more difficult in the infrared than in the visible. For stars, which have relatively well known intrinsic energy distributions, the amount of extinction is much smaller than can be found in the visible. There are highly extinguished sources, many associated with H II region/molecular cloud complexes, but these have unknown intrinsic spectra, and there is no guarantee that the dust in these regions shows the same extinction law as in less dense regions.

In spite of the difficulties, a great deal of progress has been made. The recent review by Savage and Mathis (1979) discusses the visible and ultraviolet reddening laws, as well as the infrared. This section discusses the methods for determining the infrared extinction and the results obtained in the past few years.

A. Near Infrared Extinction

The usual approach to obtaining the infrared extinction law is to observe a large number of stars so that the photometric errors of individual observations are not important. Unreddened and heavily reddened stars are observed to determine both the intrinsic colors of the stars and the reddening law. Johnson (1968) conducted the earliest of the extensive infrared studies and showed that there exists an extinction law that is closely followed in many regions of the galaxy. Johnson approximated that law with "van de Hulst curve 15," which is a theoretical extinction law for dust particles of a certain size distribution and refractive index (van de Hulst 1946). Curve 15 was derived before even the visible extinction law was well known and was meant to illustrate one possibility rather than being an attempt to model the actual extinction. Nevertheless, it fits the data surprisingly well and has often been used to estimate the extinction at wavelengths where data are unavailable or in doubt.

Johnson (1968) also suggested that there are regions in space where the extinction law differs from the normal one. That conclusion has been challenged, but more recent work (Johnson 1977, Breger et al. 1981) verifies abnormal extinction for at least the region near the Orion nebula and for certain individual stars (Savage and Mathis 1979). Part of the problem of establishing that an abnormal extinction law exists is that circumstellar dust emission from heavily reddened stars can reduce the apparent amount of extinction, generally by larger amounts at longer wavelengths. More recent studies (Lee 1970, Schultz and Wiemer 1975, Sneden et al. 1978) have attempted to exclude stars with circumstellar emission from the reddening determination, but they may not always have succeeded because there is no unambiguous indication of circumstellar emission.

Comparison of various extinction studies is often difficult because of the way the amount of extinction is normalized. A study of M stars (Lee 1970) appears at first glance to differ greatly from a similar study of O stars (Schultz and Wiemer 1975), but much of the difference turns out to be due to the different effective wavelengths of the B and V filters (Lee 1968, 1970) for the two types of stars having led to a different normalization. There is still a difference in the results, but a much smaller one than first appears.

Figure 1 shows various extinction results combined. In such a plot, both the origin and the scale of the ordinate must be determined

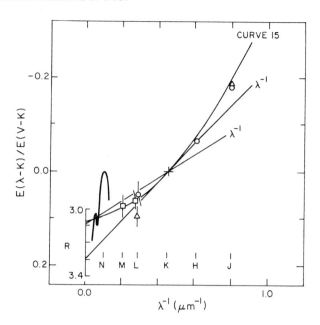

Figure 1. Wavelength dependence of interstellar extinction. The circles denote data from Lee (1970), the triangles data from Schultz and Wiemer (1975), and the squares data from Sneden et al. (1978). The data were normalized to E(V-K) and at K. The curve indicates schematically the locations and widths of the silicate features; the text should make clear that their relationship to the near infrared extinction is not well established. The lines indicate 3 extinction laws that have been suggested in the literature. A $\lambda^{-1.8}$ law has also been suggested; it is indistinguishable from curve 15 in this plot. The intercept of each extinction curve with the inset vertical axis indicates the value of A(V)/E(B-V) for that curve, provided E(B-V)/E(V-K) = 0.36 (Johnson 1968).

for each data set. For this figure, the scales were set so that the amount of extinction at V and K agree for all of the data sets. There are many additional studies that contain only infrared data and cannot be normalized the same way. Two typical examples (Jones and Hyland 1980, Tapia 1981), although not shown in the figure, agree with the data shown if J and K are used for the normalization. Van de Hulst's curve 15 is also shown in figure 1, along with 2 lines showing λ^{-1} laws. (It is known that λ^{-1} is not appropriate at shorter wavelengths, so normalization at V is not useful. Thus any straight line through the K point is a possible λ^{-1} extinction law.) One of the lines shown was chosen to pass through the 1.65-μm point, while the other passes through the point at which van de Hulst's curve 15 extrapolates to 0. Clearly the stellar data cannot distinguish among the various laws for $\lambda > 2.2$μm.

The Y intercept of figure 1 is closely related to the ratio of total to selective extinction, normally denoted R and parameterized by A(V)/E(B-V) (Johnson 1968). In order to determine R from the extinction curve, one must extrapolate from the longest wavelength measured to infinite wavelength, where the dust extinction is presumably zero. The difference between a curve 15 and a λ^{-1} form for the extinction law can make a significant difference in the value of R obtained. More direct measurements of R can be made for objects for which the actual visual extinction can be determined. Recent determinations (Sandage 1975, Turner 1976, Hawley and Duncan 1976) range from 3.2 to 3.4, but with fairly large uncertainties. The inset vertical axis of figure 1 is labeled with the value of R corresponding to the Y intercept of each reddening curve.

An extinction law can be determined any time the intrinsic spectrum of a source is known, not just from stars. H II regions are becoming increasingly important for this purpose, because the intrinsic ratio of the hydrogen lines is well known (Brocklehurst 1971, Giles 1977). Smith et al. (1981) have found that one planetary nebula gives a 1.6 to 2.5μm extinction law consistent with curve 15, but longer wavelength results will be difficult to obtain because the extinction is relatively small. Compact H II regions often suffer comparatively large extinctions, but as mentioned above, there is no assurance that the extinction law is the same in the dense molecular medium that often surrounds compact H II regions as in the diffuse interstellar medium. The meager data now available include few regions (Herter et al. 1981a,b, Willner and Pipher 1983) but are consistent with an extinction law similar to that in figure 1. These early studies also suggest an apparent optical depth at 4.05μm that is half or more of that at 2.17μm. Interpretation of extinction measurements of extended objects is complicated by the fact that spatially varying extinction causes the apparent extinction law to be flatter than the true one (Natta and Panagia 1982). This effect could account for the relatively large apparent 4.05-μm extinctions. Determining the extinction curve between 2 and 8 microns is clearly a worthwhile project for the future.

B. Silicate Extinction

Near 10 and 18μm, there is additional interstellar extinction attributed to silicates in the interstellar grains. The characteristics and identification of these absorption features are well discussed in the review by Merrill (1979). The amount of silicate absorption is often computed by assuming a particular wavelength dependence for the absorption, with the dependence used based on silicate emission features. Silicate emission and circumstellar absorption are discussed in section IV.

1. At 10μm. While the shape of the 10-μm absorption is well known and seems not to differ measurably for different sources, the total amount of absorption is impossible to measure for most sources because the underlying intrinsic spectrum is unknown. Although a number is often given for the "optical depth at 9.7μm" $\tau(9.7)$, what has in fact

been measured is a color excess, typically between 9.7μm and the average of 8 and 13μm. An assumed extinction law, i.e. the ratio of absorptions at 8, 9.7, and 13μm, is used to translate the measured color excess to an optical depth, but there is no independent knowledge of the actual total extinction. The usual assumption is that there is finite extinction at 8 and 13μm; the alternative assumption of zero extinction at those wavelengths leads to a value of $\tau(9.7)$ that is about 1.4 times smaller.

Even the color excess describing the depth of the 9.7-μm absorption can be reliably related to shorter wavelength interstellar extinction for only two sources: a B supergiant star in Cygnus and late type stars near the galactic center. In the former case (Rieke 1974, Gillett et al. 1975), the visual extinction is relatively well determined, and the ratio $A(V)/A(9.7)$ is 15 (for the usual assumption about the shape of the silicate absorption). In the second case (Becklin et al. 1978) the sources cannot be detected visually, so what is really measured is the 1.65-2.2-μm color excess. If this is translated to $A(V)$ with a standard extinction curve, the ratio of $A(V)/A(9.7)$ is 9. (The assumption of a standard extinction curve is not the source of the discrepancy, because the ratio of near infrared to 10μm color excesses is also different for the two cases.) Observations in the direction of 2 H II regions (G333.6-0.2 and W3/IRS 1) favor the upper and lower values respectively (Aitken 1981, Hackwell et al. 1978), but in both cases the silicate extinction is small and uncertain due to the unknown shape of the underlying emission spectrum. Moreover, both regions are spatially complex, and one may be observing physically distinct components at different wavelengths. Observations of the Br-α line towards several compact H II regions (Lester 1979) may tend to favor the higher value, but the translation from Br-α to near infrared or visual extinction is uncertain. Accurate measurements of radio flux density, Br-α, Br-γ, and $\tau(9.7)$ in a sample of H II regions should be possible; such measurements would provide much needed information on the various extinction ratios, provided the effect of spatially varying extinction (Natta and Panagia 1982) can be properly accounted for.

2. At 18μm. Both the shape and the relative optical depth of the 18-μm silicate feature are poorly known, because there is nearly always underlying emission in the feature. Probably the best determination is for the galactic center (McCarthy et al. 1980), where the 18-μm absorption seems to be about half as strong (in peak optical depth) as the one at 9.7μm. Actually, the main uncertainty in this ratio comes from the lack of a 10-μm spectrum with the same beam size. Polarimetry (Knacke and Capps 1979) gives a similar result if the polarization to extinction efficiency ratio of the grains is the same at the two wavelengths. The circumstellar emission from oxygen rich stars is also consistent with this ratio (Forrest et al. 1979).

Figure 1 shows a very qualitative indication of the locations and widths of the two silicate features. In view of the discussion above, it should be clear that it is presently impossible to relate the silicate absorption accurately to the near infrared extinction.

C. Far Infrared Dust Emissivity

Beyond 20µm, dust is observed in emission rather than in absorption. The emission depends on both the dust temperature and the emissivity, and it is difficult to separate the two because a higher temperature will have the same effect as a larger emissivity. Temperature variations within the observing beam may further complicate an emissivity determination. The problem is especially serious in the far infrared (20 to 200µm), but less so at longer wavelengths where the dust emits according to the Rayleigh-Jeans law and the temperature is less important. Mezger et al. (1982) present a qualitative picture of our knowledge of emissivities beyond 20µm.

The dust temperature and emissivity cannot be determined separately from far infrared observations alone. Attempts have been made to use visual observations of a reflection nebula (Harvey et al. 1980) or radio observations of a molecular cloud (Smith et al. 1979) to determine the column density of the dust. A fundamental assumption is that the far infrared emitting dust is the same as the dust whose column density is deduced from the other observations. The dust temperature is estimated from the color temperature, and the results are that the ratio of visual to infrared optical depths is about 500 at 100µm and 5000 at 900µm. These values are, however, very uncertain, and they are not both consistent with the emissivity function described below.

The variation of emissivity with wavelength has been determined for a few objects and is better known than the emissivity itself. Beyond 300µm the slope of the emissivity law can be deduced unambiguously provided all dust within the beam is hot enough to emit according to the Rayleigh-Jeans law. The emissivity seems to drop at least as steeply with increasing wavelength as $\varepsilon \propto \lambda^{-2}$ (Keene 1981, Gezari 1982, Righini et al. 1976). Observations of two thermally emitting galactic nuclei (Elias et al. 1978) are consistent with a shallower slope, but the beam necessarily covers a large linear size within which the assumption of no temperature variation undoubtedly fails.

At shorter wavelengths, a shallower dependence such as $\varepsilon \propto \lambda^{-1.3}$ has been suggested (Pipher et al. 1978, Herter et al. 1979, Thronson and Harper 1979, Erickson and Tokunaga 1980). There are differences among different sources and even within the same source observed with different beam sizes, however, so the emissivity law can hardly be considered securely known. In view of the difficulty of determining the temperature and the effects of temperature variations, considerable uncertainty is to be expected. Part of the uncertainty will be removed by observations with high spatial resolution if they ever become possible.

The above results were all obtained for dust in regions that are well mixed parts of the interstellar medium. It should be remembered that particular objects that have generated their own dust may have different emissivity laws (Moseley 1980), just as such objects do at shorter wavelengths. Emission features are discussed in section IV.

D. Other Absorption Features

There are a few other absorption features that are thought to be characteristic of the general interstellar medium because they occur in the line of sight to the galactic center, which is thought to suffer mostly the standard interstellar absorption (Becklin et al. 1978). The best known is a resolved feature near 3.39μm (Willner et al. 1979b, Wickramasinghe and Allen 1980), which has also been seen towards an OH source. The peak absorption at 3.4μm is 0.03 $\tau(9.7)$ in the case of the galactic center. Several other features between 3.0 and 3.86μm have been reported for one galactic center source (Allen and Wickramasinghe 1981). Published spectra (Willner and Pipher 1982) for 3 other galactic center sources confirm only a broad dip centered near 2.95 or 3.00μm in addition to the band near 3.4μm. Further study of these features in sources lacking 3.1μm ice absorption is necessary. Such sources generally are ones suffering very low extinction, so very high signal to noise is required to detect the bands. The difficulty is illustrated by the spectrum (Gillett et al. 1975) of Cygnus OB2 #12, where the 3.4-μm band is at best marginally present.

III. MOLECULAR CLOUD ABSORPTION FEATURES

If spectral features are observed in only limited regions of space, they are considered characteristic of grains that formed or have been processed in that region. A well known example is the case of dust condensation in circumstellar shells. At the relatively high temperatures and densities of the circumstellar environment where dust condenses, carbon monoxide is the most stable form of both carbon and oxygen. As a result, the less abundant of the two elements is completely associated with the other and is not available to form grains. Circumstellar shells thus can be uniquely characterized as carbon rich or oxygen rich, and the infrared dust signatures reflect this dichotomy (Merrill 1977).

The carbon or oxygen rich dichotomy that applies to circumstellar shells may not apply to other environments. In particular, in molecular clouds the density is much lower, and there is no reason to expect CO to be completely associated. Indeed observations (Knapp et al. 1976, Dickman 1978) indicate that it is not, and both carbon and oxygen should be available to form or be processed with dust grains. Furthermore, the interstellar medium, including molecular clouds, presumably contains a mixture of grains formed in many different environments. Such grains include types that formed in both oxygen and carbon rich environments and perhaps grains that formed in places that cannot be characterized as either.

Unlike the case for emission features, a material producing an absorption feature must constitute a substantial fraction of the material in the line of sight. Such features are thus diagnostic of the composition of the material in the line of sight. Often the material may be in a molecular cloud, in which case infrared observations can be compared to radio molecular emission observations. Such comparisons may be misleading, however, because the two types of observation do not sample the same material. The absorption features sample the line of sight to the source whose light is being absorbed, and if that source is also a region of enhanced density, the regions near the source will be weighted heavily in the observations. The emission observations, on the other hand, sample a large volume with nearly equal weight. If the emission line happens to be optically thick, high density regions may even receive lower weight than the rest of the cloud.

TABLE 1

OBSERVED ABSORPTION FEATURES

λ (μm)	Identification	Where observed
~9.7	Silicates	Any long enough path – stars,
~18	"	galactic center, H II regions, etc.
3.08	H_2O Ice	Molecular cloud sources + 1 c.s. shell
6.0	"	Molecular cloud sources + 1 c.s. shell
12.5	"	1 c.s. shell – not mol. cloud sources
42	"	Molecular cloud source
2.95	NH_3 Ice	Mol. cloud sources – wing of 3.08-μm band
3.3–3.6	NH_3+H_2O Ice	Mol. cloud sources – wing of 3.08-μm band
6.8	?	Molecular cloud sources
4.67	CO Ice	A few mol. cloud sources, not most
4.61	?	ratio of 4.67/4.61 not constant
3.9	?	1 source

Table 1 gives a summary of all infrared absorption features that have been attributed to dust. Details are given below for features other than the silicate feature, which was discussed in the previous section.

A. Water Ice

There is one infrared source that undoubtedly displays absorption features of water ice. It is usually referred to as OH 0739-14 and is a star undergoing rapid mass loss. The most recent spectral study (Soifer et al. 1981) found three absorption features at 3.1, 6.0, and near 12μm matching the laboratory spectrum of amorphous water ice. The source is an OH maser, as indicated by the designation, and there is no reason to believe that ice cannot condense in the outer part of the circumstellar shell. In fact, the puzzle is why this source is so far unique. No other star that is thought to be evolved is known to show ice absorption, and the shape of the 3.1-μm feature and the presence of the 12-μm feature are unique among all known infrared spectra.

An absorption band at 3.1μm is commonly seen in sources that are embedded in or lie behind molecular clouds (Merrill 1979, Willner et al. 1982). No such absorption is seen toward a heavily reddened star not behind a molecular cloud (Gillett et al. 1975) or toward the galactic center (Soifer et al. 1976). The band must therefore be due to some material that only exists within molecular clouds.

The wavelength of peak absorption of the molecular cloud 3.1-μm band agrees with that of the ice band seen in OH 0739-14. The short wavelength wings are hard to measure, but they probably agree also. Unlike the ice band in OH 0739-14, the molecular cloud band has a long wavelength wing extending to about 3.6μm. The molecular cloud sources also show an absorption band at 6.0μm indistinguishable from the band in OH 0739-14 and correlated in depth with that of the 3.1-μm band. At 12μm, the molecular cloud sources have no band corresponding to the one in OH 0739-14.

The 3.1 and 6.0-μm bands are usually identified as water ice because of the wavelength agreement of the absorption peaks with the laboratory spectra and because there is no other likely candidate. One problem with identifying the 3.1-μm absorptions as due to water ice is that the wing at 3.3 to 3.6μm is unexpected. There seem to be two possibilities: either another absorber is required at these wavelengths or very large ice particles, which would scatter efficiently (Leger et al. 1983) at these wavelengths, are needed. The 3.4-μm absorption feature discussed in section II.D may contribute to the absorption, but the ratio of 3.4-μm absorption to silicate absorption would have to be larger than in the general interstellar medium.

There are several pieces of evidence that there is indeed another absorber than water ice between 3.3 and 3.6μm. The first is that if the laboratory ice absorption spectrum is subtracted from a typical molecular cloud spectrum, the shape of the residual absorption (Whittet et al. 1983) resembles the absorption of a galactic center source. Many molecular cloud sources show a similar shape for the 3.1 to 3.6-μm absorption, having an inflection in their spectra at 3.30μm (Willner et al. 1982). Furthermore, in one sample of sources the 3.4-μm optical depth is better correlated with $\tau(9.7)$ than is the 3.1-μm optical depth. Finally, if scattering were important in the wing of the absorption feature, one would expect the ratio of polarization to extinction to be different at wavelengths where scattering is and is not important (Hagen et al. 1983). The wavelength dependence of the polarization of the Becklin-Neugebauer object has been measured (Kobayashi et al. 1980), and neither the ratio of polarization to extinction nor the position angle of the polarization changes significantly between 3.1 and 3.4µm.

One additional problem for identification of the 3.1 and 6.0-µm absorptions as water ice is the absence of the absorption feature near 12µm that is seen in laboratory spectra (Leger et al. 1983). If any such feature is present, it must have an optical depth less than 1/15 of $\tau(3.1)$ (Willner et al. 1982). Only if laboratory work suggests plausible physical conditions under which the 12-µm feature can be this weak can the water ice identification for the molecular cloud 3.1 and 6.0-µm absorptions be accepted. Furthermore, the conditions suggested must not apply to OH 0739-14.

Another band that may be due to water ice is a weak absorption feature at 42µm observed in only one object (Erickson et al. 1981). Although the feature is only about 5% deep, it appears in several independent data sets and is probably real. Similar observations of other objects that show 3.1- and 6.0-µm absorption at shorter wavelengths are much to be desired.

It has recently been shown (Knacke et al. 1982) that some molecular cloud sources have additional absorption near 2.97µm. This band is attributed to ammonia ice on the basis of its wavelength, but the band has been measured in too few sources to make strong statements about any correlation between its strength and that of other bands. A mixture of ammonia and water ice may also explain the 3.3- to 3.6-µm wing of the 3.1-µm band (Knacke et al. 1982, Allamandola 1983).

Studies of two particular molecular clouds (Harris et al. 1978, Whittet et al. 1982) have shown that the 3.1-µm band appears only when the extinction exceeds some critical value. The value required is lower in the Taurus Molecular Cloud than in the ρ Ophiuchi molecular cloud. The scale size for the clumps of long-chain hydrocarbon molecules seen in the TMC is about 0.1 pc or less (Avery 1980), smaller than the typical scale size in the ρ Oph cloud (Harris et al. 1978). The data for these two clouds are consistent with the conjecture by Harris et al. that 3.1-µm absorption appears at a critical value of cloud density, or equivalently extinction per linear distance. The criterion for the appearance of 3.1-µm absorption is unlikely to be total extinction alone, because the feature has recently been detected in a path with only 2.5 magnitudes of visual extinction (Goebel 1983). The path crosses the edge of a molecular cloud, where ice is evidently present.

B. Carbon Monoxide

The fundamental absorption of the carbon monoxide molecule is at 4.67µm, and a number of objects have absorptions near this wavelength. Some of these have now definitely been identified as solid and others as gaseous CO (McGregor et al. 1983). A broad feature is centered near 4.62μ and a narrower feature is centered near 4.67μm. The ratio of the feature depths is different in different objects, and with the limited sample available, the depths and ratios of the features appear better correlated with the ice/silicate ratio than with either the ice or silicate optical depths individually. Gas phase CO is probably present in at least some of the lines of sight that show the solid CO.

C. Unidentified Feature at 6.8μm

This absorption feature appears in the spectra of a large number of sources seen through molecular cloud material (Willner et al. 1982). It is not seen in the spectrum of OH 0739-14 and so is presumably not due to water ice. The relative abundance of the species producing the 6.8-μm feature is correlated with the relative abundance of the species producing the center and short wavelength edge, but not the long wavelength wing, of the 3.1-μm ice absorption.

IV. DUST EMISSION FEATURES

Emission features can be seen only when the optical depth in the continuum is no more than of order unity. The feature is seen where the emissivity and thus the optical depth are higher than at nearby wavelengths. The silicate feature is one of the most ubiquitous examples.

Even minor grain constituents may be seen in emission in the right circumstances. At short infrared wavelengths and low temperatures characteristic of many infrared objects, the Planck function is a strong function of temperature. Any grain constituent that reaches an equilibrium temperature hotter than average will emit a disproportionate share of the radiation. If such a constituent radiates in a relatively narrow spectral bandwidth, that band will appear in emission even if the abundance of the constituent is small. Thus the appearance of emission features that dominate the spectrum does not necessarily imply that the constituent that produces them is a major contributor to the mass.

Table 2 gives a summary of all of the infrared emission features that have been attributed to dust. Details are given below.

A. Oxygen Rich Features - Silicates

Both silicate emission and absorption are commonly seen in the spectra of cool, oxygen rich stars (Merrill 1977, Forrest et al. 1975). The 10-μm feature is seen in emission in stars with optically thin circumstellar shells and in absorption in stars with thick shells. The 18-μm feature is seen in emission. A possible emission feature near 33μm has also been reported on the basis of ground based broad band photometry (Hagen et al. 1975). No such feature was seen in the one candidate star examined at higher spectral resolution, but there are 3

TABLE 2

OBSERVED EMISSION FEATURES

λ (μm)	Identification	Where observed
~9.7	Silicates	O>C c.s. shells, H II regions, p.n., etc.
~18	"	O>C c.s. shells
33	"	O>C c.s. shells - not confirmed
11.5	Silicon Carbide	C>O c.s. shells, planetary nebulae
25- ~45	?	C>O c.s. shells, planetary nebulae
3.28, 3.40, 6.20, 7.7, 8.6, 11.29	?	Ionization fronts
3.28	?	IC 418 - other features missing
3.41, 3.43, 3.53	H_2CO	2 pre-main sequence stars
4.5	CO or C_3	Nova before dust formed

more stars that should be checked.

While the shape of the circumstellar 10-μm absorption feature is similar to that of the interstellar absorption feature, circumstellar emission features often appear to be narrower than absorption features (Merrill 1979). This has been explained as the absence of a featureless absorber in the circumstellar shells while such an absorber occurs in the interstellar medium. Such an absorber is seen around carbon rich stars and presumably is mixed with the silicate dust in the interstellar medium. Why stars with thick, absorbing silicate dust shells show the same spectral shape as the interstellar feature is not explained, but it may be that the observed spectrum is simply not sensitive to neutral absorption. This could be tested with numerical models.

The shapes of both the 10- and 18-μm emission features in a few stars may differ from the usual shapes (Forrest et al. 1975, 1979). It is not yet established whether the spectral differences are due to different grain properties, radiative transfer effects of different density or temperature structures in the circumstellar shells, or differences in the underlying photospheric energy distributions.

In addition to stars, silicate emission is seen from some planetary nebulae (Aitken and Roche 1982) and from some nearby H II regions such as the Orion Trapezium. In this object, the feature may have been diluted by the featureless dust component hypothesized to occur in the interstellar medium, because the contrast between the emission feature and the continuum is less than in many circumstellar shells that hypothetically represent pure silicate emission (Aitken et al. 1980).

B. Carbon Rich Features

There are two spectral features observed from circumstellar shells around stars that are known to be carbon rich, i.e. the carbon abundance exceeds the oxygen abundance. The first feature to be discovered (Treffers and Cohen 1974) is a broad emission between 10.3 and 12.7μm. The feature will be referred to as the "11-μm feature". It is narrower and centered at a different wavelength than the silicate feature discussed above, and the two features are easily distinguished in low resolution spectra. The spectra of several cool stars with 11-μm emission are shown in the review by Merrill (1977). One carbon star with the thickest known dust shell may show the 11-μm feature in absorption (Jones et al. 1978). The 11-μm feature is identified as due to silicon carbide on the basis of its wavelength and condensation models. Carbon stars with circumstellar shells also show evidence of considerable dust that has no discrete spectral features (Merrill 1977). The two constituents appear to have varying ratios in different sources and thus are probably physically distinct.

A feature similar to the 11-μm feature but extending to a slightly longer wavelength has been seen in the spectra of some planetary nebulae (Willner et al. 1979a, Aitken et al. 1979b, Aitken and Roche 1982). The nebulae where the feature has been seen are symmetric, extended, classical-appearing planetaries. Where there are good determinations, the carbon abundances in the nebulae are higher than the oxygen abundances, but the central stars do not necessarily show evidence of being carbon rich. In contrast, planetary nebulae that show silicate emission tend to be young and have high densities, and their carbon abundances are lower than their oxygen abundances.

Another emission feature is observed at longer wavelengths (Forrest et al. 1981). It begins at 23.5μm and extends to between 40 and 50μm, although the long wavelength limit is uncertain (Herter et al. 1982, Moseley and Silverberg 1983). The feature has been seen in carbon stars with a wide range of dust shell thickness and in 3 planetary nebulae, all of which show the 11-μm feature.

C. Family of 6 Unidentified Bands

Six emission bands were first found in the planetary nebula NGC 7027, although at different times by different investigators. Subsequent work has shown that the bands occur together in a variety of different kinds of objects. The bands occur at wavelengths of 3.28, 3.4, 6.2, 7.7, 8.6, and 11.3μm and have widths of about 5% of the band wavelength, with the 7.7-μm band being about twice that broad. The properties of the various bands has recently been reviewed (Aitken 1981), and here I will merely attempt to summarize the results and include more recent observations.

One key question is whether the bands always occur together or can occur singly or as smaller sets. This is actually a much more difficult question to answer on the basis of existing data than it might seem. While there are a number of instances (Aitken 1981) where bands apparently are missing or where the ratios of the various bands are different from the ratios in NGC 7027, closer examination may reveal that the observations are not clearcut. The typical problems include: 1) measurements of different spectral regions with different beam sizes or in slightly different positions and 2) the presence of a strong continuum that reduces the feature contrast and makes measurements of weak features unreliable. A third problem applies only to the 3.28-μm feature: the hydrogen Pfund δ line at 3.30μm may contribute substantial flux as in the case of NGC 6572 (Willner et al. 1979a). When these problems are taken into account, the 6 features appear very often to occur together and form a family.

Having hypothesized that the 6 features indeed form a family, it is necessary to investigate any cases where any of the features are expected but absent. The examples considered here must necessarily be illustrative rather than exhaustive. The earliest suggested case is the planetary nebula IC 418, which shows emission at 3.28μm but in none of the other features (Russell et al. 1977). None of the effects mentioned above seems to be responsible for this case. On the other hand, initial observations of the H II region G75.84+0.4 (Pipher et al. 1979) showed the 3.3 and 3.4-μm features but not the ones at 8.6 and 11.3μm. Additional observations (Herter et al. 1981a) have shown that the 6.2 and 7.7μm features are present. The continuum flux density is a strong function of beam size. The initial 8 to 13μm spectrum was taken at a different position than the 2 to 4μm spectrum, and it seems likely that a spectrum at the right position would reveal the 8.6- and 11.3-μm features. AFGL 3053 is a region where the features all appear, but the 6.2 and 7.7-μm features seem anomalously strong (Aitken 1981). Here, however, the 4 to 8-μm spectrum was taken with a substantially larger beam than was used for the other spectral regions. The case of HD 44179 illustrates the problem of low contrast for the 3.4-μm feature. If the feature fluxes are normalized to that of the 6.2-μm feature, the 3.28-μm feature is a factor of 3 stronger than in NGC 7027. A small emission bump can be clearly seen near 3.4μm (Russell et al. 1977), and although it rises only 1 or 2% above the continuum, its relative flux about as great as that of the 3.4-μm feature in NGC 7027. The star HD 97048 shows unique spectral features and is discussed below, but the best evidence indicates that the family of spectral features is present in near-normal ratios.

Considering the available observations, it seems that any assertion that the 6 features listed above do not form a family still has the burden of proof. This is not to say that the features all are due to a single material, but they are probably due to materials that form and are destroyed together. In this view, the 3.3-μm feature seen alone in

IC 418 may either be completely unrelated to the 3.28-μm feature seen with the rest of the family, or it may be that IC 418 presents a sufficiently unusual environment that only a single one of the normal combination of materials is present. Observations with higher spectral resolution might help to decide between these possibilities.

The ingredients necessary to produce the emission features seem to be dusty neutral material and ultraviolet radiation. The features are seen with the largest equivalent width known in an ionization front just outside the ionized region of the Orion nebula (Aitken et al. 1979a, Sellgren 1981). The energy distribution of the ultraviolet radiation does not seem to matter; emission features are seen from NGC 7027, which has a central star temperature >125000 K, and HD 44179, which is of spectral type B9-A0 (Cohen et al. 1975). This comparison may be a bit misleading, because few hydrogen ionizing photons are likely to escape from an H II region. Nevertheless, a broad range of wavelengths, perhaps even including some in the visible part of the spectrum, are apparently capable of exciting the features (Aitken and Roche 1983). A number of stars show 3.3-μm emission (Allen et al. 1982), and it would be interesting to search for CO or other molecular emission from such objects to see to what extent molecular gas is necessary.

Although it would be useful to characterize the grains associated with the family of emission features as either oxygen or carbon rich, that cannot now be done. The features occur mostly in objects associated with interstellar rather than circumstellar molecular clouds. As discussed above, such material cannot be uniquely characterized as oxygen or carbon rich. Many planetary nebulae that show the features are known to be carbon rich in the gas, but M1-11 is the only nebula that unambiguously shows both the emission features and the 11-μm feature characteristic of carbon stars (Aitken and Roche 1982), thus implying that the dust is also carbon rich. No objects having both feature and silicate emission from the same spatial region are known, but the star HD 44179, which shows feature emission, has an oxygen rich photosphere. If, however, the star is a pre-main sequence object, the dust is presumably a concentration from the interstellar medium rather than having been manufactured in the circumstellar shell. The possibility that the binary companion is carbon rich also cannot be ruled out.

The peak wavelength of the 3.28-μm feature has been suggested to be different in different objects (Allen et al. 1982). Within a single data set, where wavelength calibration is less of an issue, this conclusion rests on a single object, and the wavelength difference from an average of 3 other objects is only 0.3 of a resolution element. The average peak wavelength is 3.293μm, compared to 3.289 ± 0.005μm from higher resolution observations of two objects (Tokunaga and Young 1980) and 3.27 ± 0.01μm from lower resolution data. There are a number of narrow atmospheric absorptions shortwards but not longwards of the peak of the 3.28μm feature, and these complicate any determination of the peak wavelength. Higher spectral resolution observations are particularly difficult to correct. Finally, different continuum slopes may shift the position of peak flux density. The available evidence thus does not seem sufficient to confirm differences in the peak wavelength of the feature.

Finally, the question of the excitation of the features has been well reviewed by Aitken (1981). One further observational point that may be relevant is that in some sources that emit the features, the continuum suggests rather high grain temperatures, above 1000 K in at least one case (Cohen et al. 1975). The possible excitation mechanisms appear to be ultraviolet fluorescence and thermal emission from abnormally small, thus abnormally hot, grains. The fluorescence hypothesis requires photon emission efficiencies above 1 in some cases (Dwek et al. 1980, Aitken and Roche 1983), reasonable efficiencies must be maintained for high grain temperatures, and the insensitivity to the ultraviolet energy distribution must be explained. On the other hand, the thermal hypothesis should admit the possibility of emission by cool grains that emit at, say, 11.3μm but not at 3.28μm. No such case has been observed.

D. 3.5-μm Feature in HD 97048

Blades and Whittet (1980) found a 3.5-μm emission feature, together with features at 3.3 and 3.4μm, in the spectrum of the pre-main sequence star HD 97048. Subsequent observations (Baas et al. 1983) have shown that the feature is centered at 3.53μm and has a wing at shorter but not longer wavelength. The 3.4-μm feature is double, with blended peaks at 3.41 and 3.43μm, and a shoulder at 3.40μm. At longer wavelengths, the 8.6- and 11.3-μm features are present (Aitken and Roche 1981), as is probably the long wavelength wing of the 7.7-μm feature. (The 4- to 8-μm spectrum has not been observed.) Thus the entire family of features discussed above is likely to be present.

On the basis of wavelength, either the shoulder or the 3.41-μm peak, but not the 3.43-μm peak, can be identified with the 3.4-μm feature in the family discussed above. Identification with the weaker shoulder makes the ratios of the features in the family consistent with their ratios in other objects. An attempt should therefore be made to identify species producing emission peaks at 3.41, 3.43, and 3.53μm. Baas et al. (1983) have suggested formaldehyde ices embedded in cold grain mantles. They have also shown that if this identification is correct, the excitation of the features is likely to be fluorescent rather than thermal.

In spite of a fairly extensive search (Allen et al. 1982), including spectra of many apparently similar sources, the emission features seen in HD 97048 remain almost unique. Weak features are seen in the spectra of Elias 1, a source in the Taurus molecular cloud (Elias 1978), and TY CrA (Whittet et al. 1982). Both sources, like HD 97048, illuminate reflection nebulae. A transient feature at 3.5μm was seen in the spectrum of Nova Cygni 1975 (Grasdalen and Joyce 1976), but at that early stage of the outburst, the physical conditions in the envelope indicated by the infrared spectrum were much different than those in the reflection nebula sources. Evidently very special conditions are needed to produce these features.

E. 4.5-μm Feature in Nova Vulpecula 1976

The observations of this feature are reviewed by Merrill (1977). The feature disappeared just prior to the formation of infrared featureless dust grains. Possible identifications for the featureless grains are graphite or amorphous carbon. One explanation for the feature is C_3 emission, while another possibility is emission in the fundamental band by gas phase CO. Emission by CO in the overtone band at 2.3μm was seen at about the same time by Ferland et al. (1979), and they inferred that the fundamental band emission should be strong enough to account for the observed emission feature.

V. CONCLUSIONS

Infrared astronomers have now identified a number of spectral features that must be attributed to dust, although such an attribution may be in doubt in a few cases. The situation for normal stars with circumstellar shells is satisfactory, but only by comparison with the situation for other objects. Even for the stars, not all of the spectral features can be identified with particular grain constituents. There is also insufficient knowledge of the actual dependence of emissivity on wavelength. Studies of possible differences in the emissivity spectra of different objects have just begun, and the causes of such differences are still unknown.

Understanding of the features that are not seen in circumstellar shells is still rudimentary. Some features, such as water ice, have been identified with a reasonable degree of confidence, but for many others identifications are lacking. The types of regions in space that produce some features are becoming known, but for many features only a handful of cases are known.

Improved understanding of interstellar dust will require improvements in several areas. 1) More laboratory spectra are needed, as is a better understanding of the detailed origin of particular spectral features. Evidence is accumulating that the details of dust processing are important, as well as simply the bulk composition. It will be impossible to take spectra of all possible constituents in all possible mixtures with every possible amount and type of processing, so only a detailed theory will ultimately suffice. 2) A much larger sample of celestial spectra is needed. Many conclusions are presently based on spectra of only one or two objects, and many more are needed for any confident conclusions about what conditions give rise to particular spectral features. 3) Astronomical spectra at higher resolution are needed. Even where features are resolved at low resolution, the detailed shape measured with high signal to noise may give clues to the origin of the features. Evidence of differences from object to object are especially important. In many but not all important spectral regions, such observations may be very difficult to accomplish from the ground.

ACKNOWLEDGEMENTS

It is a pleasure to thank D. K. Aitken, R. F. Knacke, P. J. McGregor, A. Natta, J. L. Pipher, and W. A. Traub for useful discussions and comments on the manuscript. Thanks also to the organizers of the Symposium for the opportunity to hear of the latest advances in this and other fields and to present this review. I also thank the Secretary of the Smithsonian Institution for financial support from the Fluid Research Fund.

REFERENCES

Aitken, D. K.: 1981, in *Infrared Astronomy*, IAU Symposium No. 96, eds. C. G. Wynn-Williams and D. P. Cruikshank (Dordrecht:Reidel), pp. 207-211.
Aitken, D. K. and Roche, P. F.: 1981, Monthly Notices Roy. Astron. Soc. 196, pp. 39P-44P.
Aitken, D. K. and Roche, P. F.: 1982, Monthly Notices Roy. Astron. Soc. 200, pp. 217-237.
Aitken, D. K. and Roche, P. F.: 1983, Monthly Notices Roy. Astron. Soc. 202, in press.
Aitken, D. K., Roche, P. F., and Spenser, P. M.: 1980, Monthly Notices Roy. Astron. Soc., 193, pp. 207-212.
Aitken, D. K., Roche, P. F., Spenser, P. M., and Jones, B.: 1979a, Astron. Astrophys. 76, pp. 60-64.
Aitken, D. K., Roche, P. F., Spenser, P. M., and Jones, B.: 1979b, Astrophys. J. 233, pp. 925-934.
Allamandola, L. J.: 1984, *Galactic and Extragalactic Infrared Spectroscopy*, XVIth ESLAB Symposium, eds. M. F. Kessler, and J. P. Phillips (Dordrecht: Reidel). This volume.
Allen, D. A., Baines, D. W. T., Blades, J. C., and Whittet, D. C. B.: 1982, Monthly Notices Roy. Astron. Soc. 199, pp. 1017-1024.
Allen, D. A. and Wickramasinghe, D. T.: 1981, Nature 294, pp. 239-240.
Avery, L. W.: 1980, in *Interstellar Molecules*, IAU Symposium No. 87, ed. B. H. Andrew, (Dordrecht:Reidel) pp. 47-58.
Baas, F., Allamandola, L. J., Geballe, T. R., Persson, S. E., and Lacy, J. H.: 1983, Astrophys. J. 265, in press.
Becklin, E. E., Matthews, K., Neugebauer, G., and Willner, S. P.: 1978, Astrophys. J. 220, pp. 831-835.
Blades, J. C. and Whittet, D. C. B.: 1980, Monthly Notices Roy. Astron. Soc.: 191, pp. 701-709.
Breger, M., Gehrz, R. D., and Hackwell, J. A.: 1981, Astrophys. J. 248, pp. 963-976.

Brocklehurst, M.: 1971, Monthly Notices Roy. Astron. Soc. 153, pp. 471-490.
Cohen, M., Anderson, C. M., Cowley, A., Coyne, G. V., Fawley, W. H., Gull, T. R., Harlan, E. A., Herbig, G. H., Holden, F., Hudson, H. S., Jakoubek, R. D., Johnson, H. M., Merrill, K. M., Schiffer, F. H. III, Soifer, B. T., and Zuckerman, B.: 1975, Astrophys. J. 196, pp. 179-189.
Dickman, R. L.: 1978, Astrophys J. Suppl. 37, 407-427.
Dwek, E., Sellgren, K., Soifer, B. T., and Werner, M. W.: 1980, Astrophys. J. 238, pp. 140-147.
Elias, J. H.: 1978, Astrophys. J. 224, pp. 857-872.
Elias, J. H., Ennis, D. J., Gezari, D. Y., Hauser, M. G., Houck, J. R., Lo, K. Y., Matthews, K., Nadeau, D., Neugebauer, G., Werner, M. W., and Westbrook, W. E.: 1978, Astrophys. J. 220, pp. 25-41.
Erickson, E. F., Knacke, R. F., Tokunaga, A. T., and Haas, M. R.: 1981, Astrophys. J. 245, pp. 148-153.
Erickson, E. F. and Tokunaga, A. T.: 1980, Astrophys. J. 238, pp. 596-600.
Ferland, G. J., Lambert, D. L., Netzer, H., Hall, D. N. B., and Ridgway, S. T.: 1979, Astrophys. J. 227, pp. 489-496.
Forrest, W. J., Gillett, F. C., and Stein, W. A.: 1975, Astrophys. J. 195, pp. 423-440.
Forrest, W. J., Houck, J. R., and McCarthy, J. F.: 1981, Astrophys. J. 248, pp. 195-200.
Forrest, W. J., McCarthy, J. F., and Houck, J. R.: 1979, Astrophys. J. 233, pp. 611-620.
Gezari, D. Y.: 1982, Astrophys. J. 259, pp. L29-L33.
Giles, K.: 1977, Monthly Notices Roy. Astron. Soc. 180, pp. 57P-59P.
Gillett F. C., Jones, T. W., Merrill, K. M., and Stein, W. A.: 1975, Astron. Astrophys. 45, p. 77-81.
Goebel, J. H.: 1983, Astrophys. J. (Letters) 268, L41-45.
Grasdalen, G. L. and Joyce, R. R.: 1976, Nature, 259, pp. 187-189.
Hackwell, J. A., Gehrz, R. D., Smith, J. R., and Briotta, D.A.: 1978, Astrophys. J. 221, pp. 797-809.
Hagen, W., Simon, T., and Dyck, H. M.: 1975, Astrophys. J. 201, pp. L81-L84.
Hagen, W., Tielens, A. G. G. M., and Greenberg, J. M.: 1983, Astron. Astrophys. 117, pp. 132-140.
Harris, D. H., Woolf, N. J., and Rieke, G. H.: 1978, Astrophys. J. 226, pp. 829-838.
Harvey, P. M., Thronson, H. A. Jr., and Gatley, I.: 1980, Astrophys. J. 235, pp. 894-898.
Hawley, S. A. and Duncan, D. K.: 1976, Publ. Astron. Soc. Pacific 88, pp. 672-676.
Herter, T., Briotta, D. A. Jr., Gull, G. E., and Houck, J. R.: 1982, Astrophys. J. 259, pp. L25-L27.
Herter, T., Duthie, J. G., Pipher, J. L., and Savedoff, M. P.: 1979, Astrophys. J. 234, pp. 897-901.
Herter, T., Helfer, H. L., Pipher, J. L., Forrest, W. J., McCarthy, J., Houck, J. R., Willner, S. P., Puetter, R. C., Rudy, R. J., and Soifer, B. T.: 1981a, Astrophys. J. 250, pp. 186-199.

Herter, T., Pipher, J. L., Helfer, H. L., Willner, S. P., Puetter, R. C., Rudy, R. J., and Soifer, B. T.: 1981b, Astrophys. J. 244, pp. 511-516.
Hulst, H. C. van de: 1946, Recherches Astron. de l'Observatoire d'Utrecht 11, pp. 1946-1949.
Johnson, H. L.: 1968, in Nebulae and Interstellar Matter, eds. B. M. Middlehurst and L. H. Aller (Chicago:University of Chicago Press), pp. 193ff.
Johnson, H. L.: 1977, Rev. Mexicana Astr. Astrofis. 2, pp. 175-180.
Jones, B., Merrill, K. M., Puetter, R. C., and Willner, S. P.: 1978, Astron. J. 83, pp. 1437-1439.
Jones, T. J. and Hyland, A. R.: 1980, Monthly Notices Roy. Astron. Soc. 192, pp. 359-364.
Keene, J.: 1981, Astrophys. J. 245, pp. 115-123.
Knacke, R. F. and Capps, R. W.: 1979, Astron. J. 84, pp. 1705-1708.
Knacke, R. F., McCorkle, S., Puetter, R. C., Erickson, E. F., and Krätschmer, W.: 1982, Astrophys. J. 260, pp. 141-146.
Knapp, G. R., Kuiper, T. B. H., and Brown, R. L.: 1976, Astrophys. J. 206, pp. 109-113.
Kobayashi, Y., Kawara, K., Sato, S., and Okuda, H.: 1980, Publ. Astron. Soc. Japan 32, pp. 295-302.
Lee, T. A.: 1968, Astrophys. J. 152, pp. 913-941.
Lee, T. A.: 1970, Astrophys. J. 162, pp. 217-238.
Leger, A., Gauthier, S., Defourneau, D., and Rouan, D.: 1983, Astron. Astrophys., in press.
Lester, D. F.: 1979, Bul. Amer. Astron. Soc. 11, p. 399.
McCarthy, J. F., Forrest, W. J., Briotta, D. A. Jr., and Houck, J. R.: 1980, Astrophys. J. 242, pp. 965-975.
McGregor, P. J., Persson, S. E. Lacy, J. H., Baas, F., Allamandola, L., Lonsdale, C. J., and Geballe, T. R.: 1982, Galactic and Extragalactic Infrared Spectroscopy, XVIth ESLAB Symposium, eds. M. F. Kessler, J. P. Phillips and T. D. Guyenne. ESA-SP-192, pp. 51-55.
Merrill, K. M.: 1977, in The Interaction of Variable Stars with their Environment, IAU Colloquium No. 42, eds. R. Kippenhahn, J. Rahe, and W. Strohmeier (Bamberg:Veroffentlichungen der Remeis-Sternwarte Bamberg Bd. XI, No. 121), pp. 446ff.
Merrill, K. M.: 1979, Astrophys. Space Sci. 65, pp. 199-214.
Mezger, P. G., Mathis, J. S., and Panagia, N.: 1982, Astron. Astrophys. 105, pp. 372-388.
Moseley, H.: 1980, Astrophys. J. 238, pp. 892-904.
Moseley, H. and Silverberg, R. F.: 1983, Bull. Am. Astron. Soc. 14, p. 893.
Natta, A. and Panagia, N.: 1982, Galactic and Extragalactic Infrared Spectroscopy, XVIth ESLAB Symposium, eds. M. F. Kessler, J. P. Phillips, and T. D. Guyenne. ESA-SP-192, pp. 79-81.
Pipher, J. L., Duthie, J. G., and Savedoff, M. P.: 1978, Astrophys. J. 219, pp. 494-497.
Pipher, J. L., Soifer, B. T., and Krassner, J.: 1979, Astron. Astrophys. 74, pp. 302-307.
Rieke, G. H.: 1974, Astrophys. J. 193, pp. L81-L82.
Righini, G., Simon, M., and Joyce, R. R.: 1976, Astrophys. J. 207, pp.

119-125.
Russell, R. W., Soifer, B. T., and Morrill, K. M.: 1977, Astrophys. J. 213, pp. 66-70.
Sandage, A.: 1975, Publ. Astron. Soc. Pacific 81, pp. 853-857.
Savage, B. D. and Mathis, J. S.: 1979, Ann. Rev. Astron. Astrophys. 17, pp. 73-111.
Schultz, G. V. and Wiemer, W.: 1975, Astron. Astrophys. 43, pp. 133-139.
Sellgren, K.: 1981, Astrophys. J. 245, pp. 138-147.
Smith, H. A., Larson, H. P., and Fink, U.: 1981, Astrophys. J. 244, pp. 835-843.
Smith, J., Lynch, D. K., Cudaback, D., and Werner, M. W.: 1979, Astrophys. J. 234, 902-908.
Sneden, C., Gehrz, R. D. Hackwell, J. A., York, D. G., and Snow, T. P.: 1978, Astrophys. J. 223, pp. 168-179.
Soifer, B. T., Russell, R. W., and Merrill, K. M.: 1976, Astrophys. J. 207, pp. L83-L85.
Soifer, B. T., Willner, S. P., Capps, R. W., and Rudy, R. J.: 1981, Astrophys. J. 250, pp. 631-635.
Tapia, M.: 1981, Monthly Notices Roy. Astron. Soc. pp. 949-965.
Thronson, H. A. Jr. and Harper, D. A.: 1979, Astrophys. J. 230, pp. 133-148.
Tokunaga, A. T. and Young, E. T.: 1980, Astrophys. J. 237, pp. L93-L96.
Treffers, R. and Cohen, M.: 1974, Astrophys. J. 188, pp. 545-552.
Turner D. G.: 1976, Astrophys. J. 210, pp. 65-75.
Whittet, D. C. B., Davies, J. K., Bode, M. F., Evans, A., and Longmore, A. J.: 1982, **Galactic and Extragalactic Infrared Spectroscopy**, XVIth ESLAB Symposium, eds. M. F. Kessler, J. P. Phillips, and T. D. Guyenne. ESA-SP-192, pp. 59-64.
Wickramasinghe, D. T. and Allen, D. A.: 1980, Nature 287, pp. 518-519.
Willner, S. P., Gillett, F. C., Herter, T. L., Jones, B., Krassner, J., Merrill, K. M., Pipher, J. L., Puetter, R. C., Rudy, R. J., Russell, R. W., and Soifer, B. T.: 1982, Astrophys. J. 253, pp. 174-187.
Willner, S. P., Jones, B., Puetter, R. C., Russell, R. W., and Soifer, B. T.: 1979a, Astrophys. J. 234, pp. 496-502.
Willner, S. P. and Pipher, J. L.: 1982, in **The Galactic Center**, AIP Conf. Proc. No. 83, eds. G. R. Riegler and R. D. Blandford, (New York:AIP), pp. 77-81.
Willner, S. P. and Pipher, J. L.: 1983, Astrophys. J. 265, in press.
Willner, S. P., Russell, R. W., Puetter, R. C., Soifer, B. T., and Harvey, P. M.: 1979b, Astrophys. J. 229, p. L65-L68.

FORMATION OF MOLECULES ON DUST

David A. Williams

Mathematics Department, UMIST, Manchester, England

ABSTRACT

The contribution to interstellar chemistry from reactions on the surfaces of dust grains is described. Results of new calculations of sticking probability and surface mobility are presented, and the implications for interstellar chemistry are briefly explored. Reactions at the surfaces of oxide grains are reviewed, and their contributions to sulphur-bearing molecules, and hydroxides are described. The disintegration of carbon grains in interstellar shocks is proposed to account for a substantial interstellar abundance of carbon chain molecules.

1. INTRODUCTION

It is widely accepted that interstellar molecular hydrogen is most effectively formed in reactions on the surfaces of dust grains. This conclusion is arrived at partly by default: for no known gas phase mechanism under the conditions in interstellar clouds can provide a formation rate adequate to compete with H_2 photodissociation. For example, direct two-body radiative association in the gas phase is strictly forbidden by selection rules. Reactions involving negative ions, $H^- + H \rightarrow H_2 + e$, are important only when the electron abundance approaches that of neutral hydrogen. Reactions of hydrides, $XH + H \rightarrow X + H_2$ do not contribute significantly because XH is of low abundance. Three body reactions require very much higher densities for them to proceed efficiently. (See Duley and Williams 1983a for a review of all these processes). On the other hand, recombinations of atoms on surfaces are known to be important in the laboratory; and simple models for recombination of H-atoms on the surfaces of interstellar dust grains give a rate coefficient for H_2 formation which is consistent with results from observations (Jura 1975a,b).

The formation of molecules other than H_2 on the surfaces of dust grains is, however, controversial. This is because gas phase models of interstellar chemistry have been remarkably successful in explaining the existence and in some cases the relative abundance of interstellar molecules, and this is achieved with a well-defined chemistry, albeit one which contains a very high proportion of unknown rate coefficients. By contrast, reactions on grain surfaces are poorly defined, especially because the nature of the grain material is not yet determined. However, since surface reactions forming molecules other than H_2 are well-known in the laboratory, it is important to decide whether grain surface reactions do contribute in any significant way to interstellar chemistry. If so, then not only are molecular abundances affected, but also the determination of interstellar parameters (such as cosmic ray ionization rate) from molecular observations.

That grains have the potential to contribute significantly is readily seen from the following simplistic model. Consider the reaction of atom X with a grain, leading to the formation of molecule XY,

$$X + Y\text{-}g \rightarrow XY + g$$

(where g represents the grain). If we assume that every collision of X with the grain leads to XY formation, then the XY formation rate is

$$F(XY) = n(X) n_g \pi a^2 V_X \quad cm^{-3} \; s^{-1}$$

where n_X and n_g are the number densities of grains and X, πa^2 is a mean cross section of the grains, and V_X a mean thermal speed of X. From extinction measurements

$$\pi a^2 n_g Q_{ext} \simeq 10^{-21} n$$

where n is the total hydrogen density and Q_{ext} is the efficiency for extinction of starlight. Assuming that $Q_{ext} \simeq 1$, and that $V_X \simeq 10^4$ cm s^{-1} then

$$F(XY) \simeq 10^{-17} n(X) n \quad cm^{-3} \; s^{-1}$$

If $X \equiv H$, then this result is close to the canonical value which is reasonably consistent with the observations. For other species, however,

$$F(XY) \gtrsim 10^{-21} n^2 \quad cm^{-3} \; s^{-1}.$$

In diffuse clouds, photodissociation is an important loss. If the photodissociation rate is $\sim 10^{-10}$ s^{-1}, then

$$n(XY)/n \sim 10^{-11} n,$$

so that $n \sim 10^3$ gives a value of this relative abundance of 10^{-8}, consistent with molecular fractional abundances in diffuse clouds. In dark clouds, reactions of molecules with ions may be the important loss route. If this occurs at a rate $\sim 10^{-9}$ n(ion), then, assuming n(ion) $\simeq$ n(e) and that n(e), the electron density, is $\sim 10^{-7}$ n, then n(XY)/n $\simeq 10^{-5}$. We conclude that surface reactions may be able to contribute to abundances in diffuse clouds, and in dark clouds are capable of modifying the molecular abundances very significantly.

The evidence in favour of gas phase reactions, particularly ion-molecule reactions, is compelling (Watson and Walmsley 1982) and - apart from the seminal molecule H_2 - it seems likely that for much of the chemistry grains merely contribute to the general network of reactions. In some special cases, however, grains may be able to supply a chemical pathway to a particular molecule difficult to form in the gas, e.g. H_2. This may be the case for NaOH (tentatively detected by Hollis and Rhodes 1982) and for sulphur-bearing molecules (Millar 1982). In addition, a number of molecules presently non-detected may, if subsequently observed, find their origin in surface reactions. Such molecules are, e.g. NH, CH_4 (which might be detectable in a deuterated form) or H_2CO in diffuse clouds (Duley, Millar and Williams 1978).

Interstellar grains affect interstellar chemistry passively, by attenuating UV and visible radiation and so shielding molecules from photodissociation. Grains also deplete from the gas various elements (such as Ca, Mg, Si, Al) which form refractory cores. On these cores molecular coatings may accrete, so removing molecules from the gas phase. In this article we take these processes for granted, and restrict our discussion to the case where grains are actively involved in interstellar chemistry, by means of reactions occurring on their surfaces. We report some recent work which helps the description of molecule formation on interstellar grains.

2. BASIC PARAMETERS

The simplistic description of molecule formation given above must be extended to be more realistic. For example, the incident atom X must stick to the grain if reaction is to occur. There will be a certain probability of sticking, dependent on the material of the grains, the nature of the atom-grain interaction, and the gas and grain temperatures. Further, if atom X has not arrived at the site where it may react with Y, then it must be able to move to that site within a time sufficiently short that adsorption has not occurred.

A comprehensive *ab initio* treatment of the sticking problem has been given recently by Leitch-Devlin (1982). The method is fully quantum mechanical, and may be applied to a variety of systems. It represents a considerable advance on earlier estimates (Hollenbach and Salpeter 1970). As it approaches the surface, the atom experiences an interaction potential, and if sticking is to occur a free-bound transition must take place in the atom-surface system, accompanied simultaneously by the excitation of a lattice phonon of energy equal to that released in the transition. The model therefore requires a representation of the atom-surface interaction potential and a knowledge of the phonon spectrum in the solid. Some illustrative results of this calculation of the sticking probability S are given in Figure 1. In the cases illustrated, the interaction potential is either weak (i.e. physisorption) or strong (chemisorption). Not all the cases in Figure 1 are realistic, but where they are, the interaction is given the appropriate characteristics.

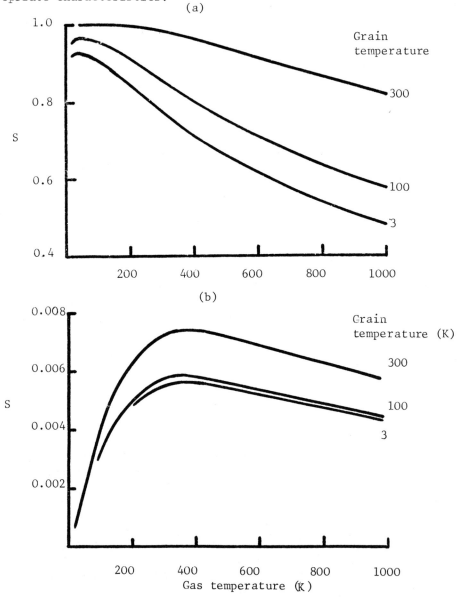

Figure 1. Sticking probabilities for H-atoms on
(a) graphite - chemisorption
(b) magnesium oxide - physisorption
as a function of gas and grain temperatures.

The results of a wide variety of cases may be summarized as follows:
(i) S increases with the grain temperature, (ii) S usually rises to a maximum and then falls monotonically as the gas temperature increases, (iii) when the interaction is strong, S is generally large, $S \sim 1$ for "soft" lattices and 0.3 for "hard" lattices. As an example, it is apparent that "hard" lattices, such as oxides, which physisorb H-atoms do not retain H-atoms, so H_2 formation does not occur on the perfect surface; sticking should, however, occur at defect sites where the interaction potential is stronger. These calculations are, of course, idealized descriptions. However, they show clearly that S depends strongly on several factors; in particular, the nature of the atom-surface interaction and the character of the lattice.

Assuming that an atom has been able to stick to the surface of a grain, then it may have to move to a site where reaction will occur. Motion of the adsorbed atom across a perfect surface is impeded by a periodic potential. If the temperature is low then motion proceeds via quantum mechanical barrier penetration. If the temperature is sufficiently high the motion is classical. The problem has been considered previously by several authors (e.g. Hollenbach and Salpeter 1970, Goodman 1978). Leitch-Devlin (1982) has given the most complete description to date; he allows for a variety of periodic potential shapes, representing these shapes as Fourier expansions. A thermal average of the group velocity for a quantum mechanical wave packet in this potential is then calculated. A representative result given in Figure 2 is for a purely sinusoidal barrier. The contours are labelled with the value of the relative mobility, u_r, which is the group velocity average, u_g, in terms of the two dimensional thermal velocity at that temperature T, thus

$$u_r = u_g / (\frac{\pi k T}{2M})^{\frac{1}{2}}$$

for an atom of mass M. The results are given in terms of the scaled barrier height, η, and scaled temperature, τ,

$$\eta = M a^2 U / (8 \pi^2 \hbar^2), \quad \tau = kT/U$$

where U is the effective potential barrier, a is the lattice period, and T is the temperature. Figure 2 shows that for small barrier heights (i.e. small η), u_r is independent of temperature, as expected. For small temperatures and large barrier heights the mobility is very small.

Consider an example in which H-atoms adsorbed on the lattice experience a potential energy barrier U to motion across the surface, and $U/k = 150$ K. If the lattice temperature is 10 K, then $\eta = 0.35$ and $\tau = 0.067$, and so from Figure 2, $u_r \sim 0.4$. Thus, H-atoms diffuse over the surface with a speed of 0.4 times their two dimensional speed on a flat surface; $u_g \sim 10^4$ cm s^{-1}. For O-atoms similarly physisorbed, the result is 10^{-38} cm s^{-1}. Thus, in comparison to H-atoms, O-atoms are relatively immobile and so hydride molecules are the likely result of such chemical reactions on inert grains, rather than molecules

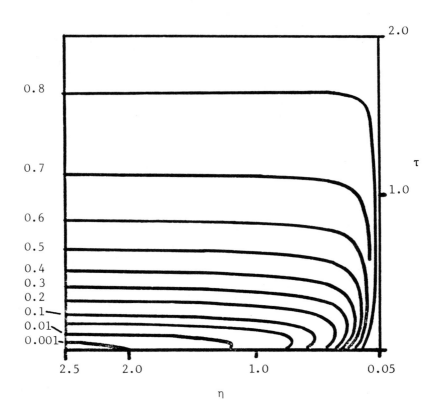

Figure 2. Mobility ratio, u_r, contours as a function of scaled barrier height, η, and scaled temperature, τ.

containing two heavy atoms. This confirms the predictions made earlier by several authors (Watson and Salpeter 1972, Pickles and Williams 1977). In the case of chemisorption, the H-atom mobility is severely reduced, but an H-atom is still able to explore the grain in the time available, and hydrides are again likely to result from this type of surface reaction.

3. COMPREHENSIVE GAS AND GRAIN MODELS OF INTERSTELLAR CHEMISTRY

Models of interstellar chemistry incorporating fairly comprehensive systems of gas phase reactions together with surface reactions forming hydrides feeding the network of gas phase reactions, have been explored (Pickles and Williams 1981). Although the grain contribution can be quite marked in some models, the uncertainty in a large number of astrophysical parameters precludes a definitive statement as to whether grains do in fact contribute in the way described above. Mann (1982) has completed a more detailed study of molecular observations in diffuse clouds and of the theoretical models of the chemistry. Diffuse clouds were chosen for study because the gas phase chemistry in them is relatively simple. Mann considered the totality of all molecular observations, not just those for a particular source. He concludes that within the observational uncertainty, there is no evidence which argues against grain contributions to the chemistry. Mann identifies the molecule NH as a very precise indicator for grain reactions in diffuse clouds. The observational upper limits for the column density $N(cm^{-2})$ for NH are given by $\log_{10} N < 11.76$ for ζ Per (Chaffee and Lutz 1977) and < 11.85 for o Per (Crutcher and Watson 1976). Models including grain chemistry for clouds of the appropriate extinction give 11.6 over a wide density range; but if grains are excluded from the chemistry, the NH abundance is a factor of 100 lower. Thus, a detection may be close if grains contribute. Mann concludes from this and a variety of other studies that, since grain contributions to the chemistry of diffuse clouds cannot be excluded, "solutions" for astrophysical parameters deduced from molecular observations cannot be unique. The possibility of surface contributions should, at this stage, continue to be entertained, and its effect on inferred astrophysical parameters explored.

4. SPECIFIC GRAIN MATERIALS

It is possible to make substantial progress in surface chemistry when specific grain models are considered. These models may give rise to a characteristic range of molecules other than the hydrides which are considered to arise on the surfaces of inert grains. We conclude this paper by referring to some recent work on particular grain models.

4.1 Oxide Grains

The case for diatomic oxide grains as an alternative or complement to silicate grains (Duley, Millar and Williams 1979) rests on the selective depletion mechanism inherent in oxides, and on the important spectral features at 2175 Å, 9.7 μm and 18 μm. Oxides are well known laboratory catalysts, and their reactivity resides in lattice defects. In the interstellar medium the $V(OH^-)_s$ site should be particularly important. Duley and Millar (1978) predicted that reactions at these sites of ions such as Na^+ (which are not absorbed into the lattice) should form gaseous NaOH. This molecule is now tentatively identified towards the galactic centre (Hollis and Rhodes 1982). Further, sulphur atoms are known to replace oxygen in oxides; reactive sites would then become $V(SH^-)_s$ sites. In interstellar clouds S may be depleted from the gas by incorporation into oxide lattices. Reactions of a variety of positive ions at these sites produce a range of sulphur-bearing molecules (Duley, Millar and Williams 1980). This proposal has been investigated by Millar (1982) who finds that the occurrence and abundance of H_2S, OCS, HNCS, CH_3SH in dense clouds may satisfactorily be explained in this way, without having to involve shock heating of interstellar gas to drive endothermic gas phase reactions (Hartquist, Oppenheimer and Dalgarno 1980).

4.2 Carbon Grains

The depletion of interstellar carbon points to the existence of carbon grains. They may absorb strongly in the vicinity of 2175 Å. Surface radicals on carbon grains have been proposed by Duley and Williams (1981) as the origin of observed IR features at 3.3, 3.4, and 11.3 μm in emission and absorption, and a variety of other possible features. Carbon grains are commonly thought to originate in the cool atmospheres of carbon rich stars, and to be carried out with the expanding outer layers, ultimately mixing with the interstellar gas. The destruction of grains occurs in interstellar shocks (Cowie 1978, Shull 1978). Laboratory evidence on the destruction of solid carbon by heating or by short duration laser pulses is that in addition to the ejection of carbon atoms, a substantial proportion of carbon chain molecules are also ejected (e.g. Berkowitz and Chupka 1964). Duley and Williams (1983b) propose that significant abundances of carbon chain molecules may arise in this way in the hot shocked gas behind the shock front. Estimates of the column density of a carbon chain molecule containing a particular number of C-atoms and terminating in a particular radical, e.g. CN, are on the order of 10^{11} cm^{-2}. This is comparable with the column densities of HC_5N and HC_7N observed in apparently diffuse cloud material in the Orion arm, towards Cas A (Bell, Feldman and Matthews 1981). Thus, chemistry in the shocked interstellar gas,(extensively modelled by Mitchell and Deveau 1982, 1983) may also have a contribution from the destruction of interstellar grains occurring in the same spatial region as that in which the gas phase reactions are enhanced.

REFERENCES

Bell, M.B., Feldman, P.A. and Matthews, H.E. 1981, Astron. Astrophys., 101, L13.

Berkowitz, J. and Chupka, W.A. 1964, J. Chem. Phys., 40, 2735.

Chaffee, F.H. and Lutz, B.L. 1977, Astrophys. J., 213, 394.

Cowie, L.L. 1978, Astrophys. J., 225, 887.

Crutcher, R.M. and Watson, W.D. 1976, Astrophys. J., 209, 778.

Duley, W.W. and Millar, T.J. 1978, Astrophys. J., 220, 124.

Duley, W.W., Millar, T.J. and Williams, D.A. 1978, Mon. Not. Roy. Astr. Soc., 186, 685.

Duley, W.W., Millar, T.J. and Williams, D.A. 1979, Astrophys. and Sp. Sci., 65, 69.

Duley, W.W., Millar, T.J. and Williams, D.A. 1980, Mon. Not. Roy. Astr. Soc., 192, 945.

Duley, W.W. and Williams, D.A. 1981, Mon. Not. Roy. Astr. Soc., 196, 269.

Duley, W.W. and Williams, D.A. 1983a "Interstellar Chemistry", London: Academic Press.

Duley, W.W. and Williams, D.A. 1983b in preparation.

Goodman, F.O. 1978, Astrophys. J., 226, 87.

Hartquist, T.W., Oppenheimer, M. and Dalgarno, A. 1980, Astrophys. J., 236, 182.

Hollenbach, D. and Salpeter, E.E. 1970, J. Chem. Phys., 53, 79.

Hollis, J.M. and Rhodes, P.J. 1982, Astrophys. J. (Letters), 262, L1.

Jura, M. 1975, Astrophys. J., 197, 575 and 581.

Leitch-Devlin, M.A. 1982, Ph.D. Thesis, Manchester University.

Mann, A.P.C. 1982, in preparation.

Millar, T.J. 1982, Mon. Not. Roy. Astr. Soc., 199, 309.

Mitchell, G.F. and Deveau, T.J. 1982, in "Regions of Recent Star Formation", R.S. Rodger and P.E. Dewdney (eds.) D. Reidel, Dordrecht, Holland, p.107.

Mitchell, G.F. and Deveau, T.J. 1983, Astrophys. J., 266, 646.

Pickles, J.B. and Williams, D.A. 1977, Astrophys. and Sp. Sci., 52, 433.

Pickles, J.B. and Williams, D.A. 1981, Mon. Not. Roy. Astr. Soc., 197, 429.

Shull, J.M. 1978, Astrophys. J., 226, 858.

Watson, W.D. and Salpeter, E.E. 1972, Astrophys. J., 174, 321.

Watson, W.D. and Walmsley, C.M. 1982, in "Regions of Recent Star Formation", R.S. Roger and P.E. Dewdney (eds.) D. Reidel, Dordrecht, Holland, p.357.

GAS PHASE CHEMISTRY IN THE INTERSTELLAR MEDIUM

William D. Watson

*Departments of Physics and Astronomy,
University of Illinois at Urbana-Champaign,
Urbana, IL 61801, USA*

ABSTRACT

   A brief discussion is given in support of <u>in situ</u>, gas phase reactions as the dominant chemical processes in the interstellar medium for molecules other than H_2. Basic aspects of the "ion-molecule" schemes in the dense cloud and diffuse clouds limits are summarized. The following topics are selected for more detailed comments -- fractionation of carbon isotopes, CH^+ and shocks, the HNC/HCN ratio, the COH^+/HCO^+ ratio, fractionation of deuterium and the electron density of interstellar clouds, and the formation of complex molecules.

I. INTRODUCTION

   A consideration of chemistry in the interstellar medium (ISM) of our own galaxy, and presumably of other galaxies, can be separated into (at least) four divisions for convenience - chemistry that occurs on the surfaces of grains, chemistry in the gas being ejected from late-type stars and its contribution to the molecules in the ISM, chemistry initiated by shock fronts propagating through the ISM, and chemistry centered around ion-molecule reactions. My topic is the last of these. Other authors in this volume will deal with surface chemistry and with shocks. I can offer some references to recent work on chemistry in stellar winds (Huggins and Glassgold 1982; Lafont, Lucas and Omont 1982; McCabe, Connon-Smith and Clegg 1979; Scalo and Slavsky 1980).

   There is no reason to doubt that the most abundant molecule in the universe - the H_2 molecule - results from reactions on the surfaces of dust grains in the ISM. I would however argue that there is good reason to doubt that surface reactions, shocks, or stellar winds are dominant for the chemistry of most of the other molecules generally observed in the ISM. Except for molecular hydrogen, there is in my opinion still no convincing mechanism for returning molecules from grain surfaces back to the gas in interstellar clouds that are at all dense. Perhaps some combination of the occasional "cleansing" of grain surfaces by shocks, etc. and low probability in situ ejection processes such as the chemical reaction itself, cosmic rays, etc. can do the job. Nevertheless, it seems out of the question that apprecible abundances of the observed molecular ions could be produced directly by surface chemistry. It seems only slightly less implausible to me that unstable isomers (HNC) and the extreme fractionation of deuterium in interstellar molecules could result from an interstellar surface chemistry. Failure to detect certain molecular species at an abundance level that would be expected from the most straightforward surface processes has been cited as evidence against major contributions from grains (NH, Crutcher and Watson 1976; NaH, Plambeck and Erickson 1982) In contrast, the possible detection of NaOH has been utilized as a point in favor of surface reactions (Hollis and Rhodes 1982). However, this is based on the absence of any explicit, gas phase proposal to date to form NaOH (but see Millar 1982) and I expect that such will be forthcoming.

The presence of unstable isomers and especially the fractionation of deuterium are compelling arguments against high temperature, near thermal environments for formation. Recognized mechanisms for isotope fractionation depend upon the effect in reactions of small zero-point energy differences between the isotopic forms of a molecule. Low temperatures ($\lesssim$ 50K) are then required so that the Boltzmann factor which is relevant for thermal processes has a value comparable with the observed enhancements of deuterium (up to about 10^4). Molecules are also observed in considerable abundance in regions of low optical depth (e.g., toward Zeta Ophiuchi) where their lifetime against photodissociation is not long enough to allow transit from a late-type star or to allow them to survive since the passage of a shock. Mass outflow and shocks certainly do make a non-negligible contribution. Molecules are observed in the material being ejected from late-type stars and lifetimes against molecular dissociation in dense ISM clouds may well be comparable to the age of the cloud. Strong shocks are observed in molecular clouds (e.g., Orion) and surely have an effect. There exists a plausible explanation for the longstanding CH^+ abundance problem in terms of shock chemistry whereas other reaction schemes seem to fail. Finally, the ability of gas-phase, ion-molecule centered reaction schemes to provide natural and quantitative explanations, although uniqueness can not be proved, for the abundances of many small molecules is a compelling argument in its favor. Notable successes include an understanding of the abundances of the key species – HCO^+, N_2H^+, HNC, OH and HD – as well as an understanding of D/H and $^{13}C/^{12}C$ fractionation in molecules. The detection of a similar, extreme enhancement of deuterium in DC_5N in TMC1 (MacLeod et al. 1981; Schloerb et al. 1981) where the long-chain hydrocarbons are common suggests to me that even the large, complex molecules in the ISM result from low-temperature, gas-phase reactions.

There are a number of reviews of the formation processes for interstellar molecules already in the literature involving this author (e.g., Watson and Walmsley 1982; Watson, 1976) and as well as those by other authors. After a brief introduction to the basic elements of in situ, gas-phase chemistry in the ISM - "ion-molecule" chemistry (Section II), I will therefore limit myself to comments on recent results relevant for certain selected topics which may be categorized as follows - $^{13}C/^{12}C$ fractionation (III), the CH^+ abundance and shocks (IV), the HNC/HCN ratio (V), the COH^+/HCO^+ ratio (VI), D/H fractionation (VII), and the formation of complex molecules (VIII).

II. BASIC ELEMENTS OF "ION-MOLECULE" CHEMISTRY IN THE ISM

Though there clearly are intermediate situations, the "dense cloud" and "diffuse cloud" limits are useful for illustrating the basic reaction schemes. Even in the dense cloud conditions, the time scale for the molecular abundances to reach steady-state is of the order of 10^6 years so that the approximation of time independence - which I will assume throughout - is also a useful limit. Photoprocesses are ignored completely in the "dense cloud" limit, whereas they are the dominant mechanism for destroying most molecules in the "diffuse cloud" limit.

The key element of "ion-molecule" schemes is the efficient transfer of the ionization of the abundant elements (hydrogen and helium) to the less abundant species (carbon, nitrogen, oxygen, etc.) with the consequent substantial increase in the rate of ionization per atom. Ions, as opposed to neutrals, are critical because activation energy barriers for exothermic chemical reactions are the exception when a positive ion is involved. When only neutrals are involved, such activation energies are common and it has been thought likely that most such reactions will have at least small energy barriers even though they may undetectable at 300K and above - the temperatures at which most laboratory studies are made. At the low temperatures of interstellar clouds, essentially any such barrier will prevent a reaction from being important. It is also the case that chemical reactions between species seem more likely to be exothermic if one partner is ionized. Since the resulting molecules are ionized, neutralizing mechanisms must be involved to produce the observed species which are mostly neutral.

A word of caution. Most chemical reactions have not been measured in the laboratory at temperatures below about 300K, though the situation has improved considerably in the past few years down to 80K for ion-molecule reactions. It has been thought that the behavior of ion-molecule reactions at very low temperatures is simple and predictable. A rate coefficient should either be Langevin or, if it is appreciably below Langevin at 300K, it must have an energy barrier that makes it completely negligible at temperatures less than about 100K. Recent studies of the astrophysically important reactions $He^+ + H_2$ (Johnsen et al. 1980) and, down to about 10K, $NH_3^+ + H_2$ (Liune and Dunn 1981) indicate that important surprises may be in store when laboratory measurements below 80K are made.

The source of primary ionizations is assumed to be the cosmic rays with energies greater than about 100 MeV/nucleon which provide an ionization rate of about 10^{-17}–10^{-18} ionizations per H-atom per second.

Dense Cloud Limit

Essentially all of the hydrogen is assumed to exist as H_2 in this case and chemical reactions are initiated by

$$\text{cosmic ray} + H_2 \rightarrow \text{cosmic ray} + H_2^+ + e \qquad (1)$$

followed by

$$H_2^+ + H_2 \rightarrow H_3^+ + H + e \qquad (2)$$

This H_3^+ is destroyed either by dissociative electron recombination

$$H_3^+ + e \rightarrow H_2 + H, \text{ or } 3H \qquad (3)$$

or by proton transfer to the most abundant molecules

$$H_3^+ + CO \rightarrow HCO^+ + H_2 \qquad (4)$$

$$+ N_2 \rightarrow N_2H^+ + H_2 \qquad (5)$$

The direct path from the presumed ionization to HCO^+ and N_2H^+ made the identification of these species in the ISM especially significant. Efforts to detect H_3^+, perhaps the most important undetected species, have not succeeded (Oka 1981) but have not yet reached the sensitivity level that is needed for middle-of-the-road predictions of its abundance.

Direct ionization of helium

$$\text{cosmic ray} + He \rightarrow \text{cosmic ray} + He^+ + e \qquad (6)$$

is perhaps more important to the chemistry because it breaks up abundant molecules

$$He^+ + CO \rightarrow C^+ + O + He \qquad (7)$$

$$+ N_2 \rightarrow N^+ + N + He \qquad (8)$$

to provide reactive atoms and ions for further reaction schemes, for example,

$$N^+ + H_2 \rightarrow NH^+ + H \qquad (9)$$

$$NH^+ + H_2 \rightarrow NH_2^+ + H \qquad (10)$$

$$NH_2^+ + H_2 \rightarrow NH_3^+ + H \tag{11}$$

$$NH_3^+ + H_2 \rightarrow NH_4^+ + H \tag{12}$$

$$NH_4^+ + e \rightarrow NH_3 + H \tag{13}$$

and

$$C^+ + NH_3 \rightarrow HCNH^+ + H \tag{14}$$

$$HCNH^+ + e \rightarrow HCN + H, \text{ and } HNC + H \tag{15}$$

The radiative association,

$$C^+ + H_2 \rightarrow CH_2^+ + h\nu \tag{16}$$

for which the reaction rate is only poorly known (see VIII), is also a key process in the initial reactions. A reaction with H_2 analogous to (11) above then contributes to producing CH_4, though a radiative association (see VIII) also is needed

$$CH_3^+ + H_2 \rightarrow CH_5^+ + h\nu \tag{17}$$

because the transfer of a single H-atom in (17) is endothermic.

The fractional ionization of the gas that results (see also VII) from a specified ionization rate in a molecular cloud is lower and more uncertain than for a primarily atomic gas due to the ion-molecule reactions. It is lower because dissociative electron recombination (e.g., as in eqn. 3) has a much larger rate coefficient than radiative electron recombination which is normally the dominant process for recombination in an atomic gas ($\approx 10^{-7}$ cm^3 s^{-1} versus $\approx 10^{-11}$ cm^3 s^{-1} at T $\approx$ 100K). It is more uncertain because the rate coefficients for dissociative electron recombination at low temperatures are open to some question (see VII), and it has been considered uncertain by many because of the possible presence of free metal atoms in the gas at near their cosmic abundance. These atoms with low ionization energies are ionized by charge-transfer from the molecular ions. They are not thought to form molecular ions or to transfer their ionization efficiently, and are thus subject to neutralization only by the slower radiative recombination. Higher ionization thus results.

The most recent numerical solution of the extensive array of rate equations that enter even when only the small molecules (up to 4 or 5 atoms) are considered is due to Graedel et al. (1982). They find that the best agreement between their calculations and observations is achieved when metal atoms are unimportant (low abundance) and when steady state has been reached.

Diffuse Cloud Limit

Cosmic ray ionization is somewhat less dominant in this case because starlight can ionize species with ionization potentials below 13.6 eV. The starlight also keeps the elements primarily in atomic form. Even for hydrogen, the atomic form is comparable in abundance with that of the molecular form. Cosmic ray ionization is essential for certain key molecules, for example.

$$\text{cosmic ray} + H \rightarrow \text{cosmic ray} + H^+ + e, \tag{18}$$

$$H^+ + D \rightarrow H + D^+, \tag{19}$$

$$D^+ + H_2 \rightarrow HD + H^+, \tag{20}$$

and

$$H^+ + O \rightarrow H + O^+, \tag{21}$$

$$O^+ + H_2 \rightarrow OH^+ + H, \tag{22}$$

...

$$H_3O^+ + e \rightarrow H_2O + H, \text{ and OH and } H_2 \tag{23}$$

The radiative association of equation (16), together with the analogous reaction involving H-atoms, converts carbon atoms into molecules. Exchange reactions produce species such as CO,

$$C^+ + OH \rightarrow CO + H^+ \tag{24}$$

Detailed numerical solutions of the relevant rate equations for the molecular abundances in diffuse clouds would seem to be more reliable than those for dense clouds because of the many fewer reactions that must be considered, because elements other than hydrogen are primarily in atomic form, and because the physical conditions in the gas are much better known. A refined chemical model exists which reproduces the observed molecular abundances, except for that of CH^+, toward Zeta Ophiuchi (Black and Dalgarno 1977). However, re-analysis of the constraints on the physical conditions for the gas toward Zeta Ophiuchi indicates that they are quite different from those used by Black and Dalgarno (Crutcher and Watson 1981; see also III). The effect of such a change in physical conditions on the predicted abundances is unclear. Study of the chemistry in diffuse clouds is valuable in itself, but also as a "laboratory" in which to gain an understanding that can be utilized in the study of the more involved chemistry of the dense regions.

III. $^{13}C/^{12}C$ FRACTIONATION

Enhancement of the abundance of ^{13}C in carbon monoxide was predicted to occur as a result of the reaction (Watson, Anicich and Huntress 1976)

$$^{13}C^+ + {}^{12}CO \rightleftarrows {}^{12}C^+ + {}^{13}CO + \Delta E \qquad (25)$$

in which $\Delta E/k \simeq 35K$. This reaction has now been measured down to about 80K in the laboratory (Smith and Adams 1980). That the observed $^{13}CO/^{12}CO$ ratio is contaminated as a result of equation (25) now seems established (Dickman et al. 1977; Langer et al. 1980).

The effect of equation (25) clearly is rather sensitive to temperature. Crutcher and Watson (1981) have shown that the competing processes are sufficiently slow that the $^{13}CO/^{12}CO$ ratio can be utilized to obtain a useful lower limit to the kinetic temperature in, at least a certain class, of diffuse clouds. The limit obtained, which is supported by an independent assessment of the $^{13}CO/^{12}CO$ ratio from ultraviolet observations (Wannier, Penzias and Jenkins 1982) and by other observational constraints, is incompatible with the Black and Dalgarno (1977) model for the "prototypical" diffuse clouds toward Zeta Ophiuchi.

Based on the assumption that an appreciable fraction of the photodissociation of CO occurs as sharp lines, Bally and Langer (1982) have shown that self-shielding can cause the abundances of the more abundant isotopic forms of CO to be enhanced at certain optical depths. A further study by Chu and Watson (1983) including the effects of equation (25) has found that the effect of any isotope-selective photodissociation of CO is more limited than indicated by the previous investigation. A separate question - about which the laboratory data do not seem to be definitive - is whether sharp resonances are a major component of the photodissociation cross section of CO.

IV. CH^+ ABUNDANCE AND SHOCKS

Since the earlist studies of the reactions of interstellar molecules (Bates and Spitzer 1951), the CH^+ abundance - first measured in about 1940 - has been a problem. As a molecular ion, CH^+ is susceptible to dissociative electron recombination

$$CH^+ + e \rightarrow C + H \qquad (26)$$

which normally has a much larger rate coefficient ($\simeq 10^{-7} cm^3 s^{-1}$) than other collision processes. Essentially all measurements of such recombinations yield rate coefficients of this magnitude, and the measurement for CH^+ down to about 0.01 eV is no exception (Mitchell and McGowan 1978). However, only one group using one technique has performed the measurement. With the measured rate coefficient for equation (26), no steady-state, chemical scheme has been reported that generates near the observed abundance of CH^+ in diffuse clouds.

Calculations have been performed which show that the observed column densities of CH^+ can be produced by shocks, propagating ($\simeq$ 10-12 km/s) through the diffuse clouds (Elitzur and Watson 1978; 1980), in which the key reaction is

$$C^+ + H_2 \rightleftarrows CH^+ + H + \Delta E \qquad (27)$$

($\Delta E \simeq 0.4$ eV). The occurence of such shocks is not inconsistent with other observational constraints and the general idea that shocks occur commonly in diffuse clouds is in the spirit of current ideas about the ISM (e.g., McCray and Snow 1979).

Collisional excitation of the rotational states of H_2 by H,

$$H + H_2 (J) \rightleftarrows H + H_2 (J') \qquad (28)$$

plays a key role in placing constraints on the shocks. The primary cooling of the hot, post-shock gas is due to the emission of radiation following the excitations of (28). Hence, the volume of the gas in which the CH^+ abundance can be influenced by a shock is determined by equations (28). Further, the tightest constraint on the shocks seems to be the directly observed abundance of the rotationally excited H_2, probably in J = 4-6. It has argued by some authors (including Hartquist and Dalgarno (1982) at this Symposium) that the collision cross sections of Green and Truhlar (1979), which were used by Elitzur and Watson (1980), are less reliable than earlier cross sections of Chu and Dalgarno (1975). I wish to re-emphasize the point made by Elitzur and Watson (1980, see page 174). Although the values for certain energies and certain J-values from Green and Truhlar (1979) are quite different from those of other recent investigators, those that are dominant in determining a total energy loss rate are quite similar. Hence, Elitzur and Watson (1980) found that the resulting energy loss rates and the structure of the shocks differ negligibly. Similarly, the actual populations of the rotational levels are also quite insensitive to the particular cross sections; for example Elitzur and Watson (1980, pg. 175) note that the $H_2(J)$ populations calculated using Green and Truhlar (1979) differ negligibly from those of Aannestad and Field (1979) who used the much earlier and more approximate cross sections of Nishimura (1968). In the case of the prototypical cloud toward Zeta Ophiuchi, comparison of the calculated $H_2(J)$ for a pre-shock density of 10 cm^{-3} and $H_2/H = 0.1$ with observations gives a shock velocity not exceeding 10 km/s. On the other hand, a shock of 12 km/s with $H_2/H = 0.01$ is found from the calculations to reproduce the column density of CH^+ toward this star. Given the large number of quantities including observational, all with some uncertainty, I do not regard such a discrepancy as meaningful. In fact, one can take the positive attitude that general agreement between the shock velocities needed for the independent CH^+ and $H_2(J)$ abundances is support for the idea of a shock formation excitation and that only a bit of fine tuning is needed to get complete consistency.

Observational support for the shock formation of CH^+ comes from studies that show the velocity of the CH^+ to be systematically shifted from that of other species (Federman 1982; see also Crutcher 1979). If CH^+ is in a hot, post-shock gas it should be redshifted by about one-quarter of the shock velocity in comparison with the cool, dense post-shock gas that is downstream.

V. HNC/HCN RATIO

The "natural" route for producing HNC and HCN in ion-molecule chemistry in the ISM is

$$C^+ + NH_3 \rightarrow HCNH^+ + H \tag{29}$$

$$HCNH^+ + e \rightarrow HCN + H, \, HNC + H \tag{30}$$

Rearrangement or isomer-selective destruction has seemed unlikely, and the HNC/HCN ratio might be expected to reflect only the branching in equation (30). Surprisingly Goldsmith et al. (1981) find it to range from about 0.03 to 0.4 in several of the warmer clouds whereas in cold clouds the ratio is greater than unity (~ 4.4 for L134). It seems that other processes must be brought to bear, the most likely of which are

$$CH_2, \, CH_3 + N \rightarrow HCN + H, \, + H_2 \tag{31}$$

and

$$NH_2 + C \rightarrow HNC + H \tag{32}$$

if these have no energy barrier at low temperatures

VI. COH^+/HCO^+

It has been recognized that formation of the isomer COH^+ may be energetically allowed in equation (4) and, if so, might reasonably be expected to be produced in comparable abundance with HCO^+ (Herbst et al. 1976). It would not be expected to be rearranged in collisions with common species (H_2, He, etc.), though it probably would be in collision with H-atoms. Recently, an extensive search has been made by Woods et al. (1983) who find $COH^+/HCO^+ \lesssim 0.01$. The absence of COH^+ can be attributed to the presence of a larger than expected abundance of atomic hydrogen in molecular clouds, though it seems more likely to be a result of a low branching ratio for the formation of COH^+ in equation (4) in light of recent laboratory experiments (Illies, Jarrold and Bowers 1982).

VII. D/H FRACTIONATION

Since the first detection of deuterium in interstellar molecules (Wilson et al, 1973), it has been found that the D/H ratio (cosmic $\approx 10^{-5}$) is greatly enhanced (x100 to x10^4 or so) in essentially all species and locations (I know of no exceptions) where the sensitivity is such that detection of the deuterated form might reasonably be expected.* The enhancement seems to be primarily a result of the reaction

$$H_3^+ + HD \rightleftarrows H_2D^+ + H_2 + \Delta E \qquad (33)$$

though

$$CH_3^+ + HD \rightleftarrows CH_2D^+ + H_2 + \Delta E \qquad (34)$$

is another exchange which is known to occur efficiently and which may contribute for some species (Watson 1976). The focus is on exchange with HD because deuterium in molecular clouds is thought to be in this form. Surprisingly, (33) and (34) seem to be the only such exchanges involving "basic" species which are rapid, at least based on measurements at 300K. Most molecules have proton affinities greater than that of H_2 and it is supposed that proton (or deuteron, actually) transfer passes along to other molecules the deuterium enhancement achieved by reaction (33). This is surely the case for DCO^+ which is formed directly from H_2D^+ through reaction (4).

Because the exchange of equation (33) must compete with equations (3) and (4), and the HD/H_2 ratio in molecular clouds can reasonably be taken to be the cosmic value, the observed DCO^+/HCO^+ provides a basis for determining useful upper limits for the CO/H_2 abundance ratio, the fractional ionization e/H_2 and the temperature of the gas which are dependent on only the ratio of rate coefficients that can in principle be measured in the laboratory (Watson 1976; 1977). Various observational studies have been made using this technique to obtain information about molecular clouds, the most recent and extensive of which is due to Guelin, Langer and Wilson (1982).

The laboratory data has also been the subject of recent investigations. That the reaction (33) proceeds rapidly at low temperatures seems clear from measurements down to ~80K (Adams and Smith 1981). There has however been some controversy about the best value for $\Delta E/k$. The first assessment based on older vibrational frequencies for H_3^+ yielded 180K (Watson 1976). Observational studies of the DNC/HNC ratio and the temperature in molecular clouds gives a best value of 240K, though with considerable uncertainty, for the $\Delta E/k$

---

* I ignore the HD molecule in diffuse clouds for which the formation rate is also "enhanced", but does not lead to abundances in excess of the cosmic abundance of deuterium.

in whatever is the exchange reaction with HD that causes the enhancement of DNC (Snell and Wootten 1979). Initial laboratory studies (Adams and Smith 1981), however, found $\Delta E/k = 90K$ for equation (33). A value this small would cause serious problems for understanding D-fractionation since, for example, DCO^+/HCO^+ seems to enhanced by about a factor of 100 in Orion A where the temperature seems to be near 50K. Further reassessment has however yielded $\Delta E/k \simeq 230K$ for equation (33) (Herbst 1982a; Smith et al. 1982) and the issue seems to be settled.

In contrast, the recombination rate for equation (3) at low temperatures is open to question. There exists a measurement with the same technique as used for the recombination study of CH^+ (Auerbach et al. 1977) which is compatible with earlier measurements using an entirely different method (Leu, Biondi and Johnsen 1973) made down to 200K. Both of these obtained a rate coefficient that is "typical" for dissociative electron recombination. Recently, theoretical arguments based on molecular orbital calculations have been presented to the effect that this recombination rate must be much smaller, in fact negligible at temperatures less than 100K in molecular clouds (Michels and Hobbs 1982). It is then presumed that the laboratory measurements are contaminated by the presence of vibrationally excited H_3^+ or have incorrectly determined collision energies. Theoretical arguments were presented some time ago, as well, for a very slow (H_3^+ + e) recombination rate (Carney and Porter 1977). Latest laboratory studies of the reaction (33) do in fact exhibit a change in slope in going to lower temperatures (Macdonald, Johnsen and Biondi 1983) that was missing in the earlier data. The issue is unresolved.

VIII. COMPLEX MOLECULES

Understanding the formation of complex molecules, e.g. the HC_nN and C_4H species observed in the cloud TMC1, is clearly a "grand" question of interstellar chemistry. However, not only are the reaction networks influenced by the same uncertainties in rate coefficients, elemental abundances, etc. as hinder the delineation of reactions to generate small molecules, but as next generation species they also depend upon the uncertain, calculated abundances of the small molecules. Some of these such as CH_4 and C_2H_2 are not directly observable. As noted in the Introduction, the detection of a high fractional abundance of DC_3N strongly suggests to me that their formation involves gas phase, in situ reactions.

I do not have the space to do justice to the various suggestions that have been made about the formation of complex molecules. I will however mention some general considerations and focus on the state of a critical physical process--radiative association.

As a molecule is "growing" by proceeding through a reaction scheme, exchange reactions with H_2 always take precedence when they have even modest rate coefficients because of the relative abundance of H_2. When an exchange reaction of some species AB^+ with H_2 is endothermic

$$AB^+ + H_2 \not\rightarrow ABH^+ + H \tag{35}$$

then there can be a competition between various, much slower channels.

$$AB^+ + e \rightarrow A + B \tag{36}$$

$$\text{rate} \sim [e] \ (10^{-6} \text{cm}^3 \text{s}^{-1}) \sim 10^{-13} n \ \text{s}^{-1}$$

$$AB^+ + H_2 \rightarrow ABH_2^+ + h\nu \tag{37}$$

$$\text{rate} = 10^{-13} n (\langle \sigma v \rangle_{rad}/10^{-13} \text{cm}^3 \text{s}^{-1}) \text{s}^{-1}$$

$$AB^+ + CD \rightarrow ABCD^+ + h\nu \tag{38}$$

$$\text{rate} = 10^{-17} n (\langle \sigma v \rangle_{rad}/10^{-13} \text{cm}^3 \text{s}^{-1})([CD]/[10^{-4} n]) \text{s}^{-1}$$

$$\rightarrow ABC^+ + D \text{ or } ABC + D^+ \tag{39}$$

$$\text{rate} \sim 10^{-14} n [CD]/[10^{-4} n] \text{s}^{-1}$$

Here, $\langle \sigma v \rangle_{rad}$ is the rate coefficient for (37) or for (38), and $n(\text{cm}^{-3})$ is the number density of molecular hydrogen. Though the relevant rate coefficients and fractional abundances vary greatly, the above scaling reflects reasonable maximum values.

It is clear that radiative association potentially plays a key role, but that the rate coefficients must be much larger than the values of $\sim 10^{-17} \text{cm}^3 \text{s}^{-1}$ that are familiar from dealing with radiative association between atoms. Laboratory studies relevant to astrophysical radiative associations are, by necessity due to their slow rates, indirect and involve 3-body association,

$$A + B + M(v) \rightarrow AB^* + M(v) \rightarrow AB + M(v') \tag{40}$$

where $M(v)$ represents some unreactive species (like He) with velocity v. Measurements of the rate coefficients for reactions of the form (40) provide a basis for assessing the efficiency with which the excited complex is formed and its lifetime. A radiative association, e.g.,

$$A + B \rightarrow AB^* \rightarrow AB + h\nu \tag{41}$$

differs from (40) in that stabilization of the complex is due to the emission of radiation instead of to the transfer of energy to M. Deducing rate coefficients for the radiative association process has been somewhat controversial. However, reasonable agreement now seems to have been reached (Bates 1980; Herbst 1980) though there exists at least order-of-magnitude, acknowledged uncertainty in the actual rate coefficients. Rate coefficients for reactions of the form of equation

(37) are found to be near $10^{-13} cm^3 s^{-1}$ in some cases at $T \leq 50$ K. When the neutral partner is slightly larger than H_2 (e.g., HCN, NH_3), rate coefficients for radiative association (equation 38) that are near the classical collision rate ($\sim 10^{-9}-10^{-10} cm^3 s^{-1}$) have been calculated. There is one system for which an actual laboratory measurement has been reported (McEwan et al 1980)--CH_3^+ + HCN. A rate coefficient of $\sim 1 \times 10^{-10} cm^3 s^{-1}$ is obtained. Studies of the key radiative association between C^+ and H_2 (eqn. 16) seem to have converged to a value $\leq 10^{-15} cm^3 s^{-1}$ at temperatures less than 50 K as the "best" value consistent with data on 3-body associations (Herbst 1982b).

Finally, it is encouraging to note that definitive laboratory tests of proposed reaction schemes for forming complex molecules in the ISM can be made, even when they eliminate possibilities as was done by Herbst, Adams and Smith (1983).

The author's research is supported by the U.S. National Science Foundation.

References

Aannestad P. A., Field G. B. (1973) Astrophys. J. **186**, L29.
Adams N., Smith D. (1981) Astrophys. J. **248**, 373.
Auerbach D., Cacak R., Caudano R., Gaily T. D., Keyser C. J., McGowan J. W., Mitchell J. B. A., Wilk S. F. J. (1977) J. Phys. B **10**, 3797.
Bally, J., Langer, W.D. (1982) Astrophys. J. 255, 143.
Bates, D. R. (1980) J. Chem. Phys. **73**, 1000.
Bates D. R., Spitzer L. (1951) Astrophys. J. **113**, 441.
Black J. H., Dalgarno A. (1977) Astrophys. J. Suppl. Ser. **34**, 405.
Carney G. D., Porter R. N. (1977) J. Chem. Phys. **66**, 2756.
Chu S., Dalgarno A. (1975) Astrophys. J. **199**, 637.
Chu Y.-H, Watson W. D. (1983) Astrophys. J., in press (Apr. 15 issue).
Crutcher R. M. (1979) Astrophys. J. **231**, L151.
Crutcher R. M., Watson W. D. (1976) Astrophys. J. **209**, 778.
Crutcher R. M., Watson W. D. (1981) Astrophys. J. **244**, 855.
Dickman R. L., Langer W. D., McCutcheon W. H., Shuter, W. L. H. (1977) in "CNO Isotopes in Astrophysics" (p. 95), ed. J. Audouze, D. Reidel Pub. Co., Dordrecht, Holland.
Elitzur M., Watson W. D. (1978) Astrophys. J. **222**, L141.
Elitzur M., Watson W. D. (1980) Astrophys. J. **236**, 172.
Federman S. R. (1982) Astrophys. J. **257**, 125.
Green S., Truhlar D. G. (1979) Astrophys. J. **231** L101.
Guelin M., Langer W. D., Wilson R. W. (1982) Astron. Astrophys. **107**, 107.
Goldsmith P. F., Langer W. D., Ellder J., Irvine W., Kolberg E. (1981) Astrophys. J. **249**, 524.
Graedel T. E., Langer W. D., Frerking M. A. (1982) Astrophys. J. Suppl. Ser. **48**, 321.
Hartquist T.W., Dalgarno A. (1982), 16th ESLAB Symposium: Galactic and Extragalactic Infrared Spectroscopy. ESA SP-192. Eds. Kessler, M.F., Phillips, J.P. and Guyenne, T.D., p. 29-31.

Herbst E. (1980) Astrophys. J. 241, 197.
Herbst E. (1982a) Astron. Astrophys. 111, 76.
Herbst E. (1982b) Astrophys. J. 252, 810.
Herbst E., Adams N., Smith D. (1983) Astrophys. J., in press.
Herbst E., Norbeck J. M., Certain P. R., Klemperer W. (1976) Astrophys. J. 207, 110.
Hollis J. M., Rhodes P. J. (1982) Astrophys. J. 262, L1.
Huggins P. J., Glassgold A. E., (1982) Astrophys. J. 252, 201.
Illies A. J., Jarrold M. F., Bowers M. T. (1982) J. Chem. Phys. 77, 5847.
Johnsen R., Chen A., Biondi M. (1980) J. Chem. Phys. 72, 3085.
Lafont S., Lucas R., Omont A. (1982) Astron. Astrophys. 106, 201.
Langer W. D., Goldsmith P. F., Carlson E. R., Wilson R. W. (1980) Astrophys. J. 235, L39.
Leu M. T., Biondi M., Johnsen R. (1973) Phys. Rev. A. 8 413.
Liune J. A., Dunn G. H. (1981) in "Proceedings of XII ICPEAC" (p. 1035), ed. S. Datz, North-Holland Pub. Co., Amsterdam.
Macdonald J., Johnsen R., Biondi M. (1983) in preparation.
MacLeod J. M., Avery L. W., Broten N. W. (1981) Astrophys. J. 251, L33.
McCabe E. M., Smith R. C., Clegg R. E. S. (1979) Nature 281, 263.
McCray R., Snow T. P. (1979) Ann. Rev. Astron Astrophys. 17, 213.
McEwan M. J., Anicich V. G., Huntress W. T., Kemper P. R., Bowers M. (1980) Chem. Phys. Letters 75, 278.
Michels H. H., Hobbs R. H. (1982) Presented at the 35th Annual Gaseous Electronics Conference, Dallas, Texas.
Millar T. J. (1983) contributed paper to this Symposium.
Mitchell J. B. A., McGowan, J. W. (1978) Astrophys. J. 222, L77.
Nishimura S. (1968) Ann. Tokyo Astron. Obs., 2nd Ser. 11, 33.
Oka T. (1981) Phil. Trans. Roy. Soc. Lond. A 303, 543.
Plambeck R. L., Erickson N. R. (1982) Astrophys. J. 262, 606.
Scalo J. M., Slavsky D. B. (1980) Astrophys. J. 239, L73.
Schloerb F. P., Snell R. L., Langer W. D., Young J. S. (1981) Astrophys. J. 251, L37.
Smith D., Adams N. (1980) Astrophys. J. 242, 424.
Smith D., Adams N., Alge E. (1982) Astrophys. J. 263, 123.
Snell R. L., Wootten A. (1979) Astrophys. J. 228, 748.
Wannier P. G., Penzias A. A., Jenkins E. B. (1982) Astrophys. J. 254, 100.
Watson W. D. (1976) Rev. Mod. Phys. 48, 513.
Watson W. D. (1977) in "CNO Isotopes in Astrophysics" (p. 105), ed. J. Audouze, D. Reidel Pub. Co., Dordrecht, Holland.
Watson W. D., Anicich V., Huntress, W. T. (1976) Astrophys. J. 205, L165.
Watson W. D., Walmsley C. M. (1982) in "Regions of Recent Star Formation" (p. 357), ed. R. S. Roger and P. E. Dewdney, D. Reidel Pub. Co., Dordrecht, Holland.
Wilson R. W., Penzias A. A., Jefferts K. B., Solomon, P. M. (1973) Astrophys. J. 179, L107.
Woods R. C., Gudeman C. S., Dickman R. L., Goldsmith P. F., Huguenin G. R., Irvine W. M., Hjalmarson A., Nyman L.-A., Olofsson H. (1983) Astrophys. J., in press.

OBSERVATIONAL CONSTRAINTS ON INTERSTELLAR CHEMISTRY

G Winnewisser

I. Physikalisches Institut, Universität zu Köln, Köln, W. Germany

I. INTRODUCTION

Ever since molecular species have been discovered in space in the late 30's and early 40's by the optical identification of CH, CH^+ and CN in absorption towards nearby hot stars, the question of molecule formation has accompanied the observational efforts. These early discoveries by Adams, Dunham, Swings, Rosenfeld and McKellar are separated by a gap of 28 years from the radio-discovery of the first polyatomic molecule (ammonia, NH_3) in interstellar space by Townes and collaborators.

The lack of sensitive and proper technical equipment (i.e. telescopes, receivers, data processing facilities etc.) certainly has contributed to the length of this gap. However, the generally adopted view at that time, that the "hostile" interstellar gas cannot support anything more complex than simple molecular fragments, seems to have been for some time at least an even more serious obstacle to the investigation of the interstellar medium, than the technical shortcomings.

The purpose of this contribution is to point out presently existing observational constraints in the detection of interstellar molecular species and the limits they may cast on our knowledge of interstellar chemistry. In Section II I try to summarize the constraints which arise from the molecular side whereas in Section III I will point out some technical difficulties encountered in detecting new species. The last section will then refer to some implications for our understanding of molecular formation processes.

II. OBSERVATIONAL CONSTRAINTS : MOLECULAR

The presently known interstellar molecules now number more than 50. They consist of 13 inorganic molecules and 41 organic molecules, about half of which are unstable compounds. With the exception of about half a dozen simple species, all have been detected by radioastronomical means. Owing to its high spectral resolution capability, spectroscopy in the microwave, millimeterwave, and adjoining submillimeter wave region has become a powerful diagnostic tool. The proper chemical identification of interstellar molecules has to meet the exacting test with available highly precise laboratory data : interstellar line frequencies can be matched to laboratory frequencies often to a few parts in a million. Thus most of the identifications are to be considered certain or in some cases nearly so.

With some exceptions spectral lines observed to date are pure rotational transitions excited essentially by collisions with H_2. Radio astronomy however provides access only to the class of polar molecules, i.e. molecular species which possess a permanent electric dipole moment. Spectra of all the molecules in the millimeter wave region are caused by electric dipole transitions, and thus symmetrical molecules without a dipole moment and allowed rotational spectra as a rule cannot be observed in the microwave-, millimeterwave region. However, the existence of several non-polar molecules in space can be inferred from known polar derivatives, e.g. the molecules CO_2, CS_2, CH_4, C_2H_2, (acetylene), C_2H_4, C_2H_6, C_2N_2 must be fairly abundant in interstellar clouds, for all the appropriate smaller functional groups have been detected.

However, most of these non-polar molecules are infrared-active and give rise to intense ro-vibrational spectra, and should be discovered this way.

The list of interstellar molecules clearly suggests that carbon is of fundamental importance in molecular clouds and thus organic chemistry is prevalent in interstellar space. It seems however to be an organic chemistry without two important classes : ring molecules and branched chains are missing to best of our knowledge.

The present limit of detection lies with molecules whose number density relative to H_2, the most abundant molecule, is of the order 10^{-10}. This limit is dependent on the molecule's size and structure, the latter determining the rotational partition function. For simple structured molecules, i.e. linear molecules - such as chains - the partition function favours their detection, whereas for large organic molecules (assymmetric top molecules) the partition function is large and the energy is spread over many low lying levels. Thus although the abundance ratio of large organic molecules (e.g. CH_3- CH_2- CN) compared to a small organic molecule (e.g. HCN) may be 1 : 100 the observed rotational line ratios however may be a factor of 10 to 20 less.

One may thus expect that with more sensitive receiving equipment more and even larger molecules than presently known can be detected. However the high spectral resolution method of molecular identification will break down once the confusion limit is reached, i.e. when the molecular line density in the spectrum together with the appropriate line width will produce a type of "continuous spectrum". This limit will be reached faster for sources with broad lines such as SgrB2, than in sources where the lines are fairly narrow.

This situation becomes then comparable to the present dilemma of trying to deduce the composition of interstellar dust, by matching the observed interstellar absorption/emission features with fundamental vibrations of known molecules or functional groups. This has been done with great care and encouraging success by Allamandola and Greenberg and is being reported here at this meeting. However since the interstellar spectra from dust intrinsically can be obtained at low spectral resolution only - and I see no possibility how this situation can be improved - the information content which can be extracted will remain limited. This is generally true in spectroscopy, and failure to fully resolve the spectra provides the basis for the confusion limit.

Interstellar molecules known to date are composed of the six elements H, C, N, O, S, and Si. In the case of Si only the diatomic species SiO and SiS are found close to stars and infrared sources embedded in molecular clouds. However, one should expect molecules made of F, Na, Mg, Al, P, Cl, and Fe. Some of these molecules have been searched for particularly as hydrides or oxides but not found.

III. OBSERVATIONAL CONSTRAINTS : TECHNICAL

Up to now the most readily observed spectral lines of interstellar molecules are rotational transitions in the millimeter wave region. However the recently discovered warm molecular clouds and the hot protostellar surroundings clearly indicate that rotational transitions and ro-vibrational transitions in the submillimeter- and far infrared regions will greatly add to our knowledge once these relatively unexplored regions of the electromagnetic spectrum become accessible to routine-type observations. For this to happen two essential obstacles have to be overcome : construction of sensitive- high resolution type receiving equipment and removal of the strongly absorbing earth atmosphere. The latter can really only be achieved by space telescopes. Once these telescopes will be in operation a rich harvest of interesting science will be expected, particularly in high resolution spectroscopy. Details of this problem have been discussed during this conference by D. Buhl.

IV. GAS-PHASE vs. SURFACE CHEMISTRY

Since the first attempts to cast the subject of interstellar chemistry into quantitative terms by Bates and Spitzer (1951) in understanding the abundance of CH and CH^+, there has been considerable discussion about the relative importance of gas phase chemistry versus catalytic processes on interstellar grain surfaces. This issue has remained the great unknown in interstellar chemistry and has assumed new importance by being shifted to the question to which extent both mechanisms contribute to the formation of larger molecules, e.g. to the formation of long chain molecules. The issue has not become easier by the fact that one has to differentiate between the chemistry which occurs in cold clouds, in shock fronts and in molecular envelopes of stars.

Ever since the early succes in understanding formation of H_2 on the surfaces of grains, surface chemistry has been hampered seriously by the lack of predictability, which is connected with the intrinsic difficulties of the process itself. Firstly there is still now a great lack of detailed knowledge of what precisely grains are made of - and this is connected with the low information content of unresolved spectra - and secondly the details of the catalytic processes remain rather uncertain and speculative, even so we note encouraging progress in understanding reactions on grain surfaces (D.A. Williams, this conference). The central problem though remains, i.e. how the product molecule could acquire sufficient kinetic energy to escape from the grain surface.

On the other hand gas-phase reactions between positive ions and neutrals have been recognized in the early 70's to be a basic process giving birth to interstellar molecules (W.D. Watson 1976). He will deal with gas phase chemistry in this volume.

In summary I may then note concerning the question of the relative importance of catalytic processes on grain surfaces versus gas phase reactions that the results for simple molecules (such as HCO^+, HCN, NH_3, ...) lend preference to ion-molecule reactions over the competing surface chemistry models. The detection of cyanopolyyne molecules with a long linear carbon backbone has reveived the discussion of how and where these molecules form : gas-phase and grain-surface reactions are held responsible, and at present it seems difficult to differentiate between the two.

In addition present molecule formation models can only include into calculations the abundances of known interstellar molecules; for all others "educated" guesses have to be made. This however is a limitation which will be overcome once appropriate measurements become available.

REFERENCES :

Allamandola, L. J. (1984) This volume
Bates, D.R., and Spitzer, L. Astrophys. J. <u>113</u>, 441 (1951)
Buhl, D., Chin, G., and Petuchowski (1984) This volume
Watson, W.D. (1984) This volume (and references therein)
Watson, W.D. Rev. Mod. Phys. <u>48</u>, 513 (1976)
Williams, D.A. (1984) This volume
Winnewisser, G. (1981), Topics in Current Chemistry <u>99</u>, 39, Springer Verlag

SECTION II: EMISSION PROCESSES AND THEIR INTERPRETATION

ATOMIC AND IONIC EMISSION PROCESSES

D.R. Flower

Physics Department, Durham University, Durham DH1 3LE, U.K.

ABSTRACT

We consider atomic processes relevant to the interpretation of infrared measurements, with special emphasis on important, recent developments.

1. BASIC EMISSION THEORY

Consider an atom with two energy levels, E_i and E_j, and let $E_j - E_i = h\nu$, where ν is the frequency of the transition. The rate coefficient for collisional deexcitation is

$$q_{j \to i}(T_k) = \int_0^\infty v_j \, Q_{j \to i} \, f(v_j) \, dv_j \qquad (1)$$

where T_k is the kinetic temperature, v_j is the initial centre of mass velocity of the projectile, $Q_{j \to i}$ is the deexcitation cross-section, and $f(v_j)$ is the relative velocity distribution, usually taken to be Maxwellian. The principle of detailed balance relates the excitation and deexcitation rate coefficients:

$$\omega_i \, q_{i \to j}(T_k) = \omega_j \, q_{j \to i} \, \exp(-h\nu/\kappa T_k) \qquad (2)$$

where h is Planck's constant and κ is Boltzmann's constant; the degeneracies of the energy levels are denoted by ω_i, ω_j.

Let the local radiation density be ρ_ν. If a black-body distribution is appropriate,

$$\rho_\nu = W \, 8\pi \, h \left(\frac{\nu}{c}\right)^3 \left(\exp(h\nu/\kappa T_r) - 1\right)^{-1} \qquad (3)$$

where W is the appropriate geometrical dilution factor and T_r is the radiation temperature. The rate of radiative excitation is $n_i B(i \to j) \rho_\nu$ and the rate of radiative deexcitation is $\left(n_j B(j \to i) \rho_\nu + A(j \to i)\right)$, where n denotes a level population and A,B the appropriate Einstein coefficients.

The atomic level populations are determined by a set of linear, time-dependent equations of the form

$$\frac{d}{dt} n_j = n_i \left(B(i \to j) \rho_\nu + n_p q(i \to j)\right) -$$
$$- n_j \left(B(j \to i) \rho_\nu + A(j \to i) + n_p q(j \to i)\right) \quad (4)$$

where n_p is the perturber number density. In (4),

$$\frac{d}{dt} = \frac{\partial}{\partial t} + \underline{u} \cdot \underline{\nabla} \quad (5)$$

where $\underline{u}$ is the flow velocity of the gas. In a stationary state,

$$\frac{\partial}{\partial t} \equiv 0 . \quad (6)$$

In the static case,

$$\frac{d}{dt} \equiv 0 . \quad (7)$$

Condition (6) is more likely to be valid than condition (7) but may not be valid, for example, behind a shock wave. In any case, the relevant dynamical time scales should be compared with the characteristic times of the important atomic and molecular processes before (6) or (7) is assumed.

If the static limit applies, the equations (4) may be written

$$\sum_i n_i a_{ij} = 0 \quad (8)$$

where we now generalise to a many level atom. The normalisation condition is

$$\sum_i n_i = N , \quad (9)$$

where N is the total atomic number density. These equations are usually referred to as "statistical equilibrium equations". In matrix form, they become

$$\underset{\sim}{n}\, \underset{\sim}{a} = \underset{\sim}{N}, \tag{10}$$

where $\underset{\sim}{n}$ and $\underset{\sim}{N}$ are row vectors and $\underset{\sim}{a}$ is a square matrix. The general solution of (10) is

$$\underset{\sim}{n} = \underset{\sim}{N}\, \underset{\sim}{a}^{-1}. \tag{11}$$

It is instructive to consider two limiting cases:

(i) radiation dominated ($n_p \to 0$)

Then,

$$n_i\, B(i \to j)\, \rho_\nu = n_j \left(B(j \to i)\, \rho_\nu + A(j \to i) \right)$$

and, using the relations between the Einstein coefficients, we obtain

$$n_j/\omega_j = (n_i/\omega_i)\, \exp(-h\nu/\kappa T_r) \tag{12}$$

i.e. a Boltzmann distribution at the radiation temperature.

(ii) matter dominated ($n_p \to \infty$)

$$n_i\, n_p\, q(i \to j) = n_j\, n_p\, q(j \to i)$$

or, using (2),

$$(n_j/\omega_j) = (n_i/\omega_i)\, \exp(-h\nu/\kappa T_k) \tag{13}$$

i.e. a Boltzmann distribution at the kinetic temperature.

The general solution (11) requires a knowledge of the matrix $\underset{\sim}{a}$ and hence of ν, $A(j \to i)$ and $Q(j \to i)$. The latter should be calculated at a number of energies sufficient for the evaluation of the integral (1). Mendoza (1) has made a remarkable compilation of atomic data relevant to regions of ionised hydrogen. Additional processes, which are important in the cooler parts of the interstellar medium, are to be considered below.

2. PHYSICAL PROCESSES IN THE EMITTING REGIONS

2.1. Electron collisional excitation

Many man-years of programme development and production work, particularly in the groups of M.J. Seaton and P.G. Burke, have yielded a large body of good quality data on electron collisional excitation of atoms and ions in thermal plasmas. The more recent calculations include the contributions of autoionising resonances, to be discussed below.

The collision strength, Ω, for a transition is related to the cross-section, Q, through

$$\pi \Omega(i,j) = k_i^2 \, \omega_i \, Q(i \to j) \,. \tag{14}$$

When atomic units are employed ($e = m = \hbar = 1$), the square of the incident wave number, k_i^2, is numerically equal to the incident electron energy, expressed in the Rydberg unit. The collision strength is a dimensionless quantity, of order unity, which is symmetric in i and j. In terms of Ω, the deexcitation rate coefficient may be written

$$q_{j \to i}(T_k) = 8.63 \times 10^{-6} \, \frac{T(j,i)}{\omega_j \, T_k^{\frac{1}{2}}} \quad (cm^3 \, s^{-1}) \tag{15}$$

where

$$T(j,i) = \int_0^\infty \Omega(j,i) \, \exp(-x_j) \, dx_j \tag{16}$$

and with $x_j = mv_j^2/(2\kappa T_k)$. If Ω is independent of v, $T = \Omega$ is independent of T_k^j. However, Ω is generally <u>not</u> energy independent. Indeed, Ω is <u>strongly</u> energy dependent in the vicinity of an autoionising resonance. This point is well illustrated by recent calculations (2) of the collision strengths for the O III $2p^2$ $^3P_J - ^3P_{J'}$ fine structure transitions. Fortunately, T, which is the quantity required in astrophysical applications, is a much more slowly varying function of T_k, and compact tabulations of $T_{j,i}(T_k)$ are possible.

2.2. Photoionisation

Mendoza (1) includes in his review a list of references to data on photoionisation cross-sections of atoms and ions.

2.3. Recombination

In low density plasmas, electron-ion recombination is a radiative process, the inverse of photoionisation. Radiative recombination and photoionisation are related through the principle of detailed balance:

$$\alpha_i(T_k) = \frac{1}{c^2} \left(\frac{2}{\pi}\right)^{1/2} (m\kappa T_k)^{-3/2} \frac{\omega_i}{\omega_+} \exp(I_i/\kappa T_k)$$

$$\times \int_{I_i}^{\infty} (h\nu)^2 a_\nu \exp(-h\nu/\kappa T_k) \, d(h\nu) \qquad (17)$$

where α_i denotes the coefficient for recombination to form the state i, ω_i and ω_+ are the degeneracies of the state i and of the recombining ion, respectively, I_i (> 0) is the ionisation potential of the state i, and a_ν is the corresponding photoionisation cross-section. If a_ν in (17) includes the contribution of autoionising resonances, no further correction to α_i is necessary to allow for the process of dielectronic recombination, which we now discuss.

The importance of dielectronic recombination in high temperature astrophysical plasmas, such as the solar corona, has been recognised since the early work of Burgess (3,4). One of the most important developments in the study of low density astrophysical plasmas in recent years has been the demonstration, by Beigman and Chichkov (5) and by Storey (6), that dielectronic recombination is important for many ions of the abundant elements C, N and O at the much lower temperatures ($T_k \approx 10^4$ K) encountered in gaseous nebulae. A complete set of calculations for C, N and O is now available (7).

Consider the process

$$X^{+m} + e \rightarrow \left(X^{+(m-1)}\right)^* \qquad (18)$$

where $X^{+(m-1)}$ is formed in an autoionising state (which lies above the first ionisation limit). From the principle of detailed balance, the rate coefficient for process (18) is given by

$$\gamma = \frac{\omega_*}{2\omega_{+m}} \left(\frac{h^2}{2\pi m\kappa T_k}\right)^{3/2} \exp(-E_*/\kappa T_k) A_a \qquad (19)$$

where $2\omega_{+m}$ and ω_* are the degeneracies of the left- and right-hand sides of (18), respectively (the factor two accounts for the spin of the initially free electron), E_* (> 0) is the energy of the auto-ionising state relative to the ground state of X^{+m}, and A_a is the autoionisation probability. In dielectronic recombination, (18) is followed by

$$\left(X^{+(m-1)}\right)^* \rightarrow X^{+(m-1)} + h\nu \qquad (20)$$

i.e. by a radiative stabilisation process. The dielectronic recombination coefficient is

$$\alpha_d = \gamma \, A_r/(A_r+A_a) \tag{21}$$

where A_r is the radiative transition probability.

Let us denote by n the principal quantum number of the state into which the electron is captured in process (18). At high (i.e. coronal) temperatures, highly excited Rydberg states (n large) are preferentially populated, and, as $A_r \sim n^{-3}$, the probability of stabilisation through a radiative transition of the outer, captured electron is small. Under these conditions, process (20) occurs through radiative transitions of the core (3,4). However, at low (i.e. nebular) temperatures, capture occurs preferentially into states of lower n, and stabilisation through radiative transitions of the outer electron must also be considered (5,6).

An informative example is the process of dielectronic capture

$$C^{2+}(2s^2 \, {}^1S) + e \to C^+(2s \, 2p(^3P^o) \, 3d \, {}^2F^o) \tag{22}$$

for which $E_* \approx 3\,300$ cm^{-1} (easily accessible at nebular temperatures: 1 cm^{-1} ≡ 1.44 K). In this case, the process of stabilisation through a radiative transition of the core,

$$C^+(2s \, 2p(^3P^o) \, 3d \, {}^2F^o) \to C^+(2s^2(^1S) \, 3d \, {}^2D) + h\nu \tag{23}$$

is spin-forbidden and has a small probability. In fact, stabilisation occurs principally through

$$C^+(2s \, 2p \, 3d \, {}^2F^o) \to C^+(2s \, 2p^2 \, {}^2D) + h\nu \tag{24}$$

as indicated in Fig. 1.

Storey (6) makes the assumption that $A_a \gg A_r$, in which case (19) and (21) give

$$\alpha_d = \frac{\omega_*}{2\omega_{+m}} \left(\frac{h^2}{2\pi m \kappa T_k}\right)^{3/2} \exp(-E_*/\kappa T_k) \, A_r \tag{25}$$

and the dielectronic recombination coefficient is determined by the probability of stabilisation through a radiative transition of the captured electron. In this case, dielectronic recombination assimilates to a radiative recombination process, proceeding via autoionising states, and, as such, is already included in equation (17), providing a_ν incorporates the contribution of the autoionising resonances to the photoionisation cross-section. However, if the condition $A_a \gg A_r$ does not hold, collisional and radiative processes must be considered simultaneously and treated in a consistent manner, i.e. the radiation field must be quantised (8).

ATOMIC AND IONIC EMISSION PROCESSES

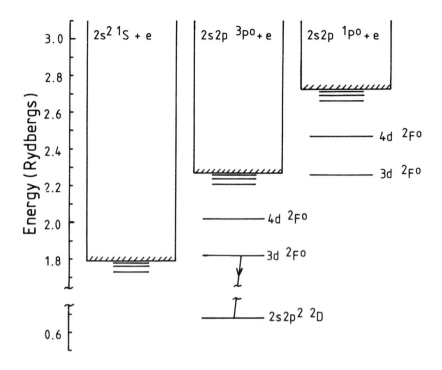

Figure 1. Illustrating the dielectronic recombination of $C^{2+}(2s^2\ ^1S)$.

The current status of calculations of recombination processes is reviewed by Storey (9), who suggests a procedure for obtaining a best estimate of the recombination coefficients for positive ions at nebular temperatures. Work in this area is continuing and further, important results might be expected during the next few years.

2.4. Charge transfer

We consider the processes

$$X^{+m} + H^0 \rightleftarrows X^{+(m-1)} + H^+ + \Delta E \tag{26}$$

and

$$X^{+m} + He^0 \rightleftarrows X^{+(m-1)} + He^+ + \Delta E \tag{27}$$

where ΔE is the energy defect. In the last few years, work in the groups of A. Dalgarno, R. McCarroll, and W. Watson has shown that charge transfer is important in the ionisation equilibrium of many abundant interstellar ions. A compilation of the results of recent calculations of rate coefficients for important charge transfer reactions of the types (26) and (27) is presented in Table 1. Although simple approximations, such as the Landau-Zener and orbiting approximations, can sometimes yield reasonable estimates of the rate coefficients for charge transfer reactions, confidence should be placed only in those results derived from quantum mechanical calculations of the relevant potential energy curves and cross sections. Even when quantum mechanical methods are used, substantial differences can subsist between the results of independent calculations, as may be seen by comparing the two sets of results in Table 1 for the reaction of C^{3+} with H^o at $T = 10^4$ K.

The charge transfer reaction

$$O^+ + H^o \rightarrow O^o + H^+ \tag{28}$$

is particularly interesting. Early work of Chamberlain (15) suggested that this reaction was effective in establishing the ionisation equilibrium between O^o and O^+ in the Cassiopeia radio source. The relevance of the reaction to the ionisation equilibrium of the interstellar medium was subsequently established by Field and Steigman (16). These authors used the classical orbiting approximation to calculate the charge transfer cross-section, assuming a statistical distribution amongst the fine structure states, ^3P_J, of the O^o product of the reaction.

More recent quantum mechanical calculations (13) have shown that the assumption that the fine structure states are populated in proportion to their statistical weights is unjustified. In Table 2, we compare the results of the quantum mechanical (13) and the classical (16) calculations. The discrepancies are substantial at low temperatures, where the quantal results are to be preferred.

At nebular temperatures ($T \approx 10^4$ K), the exact value of the rate coefficient is unimportant; it is sufficient that the rate coefficient is large ($\approx 10^{-9}$ cm^3 s^{-1}). Under these circumstances, charge transfer determines both the neutralisation and ionisation rates for oxygen, through reaction (28) and its inverse, and hence

$$\frac{N(O^+)}{N(O^o)} = \frac{8}{9} \frac{N(H^+)}{N(H^o)} \tag{29}$$

The consequences of (29) for the interpretation of the spectra of optically thick planetary nebulae were first explored by Williams (17).

Rate coefficient (10^{-9} cm^3 s^{-1})

Ion	T=10	30	50	10^2	300	10^3	3000	5000	10^4	50 000 K
C^{3+}								3.09	3.58	5.46 (10)
		1.5		1.6	1.6	1.6	1.6		1.6 (12)	
C^{4+}	3.39	3.12		2.71	2.38	2.25	2.19		2.13	3.22 (14)
N^{2+}								0.78	0.86	1.11 (10)
N^{3+}	0.35	0.30		0.25	0.25	0.43		1.54	2.93	9.47 (10)
	0.34			0.41		1.12	1.82	3.41	11.23 (14)
O^+			0.37			0.75 (13)				
O^{2+} {H^0								0.60	0.77	1.62 (10)
{He^0								0.10	0.20	0.89 (10)
O^{3+}								6.34	8.63	17.6 (10)
Ne^{3+}								4.00	5.68	13.0 (10)
Si^{2+}	1.98		1.75	1.72		2.50		4.34	5.28	7.70 (11)

Table 1. Calculated rate coefficients for charge transfer reactions with H^0 (unless otherwise indicated). Numbers in parentheses refer to the source of the data in the corresponding row.

T(K)	(a)	(b)
10	3.4	9.7
25	3.5	9.8
50	3.7	9.9
100	4.1	10.0
1000	7.5	10.2

Table 2. Rate coefficients for the charge transfer reaction $O^+ + H^o \rightarrow O^o + H^+$, in units of 10^{-10} cm^3 s^{-1}: (a) quantal calculation (13); (b) classical orbiting approximation (16).

It is clear from (26) and (27) that the influence of charge transfer as a recombination mechanism for heavy ions increases with the fractional abundance of neutral hydrogen (helium). In gaseous nebulae, the fractional abundance of neutral hydrogen increases with the effective temperature of the exciting star: at high temperatures, there is a relatively smaller flux of ionising radiation just shortwards of the Lyman limit, where the photoionisation cross-section is largest. It follows that charge transfer processes are more important in planetary nebulae than in H II regions, and more important in high excitation than in low excitation planetary nebulae. For recent reviews of charge transfer processes in the context of planetary nebulae, see McCarroll et al. (18) and Péquignot (19).

3. EXCITATION OF FINE STRUCTURE TRANSITIONS BY HYDROGEN

We consider the processes

$$C^+(^2P^o_{1/2}) + H^o \rightarrow C^+(^2P^o_{3/2}) + H^o \tag{30}$$

$$C^o(^3P_o) + H^o \rightarrow C^o(^3P_{1,2}) + H^o \tag{31}$$

and

$$O^o(^3P_2) + H^o \rightarrow O^o(^3P_{1,o}) + H^o \tag{32}$$

Cross-sections and rate coefficients for (30) have been calculated by Bazet et al. (20) and by Launay and Roueff (21). The latter authors employed a quantum mechanical, close coupling formalism, appropriate to thermal energy collisions. Launay and Roueff (22) have also calculated the cross sections and rate coefficients for (31) and (32). The rate of the reaction

$$C^+(^2P^o_{1/2}) + H_2 \rightarrow C^+(^2P^o_{3/2}) + H_2 \quad (33)$$

has been computed by Flower and Launay (23).

Processes (30-33) are important in the thermal balance of the interstellar medium. With the exception of C I $^3P_1 \leftarrow {}^3P_2$ 369 μm, the corresponding radiative transitions have been observed.

4. RECOMBINATION LINES OF HYDROGEN AND HELIUM

Previous calculations of H I, He I and He II recombination spectra by Brocklehurst (24,25) and Giles (26) have recently been extended by Hummer and Storey (27) to include additional transitions which are observed in the infrared spectra of gaseous nebulae.

5. ROTATIONAL EXCITATION OF OH IN INTERSTELLAR MASERS

Cross-sections for the rotational excitation of OH by para-H_2 have recently been computed by Dewangan and Flower (28,29). Their calculations showed that this process would produce population inversion within the Λ-doublets of the $^2\Pi_{3/2}$ ladder but not in the lower doublets of the $^2\Pi_{1/2}$ ladder.

The results of these calculations were subsequently incorporated in a radiative transfer programme incorporating the process of rotational line overlap, which is known to be important in studies of H II/OH masers (30). The principal conclusions of this study (31) were that

(i) the strong masers observed in the $^2\Pi_{3/2}$ ladder may be attributed to collisional inversion by H_2;

(ii) the masers observed in the $^2\Pi_{1/2}$ ladder are produced by "thermal" line overlaps;

(iii) the masers are effectively quenched by the far infrared radiation from local dust if the temperature of the dust is comparable to or exceeds the kinetic temperature of the gas.

Further work (32) showed that cooling of the maser regions by OH, through the process of collisional excitation of the rotational transitions followed by the emission of far-infrared photons, dominates the cooling by H_2. Indeed, the efficiency of the OH cooling process is such as to impose severe constraints on the corresponding heating mechanism (33). It would certainly seem worthwhile attempting to detect the far-infrared rotational transitions of OH in H II/OH maser sources, such as W3(OH).

ACKNOWLEDGEMENTS

The author is indebted to Dr. P.J. Storey for discussions of his work on dielectronic recombination.

REFERENCES

(1) Mendoza, C.: 1983, in "Planetary Nebulae", IAU Symposium No. 103, ed. D.R. Flower, D. Reidel (Dordrecht).
(2) Aggarwal, K.M., Baluja, K.L. and Tully, J.A.: 1982, Mon. Not. R. Astron. Soc. 201, 923.
(3) Burgess, A.: 1964, Astrophys. J. 139, 776.
(4) Burgess, A.: 1965, Astrophys. J. 141, 1588.
(5) Beigman, I.L. and Chichkov, B.N.: 1980, J. Phys. B, 13, 565.
(6) Storey, P.J.: 1981, Mon. Not. R. Astron. Soc. 195, 27P.
(7) Nussbaumer, H. and Storey, P.J.: 1983, Astron. Astrophys., to be submitted.
(8) Davies, P.C.W. and Seaton, M.J.: 1969, J. Phys. B, 2, 757.
(9) Storey, P.J.: 1982. in "Planetary Nebulae", IAU Symposium No. 103, ed. D.R. Flower, D. Reidel (Dordrecht).
(10) Butler, S.E., Heil, T.G. and Dalgarno, A.: 1980, Astrophys. J. 241, 442.
(11) Gargaud, M., McCarroll, R. and Valiron, P.: 1982, Astron. Astrophys. 106, 197.
(12) Watson, W.D. and Christensen, R.B.: 1979, Astrophys. J. 231, 627.
(13) Chambaud, G., Launay, J.M., Levy, B., Millie, P., Roueff, E. and Tran Minh, F.: 1980, J. Phys. B, 13, 4205.
(14) Gargaud, M., Hanssen, J., McCarroll, R. and Valiron, P.: 1981, J. Phys. B, 14, 2259.
(15) Chamberlain, J.W.: 1956, Astrophys. J. 124, 390.
(16) Field, G.B. and Steigman, G.: 1971, Astrophys. J. 166, 59.
(17) Williams, R.E.: 1973, Mon. Not. R. Astron. Soc. 164, 111.
(18) McCarroll, R., Valiron, P. and Opradolce, L.: 1983, in "Planetary Nebulae", IAU Symposium No. 103, ed. D.R. Flower, D. Reidel (Dordrecht).
(19) Péquignot, D.: 1983, in "Planetary Nebulae", IAU Symposium No. 103, ed. D.R. Flower, D. Reidel (Dordrecht).
(20) Bazet, J.F., Harel, C., McCarroll, R. and Riera, A.: 1975, Astron. Astrophys. 43, 229.
(21) Launay, J.M. and Roueff, E.: 1977, J. Phys. B, 10, 879.
(22) Launay, J.M. and Roueff, E.: 1977, Astron. Astrophys. 56, 289.
(23) Flower, D.R. and Launay, J.M.: 1977, J. Phys. B, 10, 3673.
(24) Brocklehurst, M.: 1971, Mon. Not. R. Astron. Soc. 153, 471.
(25) Brocklehurst, M.: 1972, Mon. Not. R. Astron. Soc. 157, 211.
(26) Giles, K.: 1977, Mon. Not. R. Astron. Soc. 180, 57P.
(27) Hummer, D.G. and Storey, P.J.: 1983, to be published.
(28) Dewangan, D.P. and Flower, D.R.: 1981, J. Phys. B, 14, L425.
(29) Dewangan, D.P. and Flower, D.R.: 1981, J. Phys. B, 14, 2179.
(30) Guilloteau, S., Lucas, R. and Omont, A.: 1981, Astron. Astrophys. 97, 347.

(31) Flower, D.R. and Guilloteau, S.: 1982, Astron. Astrophys. 114, 238.
(32) Flower, D.R., Guilloteau, S. and Hartquist, T.W.: 1982, Mon. Not. R. Astron. Soc. 200, 55P.
(33) Guilloteau, S.: 1982, Astron. Astrophys. 116, 101.

INFRARED SPECTROSCOPY OF INTERSTELLAR SHOCKS

Christopher F. McKee and David F. Chernoff
Departments of Physics and Astronomy, University of California, Berkeley, CA 94720

David J. Hollenbach
NASA-Ames Research Center, Moffett Field, CA 94035

ABSTRACT

Infrared emission lines from interstellar shocks provide valuable diagnostics for violent events in the interstellar medium, such as supernova remnants and mass outflow from young stellar objects. There are two types of interstellar shocks: In J shocks, gas properties "jump" from their preshock to their postshock values in a shock front with a thickness ≤ one mean free path; radiation is emitted behind the shock front, primarily in the visible and ultraviolet, but with a few strong infrared lines, such as OI(63μm). Such shocks occur in ionized or neutral atomic gas, or at high velocities ($v_s \gtrsim 50$ km/s) in molecular gas. In C shocks, gas is accelerated and heated by collisions between charged particles, which have a low concentration and are coupled to the magnetic field, and neutral particles; radiation is generated throughout the shock and is emitted almost entirely in infrared emission lines. Such shocks occur in weakly ionized molecular gas for shock velocities below about 50 km/s.

1. INTRODUCTION

Shocks are supersonic compressions. Since they are supersonic, the medium ahead of the shock cannot respond until the shock strikes --the shock is a "hydrodynamic surprise." Shocks compress, heat, accelerate, and increase the entropy of the gas which they encounter. They are ubiquitous in the interstellar medium (ISM) because there are many mechanisms which generate velocities in excess of the local sound speed, which ranges approximately from $10^{-0.5}$ km/s in molecular clouds to 10^2 km/s in the hot, low density component of the ISM. For example, supernova remnants have expansion velocities up to about 10^4 km/s and stellar winds up to about $10^{3.5}$ km/s, far in excess of the sound speed. Ionization fronts and cloud-cloud collisions produce velocities of 1-10 km/s, which are often supersonic in the cold components of the ISM.

Interstellar shocks play a dominant role in determining the structure of the ISM because they transmit energy from stars to the gas and fix the gas pressure (McKee and Ostriker 1977, Cox 1979). They can compress the gas to the point that it becomes gravitationally unstable, thereby inducing star formation (Elmegreen and Lada 1977). They are effective at destroying grains, thus determining the gas phase abundances in the ISM (Draine and Salpeter 1979). Finally, they may well be responsible for the acceleration of cosmic rays (e.g., Axford 1981). From the point of view of astronomers, however, shocks are important because they cause the gas to radiate, providing an invaluable diagnostic for violent events in the ISM.

Interstellar shocks of moderate velocity ($v_s < 100$ km/s) are often best observed in the infrared, both because the emission is concentrated there and because intervening dust obscures the optical radiation. Indeed, since the intensity of the shock radiation scales with the density, the brightest shocks will occur in dense gas which is so highly obscured that it will be observable only in the infrared. The strongest infrared transitions which have been observed in interstellar shocks are the rotational-vibrational transitions in H_2 (see the review by Shull and Beckwith, 1982). The presence of these lines is not sufficient in itself to establish the existence of a shock, however, since they can also be excited by ultraviolet pumping (Hollenbach and Shull 1977). Confirming evidence such as broad lines, implying high velocities, or indicators of high temperatures, such as H_2 line ratios or far infrared CO rotational emission, is also required. The best-studied shocked region is the BN-KL region of Orion (Gautier et al 1976), which is discussed below in section 4. The shock is thought to be driven by outflow from embedded young stellar objects, similar to that observed in other sources (Lada 1983). NGC 2071 is another example of such a source; in addition to exhibiting shocked H_2 (Persson et al 1981), it shows high velocity HI and CO as well (Bally and Stark 1983, Bally 1982). Infrared emission from shocks has also been observed from planetary nebulae and supernova remnants, which are at the opposite extreme of stellar evolution. The H_2 line ratios in the planetary NGC 7027 indicate shock excitation (Smith et al 1981). The supernova remnant IC443 has H_2 emission (Treffers 1979) and high velocity OH and CO (DeNoyer 1979 a,b); near infrared iron lines have been observed in MSH 15-5$\underline{2}$ (Seward et al 1983). Two galactic nuclei are known to be sources of H_2 emission, the Galactic Center (Gatley 1983) and NGC 1068 (Hall et al 1981). In each case a strong UV source is present, so it is not clear that the emission arises in a shock, although the observation of the J=16 -> J=15 CO transition in the Galactic Center (Watson 1983) suggests that at least some of the emission there arises in a shock.

In this paper we shall summarize the current understanding of interstellar shocks, with a focus on their infrared emission. The following section contains an overview of shock structure, emphasizing the distinction between classical non-magnetic shocks (termed J shocks by Draine, 1980) and shocks in magnetized, weakly ionized plasmas, in which the dissipation is due to ion-neutral collisions (C shocks). These differences are summarized in section 2.4. J shocks are the topic of section 3; they are discussed in the review by McKee and Hollenbach

(1980). Most of the work on C shocks has been done since that review, and they are discussed in section 4. The results are summarized in section 5.

2. OVERVIEW OF SHOCK STRUCTURE

2.1 J Shocks

The "classical" view of interstellar shocks was developed by Field et al (1968) and Cox (1972). Such shocks can be approximately divided into three regions: (1) a radiative precursor in which the ambient, preshock gas is heated and, perhaps, ionized by radiation from the postshock gas; (2) the shock front itself, in which the unshocked gas is accelerated and the relative kinetic energy between the unshocked and shocked gas is dissipated into heat; and (3) the post shock relaxation layer in which inelastic collisonal processes (ionization, dissociation, collisional excitation, recombination, and molecule formation) produce the shock emission and cause the gas to relax to its final downstream physical and chemical state. If the gas is substantially ionized, the dissipation in the shock front is effected by plasma instabitilities in a distance short compared to a collisional mean free path; if not, the dissipation is effected by collisions between neutral atoms or molecules in a distance of order $(n_n \sigma_{nn})^{-1}$, where n_n is the density of neutral particles and σ_{nn} the neutral particle cross section. Since the gas "jumps" from its upstream to downstream conditions across the shock front in a distance short compared to hydrodynamic and radiative length scales, such shocks are termed J shocks (Draine 1980).

In a J shock, the shock front is thin, so the physical conditions behind the shock front are related to those ahead of the front by the one-dimensional, steady state hydrodynamic equations. Conservation of mass, momentum, and energy gives the Rankine-Hugoniot relations, or jump conditions; in the absence of a magnetic field, these are

$$[\rho v] = 0, \qquad (2.1)$$
$$[p + \rho v^2] = 0, \qquad (2.2)$$
$$[v^2/2 + \gamma p /(\gamma-1)\rho] = 0, \qquad (2.3)$$

where $[x]$ is the postshock value of the variable x minus the initial value. Here the velocity v is measured in the shock frame (the frame in which all disturbances are stationary), ρ is the density, p the pressure, and γ the ratio of the specific heats. The energy conservation equation (eq. 2.3) remains valid only so long as radiative losses are negligible; the mass and momentum equations remain valid throughout the shock, provided that the steady state and one dimensional approximations are satisfied. Solution of these equations shows that the gas is supersonic ahead of the shock and subsonic behind. In terms of the isothermal Mach number

$$M = v_s/c_{s0} = (\rho_0 v_s^2/p_0)^{1/2}, \qquad (2.4)$$

where v_s is the shock velocity and ρ_0, p_0, and c_{s0} are the initial values of ρ, p, and the isothermal sound speed c_s, we have $M > \gamma^{1/2}$ ahead of the shock and $M < \gamma^{1/2}$ behind. The shock is compressive ($\rho > \rho_0$, $p > p_0$), and the compression increases monotonically with M.

Strong shocks are defined by $M \gg 1$ (i.e. $\rho_0 v_s^2 \gg p_0$), and for them the jump conditions give

$$\rho_s = \rho_0 (\gamma + 1)/(\gamma - 1) \quad \rightarrow 4\rho_0, \quad (2.5)$$

$$p_s = 2 \rho_0 v_s^2 /(\gamma + 1) \quad \rightarrow 3/4\, \rho_0 v_s^2, \quad (2.6)$$

$$kT_s = 2 \mu_s v_s^2 (\gamma - 1)/(\gamma + 1)^2 \quad \rightarrow 3/16\, \mu_s v_s^2, \quad (2.7)$$

where ρ_s, p_s, T_s, and μ_s (the mean mass per particle) are evaluated just behind the shock front; the values after the arrows are for $\gamma = 5/3$. Note that the pressure and temperature increase without limit as $v_s \rightarrow \infty$, whereas ρ_s is bounded. For $\gamma = 5/3$, the numerical value of the temperature just behind the shock front is

$$T_s = \begin{cases} 1380\ v_{s6}^2 & \text{Ionized} \\ 2900\ v_{s6}^2 & \text{Neutral Atomic} \\ 5300\ v_{s6}^2 & \text{Molecular} \end{cases} \quad (2.8)$$

where $v_{s6} = v_s/(10^6\ \text{cm/s})$.

Radiative losses in the postshock relaxation layer and excitation of H_2 vibrational/rotational levels lower the temperature of the shocked gas, and the density increases to maintain pressure balance. For $\rho \gg \rho_0$, the velocity in the shock frame $v = \rho_0 v_s/\rho$ approaches zero; then one finds $p \rightarrow \rho_0 v_s^2$. In the absence of a magnetic field, the final compression of the shocked gas is

$$\rho_f/\rho_0 = M^2 (T_0/T_f), \quad (2.9)$$

where ρ_f and T_f are the final density and temperature. In many astrophysical plasmas, however, magnetic fields limit the compression to a smaller value.

2.2 Magnetic Fields in Shocks

Magnetic fields have major effects on the structure of J shocks, although their effect on the emitted spectrum is smaller. C shocks would not exist in the absence of magnetic fields. In collisionless shocks, which are necessarily J-type, the mechanisms responsible for dissipating the relative kinetic energy of the shocked and unshocked plasmas may be completely altered by a magnetic field, as appears to be the case in the Earth's bow shock (Russell and Greenstadt 1979); this has no effect on the spectrum of the shocked gas. Cosmic rays may be accelerated with an efficiency large enough to modify the shock structure (Axford 1981), but in view of the uncertainties associated with this effect we shall not consider it here. Magnetic fields alter the jump conditions and limit the compression, which can alter the spectrum of the low temperature emission from the shock. Finally, in a partially ionized gas, the magnetic pressure may drive ionized gas ahead of the shock front in a magnetic precursor (Mullan 1971, Draine 1980). In many cases this precursor does not significantly modify the spectrum, but for sufficiently weak coupling between the magnetized, charged gas and the neutral gas the J shock structure disappears and is replaced by the C shock structure, as discussed in section 2.3.

We characterize the strength of the magnetic field by the parameter b:

$$B_0 = 10^{-6} \, b \, n_0^{1/2} \, G, \qquad (2.10)$$

with n_0 in hydrogen nuclei cm^{-3}. In terms of the upstream Alfven velocity $v_A = B_0/(4\pi\rho_0)^{1/2}$, we have

$$b = v_A / (1.8 \text{ km/s}). \qquad (2.11)$$

In interstellar clouds the quantity b is typically of order unity (see Mouschovias, 1976 and the discussion of Orion below). For $v_A > c_{s0}$, the strength of the shock is determined by the Alfven Mach number

$$M_A = v_s/v_A = (4\pi\rho_0)^{1/2} \, v_s/B_0. \qquad (2.12)$$

Under interstellar conditions, it is generally assumed that the magnetized, charged gas is tightly coupled to the neutral gas --i.e., that the magnetic field is "frozen" to the fluid. We shall analyze the validity of this assumption in interstellar shocks for the simple case in which the magnetic field $\underline{B}$ is perpendicular to the shock velocity. For a steady, planar shock Faraday's law gives

$$\underline{\nabla} \times \underline{E} = \underline{0} \quad \rightarrow \quad [\underline{E}_\perp] = 0, \qquad (2.13)$$

where $\underline{E}_\perp$ is the component of $\underline{E}$ perpendicular to $\underline{v}_s$. Under the same conditions, the current $\underline{J}$ is confined to the shock plane, since a current parallel to $\underline{v}_s$ would lead to a build up of charge on each side of the shock. The equations of motion for the charged particles then give the following simple form for Ohm's law:

$$\underline{J} = \sigma \, (\underline{E} + \underline{v}_{c\|} \times \underline{B}/c), \qquad (2.14)$$

where $\underline{v}_{c\|}$ is the component of the charged particle velocity parallel to $\underline{v}_s$ and σ is the scalar conductivity. This expression is exact for steady shocks in two cases of interest (provided the pressure tensor is diagonal): (1) a three component plasma consisting of ions, electrons, and neutrals; and (2) a plasma with an arbitrary number of species in the limit of vanishing ionization. In the latter case the conductivity is determined by the charged particle-neutral particle collision frequency and v_c is a weighted mean velocity which is close to the mean electron velocity. Combining equations (2.13) and (2.14) with Ampere's law gives

$$[\mathbf{J}/\sigma - \mathbf{v}_{c\parallel} \times \mathbf{B}] = [c \nabla \times \mathbf{B}/4\pi\sigma - \mathbf{v}_{c\parallel} \times \mathbf{B}/c] = 0. \quad (2.15)$$

We now analyze this equation on three different length scales. On the largest scale, far from the shock front, $\mathbf{B}$ approaches a constant ($\nabla \times \mathbf{B} \to 0$) and $\mathbf{v}_c$ approaches the mean fluid velocity $\mathbf{v}$, so that equation (2.15) reduces to

$$[vB] = 0 = [B/\rho] \quad (2.16)$$

with the aid of equation (2.1). Hence, on this large scale, the field is frozen to the fluid: B can increase if and only if the fluid is compressed by the same amount. This conclusion is independent of any complicated processes, such as plasma instabilities, occurring within the shock front, and it is more general than our derivation indicates, since Ohm's law reduces to equation (2.15) far from the shock front even if $\mathbf{B}$ is not perpendicular to $\mathbf{v}_s$; however, the simple result (2.16) applies only for $\mathbf{B}$ perpendicular to $\mathbf{v}_s$.

On an intermediate scale, in the vicinity of the shock front, it is possible that $\mathbf{v}_{c\parallel} \neq \mathbf{v}$ while at the same time the conductivity is high enough that $\mathbf{J}/\sigma \cong 0$. In that case the field is frozen to the charged particles but the charged particles can slip through the neutrals, so that

$$[v_c B] = 0. \quad (2.17)$$

This relation reduces to (2.16) far from the shock front, but it remains valid inside magnetic precursors and C shocks. In general, the charged particle mass flux $\rho_c v_c$ is not constant because of ionization and recombination in the shock, so one does not have $[B/\rho_c] = 0$.

Finally, on the smallest scale, we can demonstrate that the $\mathbf{J}/\sigma$ term never becomes important in interstellar plasmas, even on the relatively small length scales which occur in interstellar shocks. The ratio of the two terms in brackets in equation (2.15) is the magnetic Reynolds number (Jackson 1975)

$$R_M = 4\pi\sigma \ell_B v_c / c^2 \cong |\mathbf{v} \times \mathbf{B}/c| / |c \nabla \times \mathbf{B}/4\pi\sigma|, \quad (2.18)$$

where ℓ_B is the length scale over which B varies. For $R_M \gg 1$, the $\mathbf{J}/\sigma$ term is negligible and the field is frozen to the charged particles; for $R_M < 1$, the field diffuses through the charged particles. R_M is a minimum where ℓ_B is a minimum, corresponding to the maximum value of $\mathbf{J}$. A generous upper bound on J is $n_e e v_{Te}$, which would occur if the electrons moved through the ions at their thermal velocity v_{Te} (see the discussion of ℓ_B in Arons and Lea 1976; we assume $v_{Te} > v_s$). Ampere's law then yields

$$\ell_B \geqslant c\,B/4\pi\,n_e\,e\,v_{Te}, \qquad (2.19)$$

so that

$$R_M \geqslant 5 \times 10^6\,b\,v_6/(n_0^{1/2}\,T_{e4}), \qquad (2.20)$$

where the conductivity in a weakly ionized gas is $\sigma \cong 4 \times 10^{15}\,x_e\,T_{e4}^{-1/2}$ c.g.s. This implies that under interstellar conditions ($b \geqslant 0.1$, $n_0 \leqslant 10^8$ cm^{-3}) we have $R_M \gg 1$, even inside the shock fronts. The assumption that $v_c B$ = constant (eq. 2.17) made in all studies of interstellar shocks to date is therefore fully justified. Two caveats should be noted, however: First, the constancy of $v_c B$ does not necessarily apply on the extremely small length scales which occur in collisionless shock fronts because the electromagnetic fields are not time-steady in such fronts (Tidman and Krall 1971). Second, although the field is tied to the charged particles, it does not follow that all species of charged particles are tied to the field; the velocity v_c is very nearly equal to the mean electron velocity, and very heavy charged particles --namely, grains --can move at quite a different velocity (Draine 1980).

When the charged and neutral particles are tightly coupled, as they are almost everywhere in a J shock, the momentum jump condition (2.2) becomes

$$[p + \rho v^2 + B^2/8\pi] = 0. \qquad (2.21)$$

Just behind the shock front, $B^2/8\pi$ is at most $(\rho_s/\rho_0)^2 = 16$ times larger than the preshock value. Hence for sufficiently strong shocks ($M_A \gg 1$), the magnetic field does not alter the jump conditions across the front. Hollenbach and McKee (1979) found that equation (2.8) for the post-shock temperature with B=0 was accurate to 30% for $M_A \geqslant 6$. As radiative losses behind the shock front lead to greater compression, the terms in equation (2.21) vary as $p \propto \rho T$, $\rho v^2 \propto (\rho v)^2/\rho \propto \rho^{-1}$, and $B^2 \propto \rho^2$. Hence, at sufficiently large compressions the postshock value of the momentum flux approaches $(\rho/\rho_0)^2\,B_0^2/8\pi$. For a strong shock the initial pressure is negligible, so the initial momentum flux is simply $\rho_0 v_s^2$. Balancing these terms in accord with equation (2.21) gives the asymptotic compression

$$\rho_m/\rho_0 = B_m/B_0 = 2^{1/2}\,M_A = 77\,v_{s7}/b, \qquad (2.22)$$

where ρ_m and B_m are the final (maximum) values of the postshock density and magnetic field. For $b \cong 1$ (corresponding to a milligauss field at $n_0 = 10^6$ cm^{-3}) and $v_s \cong 100$ km/s, the compression is about two orders of magnitude.

2.3 C SHOCKS

In a partially ionized gas, the magnetic force is exerted only on the charged particles, which causes them to slip relative to the neutrals in the shock front; in terms of the above discussion of flux freezing, the charged particle velocity v_c can differ significantly from the mean fluid velocity v. In C shocks the entire shock structure is governed by this slippage, or ambipolar diffusion.

The length scale ℓ_B over which the ambipolar diffusion occurs in a shock front, whether J or C type, is determined by balancing the magnetic force on the ions with the collisional drag due to the neutrals (Spitzer, 1978)

$$B^2/8\pi\ell_B = n_i \rho (v_n - v_c) \langle\sigma v\rangle. \qquad (2.23)$$

Since the magnetic pressure gradient drives the charged particles in the upstream direction, the neutral drag on the charged particles, which varies as $v_n - v_c$, must be in the downstream direction. Hence, the charged particles must be decelerated more than the neutrals: the charged particles and magnetic fields are compressed <u>before</u> the neutrals. This can be visualized as the compressed <u>field</u> behind the shock propagating ahead as a damped hydromagnetic wave. During the compression, n_i/B remains constant, so in terms of the ion-neutral mean free path, $\lambda_{in} = 1/n_0\sigma_{in}$, the length scale ℓ_B becomes

$$\ell_B = \lambda_{in} (B/B_0)/(M_A^2 x_{i0}), \qquad (2.24)$$

where $x_{i0} = n_{i0}/n_0$ is the initial fractional ionization, and where we have set $\langle\sigma_{in}v\rangle v \cong \sigma_{in} v_s^2/2$. Note that

$$M_A^2 x_{i0} = 4\pi \rho_{i0} v_s^2/B_0^2 = M_{Ai}^2 \qquad (2.25)$$

is the Alfven Mach number relative to the ionized gas.

For a strong shock, there are three cases:
I. $M_{Ai} > 1$: J shock. In this case, equation (2.24) indicates that the field increases in a distance $\lesssim$ ion-neutral mean free path, which in turn is generally smaller than the neutral-neutral collisional mean free path λ_{nn}. Since in a weakly ionized plasma, the shock front has a thickness of at least λ_{nn}, the field will actually increase on this scale rather than on the smaller scale λ_{in}; equation (2.24) underestimates ℓ_B in this case because the relative ion-neutral drift velocity is less than the assumed value $\cong v_s$. Thin shock fronts such as this are J type. As Draine (1980) has emphasized, signals cannot propagate ahead of the shock front for $M_{Ai} > 1$, so it must be J type. Radiative losses in the shock front are negligible.

II. $M_{Ai} < 1$, $M_A > 1$, $n_0\ell_{Bm} < N_{cool}$: J shocks with magnetic precursor. The field approaches its final value B_m in a column density $n_0\ell_{Bm}$. If this is less than N_{cool}, the value required to radiate away half the relative energy of the shocked and unshocked fluids, then the charged particles and magnetic field project ahead of the neutral shock as a "magnetic precursor." Some of the relative energy is dissipated in the precursor, thereby weakening the J shock which follows the precursor.

III. $M_{Ai} \ll 1$, $M_A > 1$, $n_0 \ell_B \cong N_{cool}$: C shock. For sufficiently low M_{Ai}, especially in molecular gas which has efficient cooling and low ionization, the increase in the magnetic field can stretch out over such a large distance that the radiative losses in the precursor dissipate all the shock energy. In this case, the embedded neutral J shock, in which dissipation is effected by neutral-neutral collisions, vanishes. Instead, the density, temperature, and velocity of the gas vary continuously on a scale $\ell_B \gg 1/n_n \sigma_{nn}$, whence the term C shock, and the dissipation is effected by collisions between neutrals and the small concentration of ions. Over most of the volume of the shock front, the magnetic field is within a factor of 2 of its final value B_m [eq. 2.22], which is independent of the initial field strength B_0; this was not appreciated in early discussions of C shocks.

Historically, magnetic precursors were first analyzed by Mullan (1971). He ignored radiative losses, which is approximately valid for the atomic shocks he considered, but this prevented him from finding strong C shocks. Weak C shocks ($M_A \lesssim 2.5$) can exist even in the absence of radiation, and he did find these. Draine (1980) introduced the distinction between C and J shocks and laid the foundation for the study of the structure of C shocks. He characterized the transition from J to C shocks in terms of a critical preshock field strength B_{crit} (which is equivalent to a critical value of M_{Ai}). The value of B_{crit} depends on a number of factors, including v_s and x_{i0}, but in a weakly ionized molecular gas it could correspond to $M_{Ai} \cong 1$, or $b_{crit} \cong 5 \times 10^{-3}$ $(x_{i0}/10^{-8})^{1/2} v_{s7}$ for velocities $v_{s7} \lesssim 0.3$ which do not lead to appreciable ionization.

There are two conditions which must be satisfied for the existence of a C shock. First, as emphasized above, the shock velocity must be less than the ion magnetosonic velocity [$v_s < B/(4\pi\rho_c)^{1/2}$ for cold ions, or $M_{Ai} < 1$] in order that a precursor be able to project ahead of a jump in the neutral gas. Second, the neutral gas must remain supersonic ($v_n^2 > \gamma c_{sn}^2$) throughout the shock, where c_{sn} is the neutral thermal speed; if it did not, a J shock would occur where the gas made a transition to subsonic flow. At first sight, this seemingly contradicts the fundamental requirement that the flow behind a shock be subsonic relative to the shock front; however, it is the large Alfven velocity associated with the compressed field, rather than the thermal velocity of the neutral gas, which renders the signal velocity behind the shock large enough to satisfy this requirement. In order for the neutral gas to remain supersonic, it must remain cool (typically, $T_n < 5000$ K). Since an atomic gas with cosmic abundances is inefficient at cooling at these temperatures, strong C shocks must occur in molecular gas. Hence, C shocks are non-dissociative, in the sense that such shocks cannot dissociate most of the molecules in the gas. A corollary is that C shocks are non-ionizing; equation (2.24) shows that $\ell_B \propto x_i^{-1}$ so that if the ionization becomes too large (typically $\gtrsim 10^{-4}$) the shock becomes too thin to radiate and it becomes J type (see section 4). The contrast in the temperature variation in C shocks and J shocks is graphically illustrated in Figure 1.

To this point we have considered only the atomic and molecular consitituents of the interstellar gas. However, grains can play a major role in the structure of interstellar C shocks because they become charged and dominate the momentum transfer between the magnetic field and the neutral gas at low ionizations. If grains are dominant, then the column density of a C shock is of order

$$n_0 \Delta \ell_{sh} \cong n_0/(n_{gr} \sigma_{gr}) \cong n_0/(M_A n_{gr0} \sigma_{gr}) \cong 10^{21-22}/M_A \text{ cm}^{-2}, \quad (2.26)$$

where n_{gr0} is the initial density of grains. The value $(n_0/n_{gr0}\sigma_{gr}) = 10^{21}$ cm^{-2} is appropriate for the typical ISM (Spitzer 1978), but Draine and Roberge (1982, hereafter DR) and Chernoff et al (1982, CHM) found a value closer to 10^{22} cm^{-2} in their models for the emission from Orion. The thickness indicated in equation (2.26) is smaller than that based on charge exchange between neutrals and ions for ionizations $x_{i0} \lesssim 10^{-6.5}$ – $10^{-7.5}$, where we set $\sigma_{in} \cong 10^{-14.5}$ cm^2 in equation (2.24). The full range of effects of grains in C shocks has yet to be determined. Studies to date have neglected the inertia of the grains, an assumption which is questionable at shock velocities which are highly supersonic relative to the grains ($v_s \gg B_0/(4\pi\rho_{gr})^{1/2} \cong 10 v_A$, or $M_A \gtrsim 10$).

2.4 SUMMARY

Interstellar shocks can be classified as J type or C type. In J type shocks, the shock front --the region in which the relative kinetic energy is dissipated --is less than or of order one elastic collision mean free path in thickness, so the hydrodynamic variables suffer a non-radiative "jump" as the fluid passes through the shock front. On the other hand, in C shocks, which occur only in weakly ionized plasmas, the hydrodynamic variables change continuously over a length scale determined by the mean free path for a neutral to hit an ion or a charged grain, a length much greater than the neutral-neutral mean free path because of the low ionization. In J shocks, the temperature is a maximum just behind the shock front; at a velocity of 30 km/s, say, much of the emission is in the UV or visible. In C shocks, the temperature must be low (typically $\lesssim 5000$ K) and most of the shock energy is radiated in the infrared. Thus C shocks were designed for infrared astronomers: the emission occurs primarily in infrared molecular lines, which are readily observed in highly obscured regions and which provide a wealth of information on the physical conditions.

A complete delineation of the conditions under which shocks are J type, J type with observable precursors, or C type has yet to be made. Calculations to date show that for typical interstellar fields ($b \cong 10^6 B_0 n_0^{-1/2} \gtrsim 0.1$), shocks in molecular gas with $v_s \lesssim 50$ km/s are C type;

INFRARED SPECTROSCOPY OF INTERSTELLAR SHOCKS 113

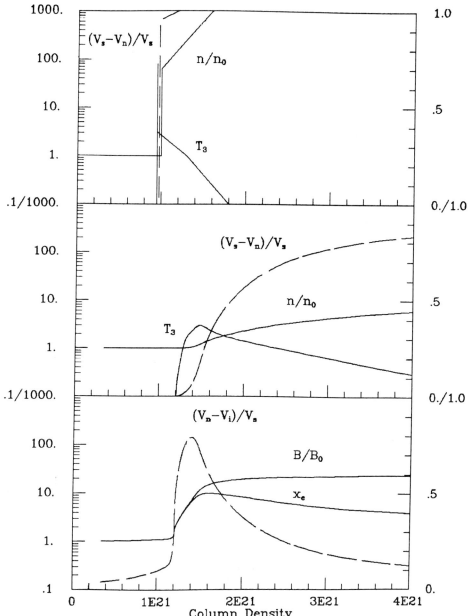

Figure 1. Comparison of J (top graph) and C (lower two graphs) shock structure. The top and middle graphs plot temperature ($T_3 = T/1000$ K), compression (n/n_0), and drift velocity ($v_s - v_n$)/v_s as functions of column density for J and C shocks with $n_0 = 10^5$ cm-3 and $v_s = 40$ km/s. (The three curves are offset slightly in the top graph for clarity.) The magnetic fields are b=.01 and b=1 for the J and C shocks, respectively. Solid lines refer to the left (log) ordinate, dashed lines to the right (linear) scale. The bottom graph plots ion-neutral slip ($v_n - v_i$)/v_s, magnetic field (B/B_0), and ionization fraction (x_e), on the same scale for the C shock.

otherwise, if the shock is strong, it is J type. It is possible to have shocks with significant magnetic precursors in atomic gas only if the shocks are weak (Mullan 1971, Draine 1980); such shocks are faint and hence difficult to observe. J shocks can be either collisional or collisionless. If the column density of shocked gas in a J shock becomes sufficiently large, as it will if the shock persists for a long enough time, then the shocked gas will cool; for $v_s \gtrsim 110$ km/s the radiation from the cooling gas will photoionize the upstream gas, ensuring that the shock is collisionless (Shull and McKee 1979). On the other hand, for $v_s \cong 60$ km/s, the pre-ionization is only about 1 % and the shock, if in a neutral gas, will be collisional. The various possible structures for interstellar shock fronts are summarized in Table 1.

Strong radiative shocks, whether C or J, convert virtually all the incident kinetic energy flux $\rho_0 v_s^3/2$ into radiation. Half this flux is directed upstream and half downstream. The total specific intensity normal to the plane of the shock, including both lines and continuum is

$$I = \rho_0 v_s^3/8\pi = 9.3 \times 10^{-8} n_0 v_{s6}^3 \text{ erg cm}^{-2}\text{s}^{-1}\text{ sr}^{-1}, \quad (2.27)$$

where $n_0 = n(H^+) + n(H^0) + 2 n(H_2)$ is the density of hydrogen nuclei in cm^{-3}. Current infrared instruments are sensitive to lines with $I \gtrsim 10^{-4}$ erg cm^{-2}s^{-1}sr^{-1}; detection of a strong line of intensity 0.1 I from a face-on shock thus requires

$$n_0 v_{s6}^3 \gtrsim 10^4. \quad (2.28)$$

For shocks in diffuse atomic gas ($n_0 \lesssim 10^2$ cm^{-3}) this requires a reasonably high velocity, $v_s \gtrsim 50$ km/s. Lower velocities are observable at the higher densities found in molecular gas. Note that the constraint (2.28) can be relaxed somewhat if there is more than one shock along the line of sight or if the shock is observed edge-on.

3. J SHOCKS

For typical interstellar magnetic fields (b $\cong$ 0.1-1) we find that shocks will be J type for $v_s > 40$-50 km/s in dense molecular gas of low ionization ($x_i \lesssim 10^{-6}$) and at lower velocities in neutral atomic gas ($x_i \cong 10^{-4}$). All shocks in ionized gas are J type. The radiation from J shocks is produced behind the shock front, and its character depends on whether it is produced before or after the shocked gas recombines.

3.1 High Temperature Structure of J Shocks: Emission from Ionized Gas

Generally, for $v_{s7} > 1$, the shock parameters plotted against column density for postshock temperatures T > 5000 K are relatively independent of the state of the ambient cloud, i.e. the preshock temperature, density and magnetic field strength and whether the ambient gas is molecular, atomic or ionic. At these high speeds the postshock temperatures exceed 10^5 K and the UV radiation produced in the shock

TABLE 1
STRUCTURE OF STRONG INTERSTELLAR SHOCK FRONTS

	J Shock $N_{sh} = n \Delta \ell_{sh} \lesssim \sigma_{nn}^{-1}$		C Shock $N_{sh} \gg \sigma_{nn}^{-1}$
Dissipation	Collisionless (Plasma Instabilities)	Collisional (Neutral-neutral)	Collisional (Ion-neutral)
Shock Thickness $\Delta \ell_{sh}$	$\ll 1/n\sigma$	$\approx 1/n_n \sigma_{nn}$	$\approx 1/n_i \sigma_{in}$
Ionization	High	Low-moderate	Low
Radiative[1]	No	No	Yes
Magnetic[1]	Usually	No	Yes
Molecular	No	If yes, usually dissociative	Yes
Grain-dominated[2]	No	No	Often

1. Refers to whether radiation or magnetic fields are important in determining the structure of the shock front; in interstellar J shocks these effects are almost always important behind the shock front provided n_0 is not too low or v_s too high.
2. Refers to whether the presence of grains has a major effect on the structure of the shock front.

will completely predissociate and preionize the preshock gas (cf. Shull and McKee 1979). The column density of hydrogen nuclei from the shock front to the cooling postshock gas at T = 10^4 K is given by

$$N_{cool}(10^4 \text{ K}) = 2 \times 10^{17} v_{s7}^{4.2} \text{ cm}^{-2}, \quad (3.1)$$

a result good to within a factor of two for $0.6 < v_{s7} < 2$ (Hollenbach and McKee 1979).

The UV radiation from this hot, ionized postshock gas not only "homogenizes" the upstream preshock gas, but it travels downstream into the cooling T < 10^4 K postshock gas, delaying recombination and molecule formation and heating the gas and dust grains. For $v_{s7} \cong 1$ the total column density of ionized gas is of order $N_{cool}(5000 \text{ K}) \cong 10^{18}$ cm^{-2}, with most of this column photoionized by the UV field and maintained at T $\cong$ 5000 K.

The normal intensity I of infrared fine structure lines in this region can be readily estimated in the limits of low and high density. At low density, each collisional excitation results in the emission of a photon, so that

$$I \cong 0.03\, v_{s7}^2 \left(\frac{\Omega}{g_1}\right)\left(\frac{n_0}{10^3 \text{cm}^{-3}}\right)\left(\frac{x_j}{10^{-4}}\right)\left(\frac{10\mu m}{\lambda}\right)\left(\frac{N_{cool}(5000\text{ K})}{10^{18}\text{cm}^{-2}}\right) \text{ erg cm}^{-2}\text{s}^{-1}\text{sr}^{-1}, \quad (3.2)$$

where Ω is the collision strength, g_1 the statistical weight of the ground state of the transition, and x_j is the fractional abundance of the emitting ion. We have assumed that the gas is not magnetically supported at 5000 K, so that the density in the emitting region is determined by the condition $2nkT \cong \rho_0 v_s^2$. On the other hand, at high density, the excited state is in LTE so that

$$I \cong 1.6 \times 10^{-4} \left(\frac{g_2}{Z}\right)\left(\frac{x_j}{10^{-4}}\right)\left(\frac{A_{21}}{10^{-4}\text{s}^{-1}}\right)\left(\frac{10\mu m}{\lambda}\right)\left(\frac{N_{cool}(5000\text{ K})}{10^{18}\text{cm}^{-2}}\right) \text{ erg cm}^{-2}\text{s}^{-1}\text{sr}^{-1} \quad (3.3)$$

where g_2 is the statistical weight of the excited state, Z is the partition function, and A_{21} the transition probability. At any density, the intensity is approximately equal to the smaller of these two expressions. For ions such as Ne II, much of the line emission originates in this recombination zone, and a necessary condition for detectability ($I \geqslant 10^{-4}$ erg cm^{-2}s^{-1}sr^{-1}) is $x_j A_{21} \geqslant 10^{-8}$, which is readily satisfied for Ne II. Emission from ions such as Si II and Fe II can arise from predominantly neutral gas at lower temperatures, and can therefore exceed the estimates in equations (3.2) and (3.3).

An important process which occurs in high-speed J shocks in the hot, $T > 10^4$ K component of the shock structure is the sputtering of dust grains (cf. Draine and Salpeter 1979). For $v_{s7} \geqslant 1$, sputtering leads to nearly cosmic gas phase abundances of most elements (including Si and Fe), a result which has important consequences on infrared line intensities.

3.3 Low Temperature Structure of J Shocks: Emission and Chemistry

<u>Low Density Regime $n_0 < 10^3$ cm^{-3}</u>. The shock structure for postshock temperatures $T \leqslant 5000$ K, unlike the hotter gas, is sensitive to the preshock density and magnetic field strength. These latter two quantities determine the postshock gas density. The gas ceases to compress at $T_m \cong 10^4\, b\, v_{s7}$ K with a density $n_m \cong 77\, b^{-1}\, v_{s7}\, n_0$ (see eq. 2.22). Above T_m, the gas pressure nT is approximately constant. The postshock density is important in the structure below 5000 K because it is here that infrared line emission generally dominates the cooling, and this cooling process can be affected by collisional deexcitation. At low postshock densities $n \leqslant 10^5$–10^6 cm^{-3}, the cooling by OI (63 μm)

dominates from 5000 K $\geq$ T $\geq$ 100 K, and the cooling column density to 100 K is given by

$$N_{cool}(100\ K) = n_0 v_s t_{cool}(OI) \cong 1 \times 10^{21} (n_0/n) v_{s7}\ cm^{-2}. \quad (3.4)$$

Assuming that the gas compression n/n_0 is limited by magnetic pressure (eq. 2.22) in this temperature range, we find

$$N_{cool}(100\ K) \cong 10^{19}\ b\ cm^{-2}. \quad (3.5)$$

The condition that OI (63μm) not be collisionally de-excited in this regime is $n_0 \leq 2 \times 10^3\ b/v_{s7}\ cm^{-3}$.

High Density Regime $n_0 \geq 10^3\ cm^{-3}$. For preshock densities $n_0 \geq 10^3$ and $v_{s7} \cong 1$, a significant change occurs in the structure between 5000 K > T > 100 K. Here, the collisional deexcitation of excited OI leads to larger cooling column densities. The heating by H_2 formation on grains (cf. Hollenbach and McKee 1979) deposits significant energy into the gas, offsets the cooling, and further increases the column density of warm gas. The formation of H_2 at warm (T > 300 K) temperatures initiates a neutral-neutral chemical reaction scheme (McKee and Hollenbach 1980) which transforms O into OH, H_2O and CO. Rotational transitions of these species and gas-grain collisions dominate the gas cooling. The dissociated hydrogen is completely recombined to molecular hydrogen by the time the gas cools to T $\cong$ 100 K and

$$N_{cool}(1000\ K) \cong 10^{20.5}\ cm^{-2},$$
$$N_{cool}(100\ K) \cong 10^{21} - 10^{22.5}\ cm^{-2}. \quad (3.6)$$

The fractional abundance of primary species at T $\cong$ 1000 K are $x(H_2) \cong 10^{-2}$, $x(O) \cong 4 \times 10^{-4}$, $x(OH) \cong 10^{-6}$, $x(H_2O) \cong 10^{-8}$ and $x(CO) \cong 3 \times 10^{-5}$.

The 6 eV -13.6 eV photons from the hot postshock gas penetrate to N $\cong 10^{21}\ cm^{-2}$ downstream, maintaining many metals in the singly ionized state. The resulting C^+, Si^+, and Fe^+ produce fine structure lines unimportant to the cooling but possibly detectable in the IR and sensitive to the gas phase abundances of these species.

3.4 Calculations of Infrared Emission from J Shocks.

Table 2 lists the results of numerical studies of J shock line intensities for fully ionized atomic gas at low density (Shull, Seab and McKee 1983) and at high density (Hollenbach and McKee 1983). The low density calculations allow for grain destruction in the shock and the final gas phase abundances relative to solar, δ, are listed for carbon, silicon, and iron; for the other elements Shull et al found $\delta > 0.75$. The high density calculations used fixed abundances with $\delta = 1$ for all elements but silicon and iron; emission from T > 10^4 K was ignored, so the intensity of Ne II (12.8μm) is a lower limit. Generally, the intensities scale roughly with abundances, except that the dominant coolants remain roughly constant.

Preliminary results for the high density cases are presented as a function of preshock density and magnetic field for a characteristic velocity of 100 km/s. The sensitivity of the results to the preshock magnetic field strength (given by the parameter b) occurs because the field determines the gas compression in the IR producing regions. Higher fields lead to less compression and the relative intensities approach the lower preshock density results.

The intensities of the infrared lines are quite small relative to the total shock intensity $\rho_0 v_s^3/8\pi$. Most of the shock radiation emerges as UV radiation, and often this radiation is absorbed by grains and reradiated as grain continuum. Shock luminosities are low; a 0.1pc shock radiates about $10^{-2} n_0 v_{s7}^3 L_\odot$. The grain temperatures remain low (Tgr < 100 K) so that grain continuum emission from shocks will be difficult to distinguish from the overall grain emission of the cloud as a whole.

Table 2 demonstrates that OI (63μm) may become observable for $n_0 \geqslant 10^2$ cm^{-3} at $v_{s7} \geqslant 1$, and at somewhat higher densities for lower velocities. For $v_{s7} \cong 1$, the OI (63μm) intensity scales with n_0 for $n_0 \leqslant 3 \times 10^4$ cm^{-3} and is a good indicator of the mass flux into the shock. The OI (63μm) from radiative shocked stellar winds ($v_w \cong 100$ km/s) may therefore probe the mass-loss rate from the star. In general these J shocks have the characteristics of weak H_2 vibrational emission (because the H_2 abundance is low at T > 1000 K), a large CO 30-29/CO 21-20 ratio when the lines are observable (because of high densities), and strong OI (63μm) emission (oxygen is a primary coolant at T $\cong$ 1000 K for most of these shocks). OH lines are strong for $n_0 \geqslant 10^4$ cm^{-3}, and Si^+ and Fe^+ lines may be strong for $n_0 \geqslant 10^4$ cm^{-3} as well, depending on their gas phase abundance. Most of the emission occurs at speeds of order v_s, since the gas is accelerated by the shock wave before it radiates the infrared.

4. C SHOCKS

4.1 Physical Conditions in C Shocks

When a C shock approaches, the ions are the first to take notice: they compress to $M_A\sqrt{2}$ times their original density (if no chemistry occurs) and, in the shock frame, decrease in velocity from v_s to $v_s/M_A\sqrt{2}$. For M_A larger than a few and for low ionization levels, the neutrals have no time to react and instead slip with respect to the field and the charged particles. The maximum slip, or drift velocity, v_d, occurs at the beginning of the shock; there $v_d = v_s (1 - 1/M_A\sqrt{2})$. Throughout the rest of the shock, the neutrals are gradually decelerated by collisions with charged particles. In C shocks the primary constituents of the plasma are not in equilibrium and have different temperatures and drift velocities. Nevertheless, the physical

TABLE 2

INFRARED EMISSION FROM J-SHOCKS

	Low n (Atomic)*			High n (Molecule Formation)*			
n_0(cm^{-3})	10	10	10^2	10^3	10^4	10^4	10^6
v_s(km/s)	40	100	100	100	100	100	100
b	0.3	0.3	0.1	0.3	0.3	3.0	0.3
N(cm^{-2})†	9.6(18)	6.6(18)	6.9(18)	1.6(19)	5.9(19)	9.1(19)	2.9(20)
n_m(cm^{-3})	9.0(2)	2.3(3)	6.8(4)	2.6(5)	2.6(6)	2.6(5)	2.6(8)
$\rho_0 v_s^3/8\pi$**	1.5(-4)	9.3(-4)	9.3(-3)	9.3(-2)	9.3(-1)	9.3(-1)	9.3(+1)
δ(C)††	0.2	0.28	0.29	1.0	1.0	1.0	1.0
δ(Si)	0.16	0.54	0.58	0.1	0.1	0.1	0.1
δ(Fe)	0.15	0.53	0.58	0.1	0.1	0.1	0.1
Hβ**	2.8(-7)	7.1(-6)	7.0(-5)	-	-	-	-
CII(158)	6.9(-7)	9.5(-7)	1.1(-6)	1.2(-5)	4.4(-6)	3.4(-6)	7.7(-7)
OI(63)	2.0(-6)	9.0(-6)	1.4(-4)	7.9(-4)	5.5(-3)	5.4(-3)	2.8(-2)
OI(145)	1.7(-7)	8.0(-7)	1.2(-5)	2.1(-5)	1.4(-4)	1.9(-4)	7.2(-4)
NeII(12.8)	3.1(-9)	9.5(-6)	1.0(-4)	3.1(-4)	1.6(-3)	2.7(-3)	4.1(-3)
NeIV(15.6)	-	2.5(-6)	2.2(-5)	-	-	-	-
SiII(34.8)	1.7(-6)	3.0(-5)	7.4(-5)	4.0(-5)	2.1(-4)	2.8(-4)	7.6(-3)
FeII(1.27)	3.4(-7)	1.9(-5)	2.6(-4)	-	-	-	-
FeII(5.0)	2.6(-7)	1.1(-5)	1.5(-4)	-	-	-	-
FeII(11.6)	4.1(-8)	1.6(-6)	2.0(-5)	-	-	-	-
FeII(15)	5.1(-8)	1.9(-6)	2.5(-5)	-	-	-	-
FeII(26)	1.2(-7)	4.0(-6)	5.3(-5)	1.2(-4)	6.8(-4)	8.0(-4)	2.1(-2)
H_2(tot)				3.0(-4)	1.5(-3)	2.0(-3)	1.2(-2)
1-0S(1)				1.3(-5)	1.1(-4)	1.0(-4)	3.1(-4)
2-1S(1)				5.4(-6)	4.7(-5)	4.7(-5)	1.4(-4)
0-0S(2)				2.9(-7)	6.7(-7)	2.8(-6)	6.1(-4)
0-0S(5)				3.8(-5)	8.0(-5)	3.7(-4)	1.5(-4)
0-0S(7)				7.0(-5)	4.0(-4)	6.8(-4)	3.9(-4)
CO(tot)				5.6(-6)	3.5(-4)	2.9(-5)	2.1(-1)
30-29				1.9(-10)	1.9(-6)	2.4(-9)	7.1(-3)
21-20				5.7(-10)	1.0(-5)	7.6(-9)	8.2(-3)
OH(tot)				4.8(-6)	6.5(-4)	2.4(-5)	3.7(-1)
$^2\pi_{3/2}$(5/2-3/2)				1.3(-7)	2.5(-5)	1.1(-6)	7.5(-3)
H_2O(tot)				4.1(-8)	9.2(-6)	7.9(-6)	2.2(0)

*Atomic: Shull, Seab, and McKee (1983); Molecule formation: Hollenbach and McKee (1983)
†Column density for T > 500 K (atomic), 10^3K (molecule formation)
**Intensities are given in erg cm^{-2} s^{-1} sr^{-1}; wavelengths are in μm
††Depletions relative to solar abundances; for the atomic case, which includes grain destruction, these are the final values.

properties can be calculated because the gradients in the transition zone, after the ion compression, are small.

The length scale ℓ_B for the ion compression is given by equation (2.24). It is worthwhile to note that it increases as the field compresses so that far downstream ℓ_B can be as much as M_A times greater than upstream. It takes a long time for the ion-neutral collisions to bring the neutrals to a complete stop with respect to the ions. This accounts for the characteristic profiles in Figure 1, in particular, for the steep initial rises and slower intermediate and final changes. It is also important because the ion compression and the steady state C shock solution develop, in a time-dependent problem, within a distance given by the upstream ℓ_B, not the much longer downstream ℓ_B.

Heating and Cooling of Charged Species (Ions, Electrons and Grains). The temperature of each species may be estimated by balancing the most important heating and cooling processes for each species. Ions are heated and cooled primarily by elastic collisions with neutrals and reach temperatures of

$$T_i = m_n v_d^2 / 3k. \qquad (4.1)$$

where m_n is the mean mass of a neutral. The electron temperature is lower than the ion temperature since the small quantity (m_e/m_n) enters in the rate of elastic heating and cooling, but not in the rate of cooling by excitation of vibrational and rotational transitions in H_2. Collisionless plasma instabilities contribute to the heating also, so that the electrons typically reach temperatures of $T_e \cong (0.1-0.2)\, T_i$. At very high temperatures, they cool also by dissociation and ionization of H_2. Grains in the shock waves are heated primarily by collisions with neutrals and are cooled by emission of far-infrared photons; typically they equilibrate at $T_{gr} < 100$ K.

Heating of Neutrals. The situation for neutrals is more complicated. Far upstream of the shock, only ions significantly heat the neutrals, but as the flow approaches the strong downstream magnetic field, heavier charged particles (grains) are slowed by the field and contribute to the heating. Grains dominate the heating if the plasma is only slightly ionized, $x_i < 3 \times 10^{-7}$, and if the electrons are sufficiently hot to charge them ($v_d > 3-5$ km/s); otherwise, ions are more important. The neutral temperature may be estimated, if charged grains are the dominant heat source, by balancing heating and cooling:

$$n_{gr}\, n\, \sigma_{gr}\, \mu_H\, |\underline{v}_n - \underline{v}_{gr}|^3 = n^2 \Lambda. \qquad (4.2)$$

where μ_H is the mass per hydrogen nucleus. The maximum temperature T_{max} occurs near the beginning of the shock where the drift velocity multiplied by the grain density is a maximum. This point occurs where the charged particles have been decelerated to a velocity $\cong 1.6\, v_s/M_A$, while the neutrals are moving at $0.8\, v_s$. The drift velocity between the neutrals and ions is

$$v_d = (0.8 - 1.6/M_A) \, v_s \equiv z \, v_s. \qquad (4.3)$$

The coupling between the grains and neutrals is governed by $\omega\tau$, the product of the grain gyrofrequency and the collisional damping time, or equivalently by

$$\zeta^2 = (\omega\tau)^2/[1 + (\omega\tau)^2]. \qquad (4.4)$$

For $\zeta \to 0$, the grains are tightly coupled to the neutrals, whereas for $\zeta \to 1$, they are coupled to the magnetic field. For $kT_n \ll m_n v_d^2$, $kT_e \cong 0.1 \, kT_i$ (eqn. 4.1), and $B = B_m$ (eqn. 2.22), Draine's (1980) results lead to

$$\zeta^2 = 2 \, [1 + (1 + 0.3 \, n_{06} \, a_{-5}^2/(v_{s6}^4 z^2))^{1/2}]^{-1}, \qquad (4.5)$$

where a_{-5} is the grain radius in units of 10^{-5} cm. His results also imply that $v_n - v_{gr} = \zeta \, v_d$ and $n_{gr}/n_{gr0} = (0.8 - z\zeta^2)^{-1}$, so that the maximum neutral temperature in a grain-dominated C shock is given implicitly by equation (4.2) as

$$[(n_{gr}\sigma_{gr}/n)_0 \, \mu_H v_s^3] \, z^3 \zeta^3/(1 - 1.25 z \zeta^2) = \Lambda(T_{max}, \, 1.25 \, n_0). \qquad (4.6)$$

The polynomial in z and ζ incorporates the effects of gas density, magnetic field strength, shock velocity, and grain size on the heating rate, which has a characteristic value $(n_{gr}\sigma_{gr}/n)_0 \, \mu_H v_s^3$. For low densities the grains are tightly coupled to the field and this factor is $M_A/4$; at high densities this factor approaches zero as ζ does. The quantity $(n_{gr}\sigma_{gr}/n)_0$ is the initial grain surface area per hydrogen nucleus, and is typically 10^{-21} cm^2 in the ISM (Spitzer 1978).

As T_{max} and v_d increase, the ionization of the gas can increase. When it rises above 3×10^{-7}, the temperature is determined by ion-neutral scattering:

$$x_i \, \langle\sigma v\rangle_{in} \, m_n \, v_d^2 = \Lambda. \qquad (4.7)$$

To estimate T_{max} requires solving for x_i, which is a complicated function of a number of exponentially sensitive reaction rates; fortunately, as we will discuss, equation (4.7) applies only in a relatively small region of parameter space.

<u>Cooling of Neutrals and Chemical Composition</u>. The cooling of the gas in a C shock depends on the the chemical composition of the predominantly neutral gas which, in turn, depends on the gas temperature. Hollenbach (1982), Hollenbach and McKee (1979), and McKee and Hollenbach (1980) have summarized the chemical reactions which are important in C shocks. The neutral chemistry is determined by the ratio $x_2 = n(H_2)/n$. For $x_2 \cong 1/2$ (molecular gas) and for an oxygen abundance exceeding that of carbon, all carbon becomes CO, the leftover oxygen becomes H_2O, nitrogen becomes NH_3, sulfur becomes H_2S, OCS, and SO, and Si goes to SiO. For $x_2 \ll 1$, chemical reactions destroy the remaining

molecules and the gas becomes atomic. For $x_2 \cong 1/4$ a significant fraction of O is in the form of OH. We note that fast ($v_s \cong 50$ km/s) C shocks incident upon molecular gas can produce quite intense OH infrared emission because of the dissociation of a fraction of the H_2 inside the shock wave. The ion chemistry, which is unimportant to the cooling of a C shock but is important to the heating, is dominated by the collisional dissociation of molecular gas ions by H and H_2 for observable C shocks ($v_s > 20$ km/s). The most abundant ion in these shocks is therefore H^+. We discuss the effects of ionization processes further in the next section.

In C shocks the primary coolants are OI, H_2, CO, OH, and H_2O (see Fig. 2). The column density of warm gas increases with density because collisional de-excitation at high densities reduces the cooling efficiency. The results are only weakly dependent on the preshock abundances of $x(O)$, $x(CO)$, $x(H_2O)$, $x(OH)$, etc., since shock chemistry rapidly equilibrates the initial distribution and since the total cooling at low temperatures and high densities depends primarily on the density of dipolar molecules.

4.2 Parameter Range of C Shocks

As discussed in section 2.3, the necessary conditions for strong C shocks are $M_A \gg 1$, $M_{Ai} < 1$, and $v_n^2 > \gamma c_{sn}^2$. The separation between C and J shocks is usually determined by the third criterion. The characteristic length scale for coupling of the field to neutrals varies as the inverse of the ionization x_i, so at low ionization levels, small increases in x_i can greatly increase the heating and ionization rates. There are a number of ionization mechanisms which become energetically feasible as the drift velocity increases. It is possible for hot, drifting ions and electrons to ionize the neutrals. Drifting grains reflect neutrals, thereby producing a high velocity population of neutrals which then ionize other neutrals. Finally, hot neutrals can ionize other neutrals. The considerable uncertainties in the forms of the ionization cross sections near threshold are ameliorated, in part, by a sudden avalanching effect: as the gas heats, it ionizes itself and destroys molecular coolants, both of which cause it to heat even more quickly. Hence, the estimated uncertainty in the ionization breakdown is about 5-10 km/s. In Figure 3 we have delineated the region in the $b-v_s$ plane in which C shocks can occur. In the parameter space beyond the ionization breakdown, the steady state solution is a J shock with precursor. These calculations were made with Draine's (1980) approximate treatment of the grain dynamics. A more accurate analysis (now underway) is needed to analyze the breakdown as $M_{Ai} \rightarrow 1$.

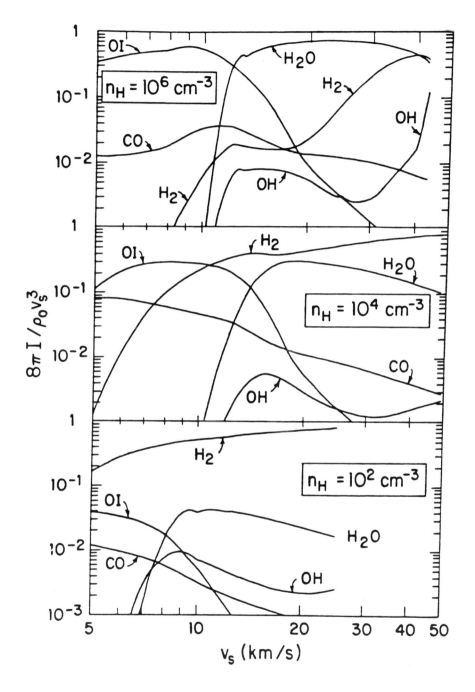

Figure 2. The major coolants in C shocks are given for $(n_0, x_e) = (10^2$ $cm^{-3}, 10^{-4}), (10^4\ cm^{-3}, 10^{-7}), (10^6\ cm^{-3}, 10^{-8})$ as a function of velocity; In each case $b = 1$. From Draine, Roberge and Dalgarno (1983).

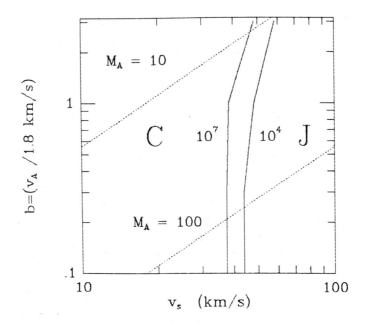

Figure 3. The allowed regime for C shocks in the $b-v_s$ plane lies to the left of the breakdown lines, which are given for densities of $n_0 = 10^4$ cm^{-3} and 10^7 cm^{-3}. Strong C shocks lie in the region $M_{Ai} < 1$ (includes the entire figure for $x_i \lesssim 10^{-6}$) and $M_A \gg 1$.

4.3 Qualitative Diagnostics

As noted previously, the major differences between J and C shocks are the following: In fast J shocks, gas directly behind the shock front reaches extremely high temperatures; molecules are generally destroyed and a substantial fraction of the shock energy emerges in optical, uv and x-ray emission lines radiated at a velocity $\cong v_s$ relative to the ambient gas. By comparison, C shocks are cooler, they radiate over a larger column density, they reach their temperature maximum (T_{max}) at lower compressions, and most of the emission occurs near the velocity of the ambient preshock gas; the shock energy is transformed into infrared line emission, principally molecular. Reference to Figure 1 shows that the highest temperatures occur when the neutrals are still nearly at rest relative to the unshocked gas; hence, for a shock advancing into stationary gas, high excitation lines should generally have narrower profiles than low excitation lines.

Computer modeling of a shock produces a 'fingerprint' which can be compared with observations to unravel the physical parameters. The input for a model consists of the parameters n_0, v_s, B, $n_{gr}\sigma_{gr}$, a, and the gas phase abundances of atoms and molecules. Before comparing the resulting "fingerprint" with observations, one must allow for extinction by dust and geometrical effects. In order to match observations with a model one would like to employ quantities which are sensitive to only a few of these parameters and insensitive to the rest. Thus, for example, far-infrared lines are insensitive to extinction, H_2 lines and line ratios are insensitive to elemental abundances, and line ratios are insensitive to geometrical effects.

As an example of the sort of qualitative diagnostics which are possible, consider Figure 4 which plots the CO 34-33/21-20 versus H_2 0-0 S(2) intensity ratio and whose purpose is to distinguish C shocks from J shocks. We have calculated the emission from a variety of C and J shocks. The S(2) line is excited in gas of 700 K or more; it gives a rough estimate of the column density of H_2 molecules hotter than that temperature. As a diagnostic, the 12 micron S(2) line has the added advantage that it is affected less than the vibrational lines by the

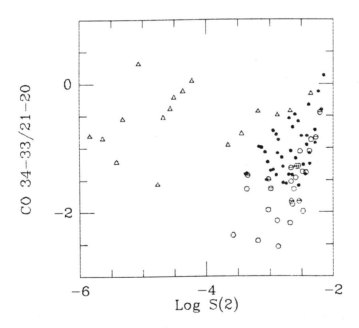

Figure 4. The intensity of H_2 v=0 S(2) emission line at 12.3 μm versus the CO line ratio 34-33/21-20, is given for J shocks (triangles) with densities 10^4-10^6 cm^{-3}, velocities 60-100 km/s, and b=0.01; and C shocks (circles) with densities 10^4-10^7 cm^{-3}, velocities 20 km/s-breakdown, and b=0.3-3. The intensity of the 1-0S(1) line in units of erg cm^{-2}s^{-1}sr^{-1}, I, is coded by the fill of the symbol: solid for $I > 10^{-3}$, open with bar for $10^{-3} > I > 10^{-4}$, and open for $10^{-4} > I$. For $n_0 = 10^6$ cm^{-3} J shocks lie close to C shocks in the figure but are clearly distinguishable by their weak 1-0S(1) emission.

degree of extinction to the source. The CO line ratio is sensitive to both temperature and density variations: higher densities bring the upper CO levels into LTE and raise the ratio. Observable C shocks generally have larger S(2) intensities (the heat is distributed over greater column density of warm gas) and for a given preshock density have lower CO 34-33/21-20 ratios (the gas density is lower) than observable J shocks. In high density ($n_0 > 10^6$ cm-3) dissociative J shocks, H_2 reforms at temperatures above 500 K and the released energy (4.48 eV per molecule) is thermalized. The gas remains warm over a long column, and so these shocks have large S(2) intensities also. Such J shocks may be distinguished from C shocks which lie nearby in Figure 4 by their very low levels of high temperature H_2 emission, such as 2 micron 1-0S(1) line.

Draine, Roberge, and Dalgarno (1983) have calculated detailed emission intensities and line ratios for H_2 and OI for fixed values of n and B with a shock velocity from 5 to 40 km/s. These are approximate calculations in that they include minimal chemistry, no ionization processes, and steady state grain dynamics. Their results are shown in Figure 5. The picture gives some idea of the variety of lines emitted and the possibilities for 'fingerprinting' a particular source.

4.4 C Shocks in Orion

Supersonic gas motions, hot molecular gas, and large-scale outflows have all been observed in the BN-KL region of Orion. The outflows center on IRc2, a luminous embedded protostar. Within a projected distance of 2×10^{17} cm from IRc2, a number of infrared H_2 lines indicating gas temperatures of $\simeq$ 2000 K are seen (Nadeau and Geballe 1979, Nadeau, Geballe, and Neugebauer 1982, Scoville et al 1982). The emission and the line profiles which show supersonic wings extending to $\pm$ 100 km/s are highly suggestive of shocks. In addition, high J rotational transitions in CO (Goldsmith et al 1981, Storey et al 1981, Watson 1982) and the ground state OH rotational transition (Storey, Watson, and Townes 1981) have been seen.

An instructive example of the inherent uncertainties in shock modeling comes from comparing two theoretical shock models of the BN-KL region of Orion which "match" the observations yet arrive at somewhat different values for the fundamental physical parameters n_0, b, x(CO), and A_v. The results of Draine and Roberge (1982 DR) and CHM are: (DR, CHM) $n_0 = (7 \times 10^5, 2 \times 10^5)$cm^{-3}, B = (1.5, 0.45)mG, v_s = (38, 36) km/s, x(CO) = $(7 \times 10^{-5}, 3 \times 10^{-4})$, and A_v = (4., 2.5)mag at 2µm.

DR attempted to match the H_2 lines emitted at Peak 1, north of BN. Here Knacke and Young (1981) observed large intensities of high pure rotational lines of H_2 (S(9) -S(15)), and DR infer high values of n_0 to insure LTE populations at 2000 K. CHM averaged all intensities over the entire shock region since the CO beam does not resolve the region and they therefore ignored these high J H_2 transitions, which are localized

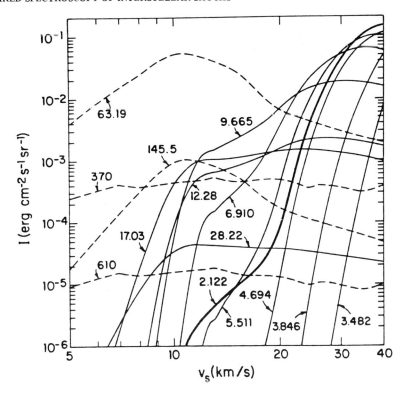

Figure 5. H_2, CI (370, 609 μm) and OI (63, 145 μm) line intensities for $n_0 = 10^6$ cm^{-3} and b = 1 as a function of shock velocity (from Draine, Roberge, and Dalgarno 1983).

and incompletely mapped. The high values of n_0 used by DR increase the CO 30-29/21-20 ratio to unacceptably high levels; however, they suggest physical arguments that the excitation cross sections for high J transitions in CO are incorrect. Recent calculations by Green and Chapman (1983) do not support this contention. By ignoring the H_2 emission from the higher rotational levels at Peak 1, CHM are able to match the CO observations with a lower density ($n_0 \cong 2 \times 10^5$ cm^{-3}) and the standard CO excitation rates (McKee et al 1982). These two constraints are illustrated in Figure 6 where the H_2 ratio 2-1 S(1)/1-0 S(1) --a measure of temperature --is plotted versus the CO ratio 21-20/30-29 --a measure of density (CHM). The H_2 lines are nearly the same wavelength and so are relatively unaffected by the unknown extinction; the CO lines are optically thin, and their ratio is independent of the CO abundance.

Since the shock intensity varies as $n_0 v_s^3$ and both DR and CHM find a similar value of v_s, the higher value of n_0 used by DR means that they must assume larger values of A_v in order that the 2 μm H_2 absolute intensities match those observed. Recent observations (Scoville et al

1982) support the lower extinctions found by CHM. In addition, DR must assume more extinction at 12 μm than do CHM. CHM were driven to higher values of CO abundance because of their lower densities. The CO lines are unattenuated by grains and therefore the absolute intensities of CO relative to extinction-corrected H_2 emission give a measure of the CO abundance. The abundance found by CHM is greater than that observed elsewhere in the ISM.

A noteworthy feature of the shock models common to both CHM and DR, but not emphasized in those papers, is the low value of $(n_{gr} \sigma_{gr}/n)_0 \cong 10^{-22}$ cm^2 required to produce viable shock models. Although grain mantles are destroyed in these C shocks, this value of grain area per hydrogen atom is nearly an order of magnitude lower than that appropriate to diffuse clouds, where grains may also be free from mantles. These low values of $(n_{gr} \sigma_{gr}/n)_0$ were required in order to spread the heating over a larger column density (see eq. 2.26), thereby keeping T_{max} at an acceptable level.

Overall, the differences between the models are relatively minor, and the models are remarkably successful in accounting for the strengths of the large number of molecular lines which have been observed in Orion. Both models require an ambient magnetic field of order a milligauss, large enough to be dynamically significant. Both predict large column densities of high velocity HI, which may have been seen in NGC 2071 (Bally and Stark 1983). CHM suggest that emission from J shocks may also be present, either from the shock which decelerates the outflow from the central source or in regions where the velocity of the shock in the ambient gas is larger than average.

5. SUMMARY

Violent events are common in the ISM, accompanying both the birth and death of stars. Such events generate strong shocks, which accelerate and heat interstellar gas. Shocks in ionic or atomic gas, or fast ($v_s \gtrsim 50$ km/s) shocks in molecular gas, are J type: the heating is abrupt and much of the emission occurs in the visible and UV, although strong infrared lines from OI, NeII, FeII, and various molecules are also produced (Table 2). Slower shocks in weakly ionized molecular gas are C type, and in such shocks virtually all the shock energy is radiated in infrared emission lines, primarily molecular. Shocks in molecular gas of relatively high ionization ($x_i \cong 10^{-4}$) near HII regions are likely to be intermediate between C and J type.

Infrared spectroscopy of interstellar shocks is vital, both to penetrate the veil of dust which obscures most of the Galaxy from visible observation and to study shocks in molecular gas. In particular, C shocks afford perfect targets for infrared spectroscopists (see Fig. 6). Infrared studies of the BN-KL region have provided dramatic evidence for energetic outflows from the young stars in that region, with an integrated energy output estimated to exceed 5×10^{47} erg (CHM). Shock models of Orion can account for the intensities of the

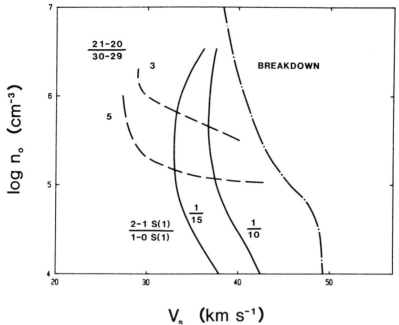

Figure 6. The observational constraints on preshock density and shock velocity for C shocks in Orion for b=1. The contours of constant CO and H_2 line ratios give the uncertainty in the observations; theoretical models are forced to lie within the intersection of the pairs of lines.

many infrared emission lines observed there (DR, CHM), but only if the magnetic field is large enough to be dynamically significant. Further work, both observational and theoretical, is necessary to understand the line profiles. Infrared studies of supernova remnants have been less informative so far, but this could change. The supernova rate in the Galaxy is estimated to be about $1/30$ yr^{-1}, yet the youngest known SNR is Cas A, which is 300 yr old. It is quite possible that heretofore unseen young SNRs will be discovered by observation of infrared line emission from the shocks which they drive into the ISM.

ACKNOWLEDGMENTS

CFM thanks ESA for the opportunity to participate in this Symposium. The research of CFM and DFC is supported in part by NSF grant AST 79 23243. DFC gratefully acknowledges the support of the California Space Institute as well. DJH acknowledges support under NASA RTOP 188-41-53-06-10.

REFERENCES

Arons, J., and Lea, S.M. 1976, Ap.J. 207, 914.
Axford, W.I. 1981, in Proceedings of the 17th International Cosmic Ray Conference, Paris, 12, 155.

Bally, J., and Stark, A.A. Ap.J. (Letters) 266, L61.
Chernoff, D.F., Hollenbach, D., and McKee, C.F. 1982, Ap.J. (Letters) 259, L97.
Cox, D.P. 1972, Ap.J. 178, 143.
Cox, D.P. 1979, Ap.J. 234, 863.
DeNoyer, L.K. 1979a, Ap.J. (Letters) 228, L41.
DeNoyer, L.K. 1979b, Ap.J. (Letters) 232, L165.
Draine, B.T. 1980, Ap.J. 241, 1021.
Draine, B.T., and Roberge, W.G. 1982, Ap.J. (Letters) 259, L91.
Draine, B.T., Roberge, W.G., and Dalgarno, A. 1982, Ap.J. 264, 485.
Draine, B.T., and Salpeter, E.E. 1979, Ap.J. 231, 438.
Elmegreen, B.G., and Lada, C.J. 1977, Ap.J. 214, 725.
Field, G.B., Rather, J.D.G., Aanestad, P.A., Orszag, S.A. 1968, Ap.J. 151, 953.
Gatley, I. 1984. This volume.
Gautier, T.N., Fink, U., Treffers, R.R., and Larson, H.P. 1976, Ap.J. (Letters) 207, L129.
Goldsmith, P.F., Erickson, N.R., Fetterman, H.R., Clifton, B.J., Peck, D.D., Tannenwald, P.E., Koepf, G.A., Buhl, D., and McAvoy, N. 1981, Ap. J. (Letters) 243, L79.
Green, S., and Chapman, S. 1983, Chemical Physics Letters (submitted).
Hall, D.N.B., Kleinmann, S.G., Scoville, N.Z., and Ridgway, S.T. 1981, Ap.J. 248, 898.
Hollenbach, D. 1982, Annals of New York Academy of Sciences 395, 242.
Hollenbach, D.J., and McKee, C.F. 1979, Ap.J. Suppl. 41, 555.
Hollenbach, D.J., and McKee, C.F. 1983, in preparation.
Hollenbach, D.J., and Shull, J.M. 1977, Ap.J. 216, 419.
Jackson, J.D. 1975, Classical Electrodynamics, 2nd ed. (New York: Wiley), p. 473.
Knacke, R.F., and Young, E.T. 1981, Ap.J. (Letters) 249, L65.
Lada, C.J. 1984. This volume.
McKee, C.F., and Hollenbach, D.J. 1980, Ann. Rev. Astr. Ap. 18, 219.
McKee, C.F., and Ostriker, J.P. 1977, Ap.J. 218, 148.
McKee, C.F., Storey, J.W.V., Watson, D.M., and Green, S. 1982, Ap.J. 259, 647.
Mouschovias, T.C. 1976, Ap.J. 207, 141.
Mullan, D.J. 1971, M.N.R.A.S. 153, 145.
Nadeau, D., Geballe, T.R., and Neugebauer, G. 1982, Ap.J., 253, 154.
Nadeau, D. and Geballe, T.R. 1979, Ap.J. (Letters), 230, L169.
Persson, S.E., Geballe, T.R., Simon, T., Lonsdale, C.J., and Baas, F. 1982, Ap.J. (Letters) 251, L85.
Russell, C.T., and Greenstadt, E.W. 1979, Space Sci. Rev. 23, 3.
Scoville, N.Z., Hall, D.N.B., Kleinmann, S.G., and Ridgway, S.T. Ap.J., 253, 136.
Seward, F.D., Harnden, F.R., Murdin, P., and Clark, D.H. 1983, Ap.J. (in press).
Shull, J.M., and Beckwith, S. 1982, Ann. Rev. Astron. Astrophys. 20, 163.
Shull, J.M., and McKee, C.F. 1979, Ap.J. 227, 131.
Shull, J.M., Seab, G., and McKee, C.F. 1983, Ap.J. (submitted).
Smith, H.A., Larson, H.P., and Fink, U. 1981, Ap.J. 244, 835.

Spitzer, L. 1978, Physical Processes in the Interstellar Medium (New York: Wiley).
Storey, J.W.V., Watson, D.M., and Townes, C.H. 1981, Ap.J. (Letters) 244, L27.
Storey, J.W.V., Watson, D.M., Townes, C.H., Haller, E.E., and Hansen, W.L. 1981, Ap.J. (Letters) 247, 136.
Tidman, D.A., and Krall, N.A. 1971, Shock Waves in Collisionless Plasmas (New York: Wiley), 175 pp.
Treffers, R. 1979, Ap.J. (Letters) 233, L17.
Watson, D.M. 1982. Ph.D. Thesis University of California, Berkeley.
Watson, D.M. 1984. This volume.

INFRARED HYDROGEN EMISSION LINES FROM H II REGIONS AND 'PROTOSTARS'

C.G. Wynn-Williams

University of Hawaii

INTRODUCTION

The first infrared hydrogen line observation of an H II region was Hilgeman's (1970) study of the Brackett-γ emission from the Orion Nebula. Subsequent studies of H II regions (see, for example, review by Wynn-Williams and Becklin 1974) and of the Becklin-Neugebauer source in Orion (Penston, Allen and Hyland 1971) over the next few years generally confirmed the expectation that hydrogen recombination lines were to be found in H II regions but not in "protostellar" infrared sources. This tidy differentiation was blunted by Grasdalen's (1976) discovery of Brackett-α emission in the Becklin-Neugebauer object. Subsequently Brackett-α or Brackett-γ emission has been detected in a significant fraction of all known compact sources in molecular clouds (see, for example, review by Wynn-Williams 1982). Recently, it has become clear that the physical conditions in the emitting regions of the compact "protostellar" sources are quite unlike those in compact H II regions and more resemble those in an outflowing extended stellar atmosphere. This distinction is reflected in the organization of this review; Section II covers the observations and theory of infrared emission from compact H II regions while Section III comprises a discussion of how these ideas are being modified in the cases where the lines are formed in stellar winds.

II. HYDROGEN EMISSION LINES FROM COMPACT H II REGIONS

The wavelengths of the atomic transitions of hydrogen are given by the equation

$$\lambda = 0.0912 \left[\frac{1}{n'^2} - \frac{1}{n^2} \right]^{-1} \mu m,$$

where n and n' are the principal quantum numbers of the upper and lower level of the transition. All transitions with n' $\geq$ 3 lie in the infrared or radio parts of the spectrum. Table 1 shows the wavelengths of some of these lines, together with an idea of their relative flux in a high-density, optically thin H II region at a temperature of 10^4 K. The most frequently observed infrared hydrogen lines are the Brackett-α and Brackett-γ lines, which have the advantage of being fairly strong, and of lying in clean atmospheric windows at wavelengths well separated from the visible. The strong Paschen-α line at 1.875 μm coincides with a telluric H_2O feature, making it hard to use except in redshifted extra-galactic objects (e.g. Neugebauer et al. 1980).

Table 1. Wavelengths (μm) and Emission Coefficients* of Selected Hydrogen Recombination lines at 10^4 K Based on Seaton's b_n Values.

	Balmer (n' = 2)	Paschen (n' = 3)	Brackett (n' = 4)	Pfund (n' = 5)	Humphreys (n' = 6)
α (n = n' + 1)	0.656 (330)	1.875 (34)	4.051 (7.0)	7.46 (2.1)	12.37 (0.78)
β (n = n' + 2)	0.486 (120)	1.282 (18)	2.625 (4.5)	4.65 (1.5)	7.50 (0.59)
γ (n = n' + 3)	0.434 (62)	1.094 (11)	2.166 (3.0)	3.74 (1.1)	5.91 (0.44)
δ (n = n' + 4)	0.410 (36)	1.005 (7.1)	1.945 (2.1)	3.30 (0.77)	5.13 (0.33)
Series Limit (n = ∞)	0.365	0.820	1.458	2.28	3.28

* Emission coefficients (units at 10^{-27} erg cm^3 s^{-1}) are in parentheses.

Calculation of the intensities of the hydrogen recombination lines has been described by Menzel (1937), Seaton (1960), Osterbrock (1974) and many others. The intensity $I_{nn'}$ (erg cm^{-2} s^{-1}) of the n-n' transition is given by

$$I_{nn'} = \frac{\gamma_{nn'}}{4\pi} \times (3 \times 10^{18} \text{ EM}),$$

where EM is the emission measure in units of pc cm^{-6} and $\gamma_{nn'}$ is an emission coefficient given by the expression

$$\gamma_{nn'} = 1.42 \times 10^{-16} \left(\frac{g_{nn'}}{n^3 n'^3}\right) b_{nn'} T_e^{-1.5} \exp(157800/n^2 T_e).$$

T_e is the electron temperature, $g_{nn'}$ is the Gaunt factor for the transition, and $b_{nn'}$ is the weighted non-LTE departure coefficient for the upper level. The Gaunt factor, which may be calculated from the formula given by Burgess (1958), is in the range 0.75 to 0.90 for most infrared lines.

The $b_{nn'}$ coefficients present a greater problem. In the approximation considered by Seaton (1960), the various ℓ states of each n level are populated according to their statistical weights. This approximation, which is valid in the high electron density limit, yields a single b_n value for each n level, independent of n'. Seaton's calculations were used to calculate the intensities shown in Table 1. A more sophisticated approach, followed by Clarke (1965) and Brocklehurst (1971) calculates the departure coefficient b_n for each ℓ state under specified conditions of electron temperature and density, then derives the effective $b_{nn'}$ value using only the permitted ℓ-ℓ' transitions. Some idea of the possible variation in the ratio of Bα to Bγ, and in the ratio Bγ to 6 cm free-free radio continuum fluxes is shown in Tables 2 and 3. Table 2 is based on the work of Giles (1977), while Table 3 is based on data cited in the Appendix of Wynn-Williams et al. (1978).

Table 2. Ratio of Bα/Bγ Intensities as a Function of Electron Temperature and Density

Electron Density	5,000 K	10,000 K	20,000 K
0	3.29	2.97	2.71
10^4 cm^{-3}	3.07	2.83	2.65
∞	2.24	2.35	2.45

b) <u>Recombination Lines and Extinctions</u>

In many observed emission line objects neither the Brackett-line ratios nor the infrared to radio flux ratio lie in the ranges (Tables 2 and 3) expected for plausible physical conditions. Two important causes of deviations are free-free self-absorption at radio wavelengths and dust extinction at infrared wavelengths. Figure 1 illustrates a graphic method of testing whether or not the observed line strengths in a parti-

Table 3. Ratio of $B\gamma$/6-cm Intensities as a Function of Electron Temperature and Density* (units are 10^{11} Hz)

Electron Density	5,000 K	10,000 K	20,000 K
0	14.8	8.8	5.0
∞	12.2	7.8	4.8

* $N(He+)/N(H+) = 0.1$

ular object are as expected for an H II region lying behind a screen of dust with an assumed extinction law. The flux of a particular emission line is plotted vertically as the equivalent 6-cm free-free flux density calculated assuming no dust absorption, and, in this case, Seaton's (1960) b_n values for 10,000 K. The horizontal axis is the normalized extinction coefficient A_λ/A_V of interstellar dust at the wavelength (λ) of the line. The point on the vertical axis is the measured radio flux density. The advantage of plotting hydrogen emission line data in the way shown in Figure 1 is that if the intrinsic strengths of the emission lines obey the assumed recombination theory, and if the foreground dust obeys the assumed extinction curve, then the points in Figure 1 should all be on a straight line, from the gradient which can be derived the extinction, and the vertical intercept the dereddened flux.

The data in Figure 1, which are all taken from the work of Lester et al. (1983) for G45.13 + 0.14A and K3-50 and Simon et al. (1983) for NGC 6334 IRS1 and GL 490, show a variety of qualities of fit to the model. The sources NGC 6334 IRS1 and G45.13 + 0.14A obey the straight line laws reasonably well, given the presence of some free-free self-absorption plus uncertainties in the interstellar reddening curve and in the electron temperature and density. The line strengths in K3-50, on the other hand, cannot be fitted to a single extinction curve, as was pointed out originally by Persson and Frogel (1974). Figure 1 demonstrates, however, that the emission in K3-50 could be modelled by a combination of a 6 Jy source behind 23 magnitudes of extinction plus a 0.15 Jy source behind 7.3 magnitudes of extinction. Models which include scattering (Harris 1975) or an inhomogeneous extinction layer (Wynn-Williams et al. 1977) are also plausible, though.

The final source, GL 490, is an example of an object for which the data do not even remotely satisfy the model; objects such as these are much better described by outflow models such as those discussed in the next section.

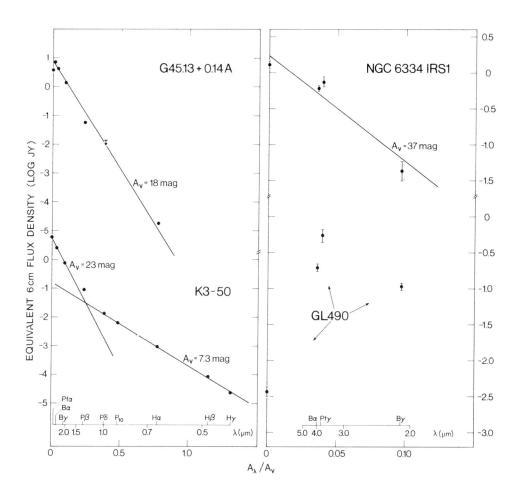

Fig. 1. Hydrogen line intensities as a function of extinction for four infrared sources. The extinction curve is based on the data of Osterbrock (1974, p. 171) at $\lambda < 2$ μm and Becklin et al. (1978) at $\lambda > 2$ μm.

A second problem with the interpretation of the infrared emission lines from compact infrared sources is the fact that in several cases the rate of ionization that is implied greatly exceeds that expected from a main-sequence star of the same luminosity. Thompson (1982) has proposed a model in which this excess ionization is produced by a hot circumstellar disk. An alternative model, in which the extra ionization arises by absorption of Balmer continuum photons (Simon et al. 1983) is discussed in Section III.

III. HYDROGEN EMISSION LINES FROM OUTFLOWING WINDS

In the last four years much evidence has accrued that many compact infrared sources in molecular clouds and other young stellar objects are expelling matter rather than accreting it (see, for example, review by Wynn-Williams 1982). The evidence for these outflows includes the proper motions of H_2O maser spots (Genzel et al. 1981), the bipolar outflows seen in CO and other microwave molecular lines (see review by Lada elsewhere in this volume), and the free-free radio spectrum (Wright and Barlow 1975, Panagia and Felli 1975, Felli and Panagia 1981). The idea that the infrared hydrogen lines seen from some sources might be formed in outflow regions came partially as a result of the failure of simple recombination theory to account for the line strengths, and partly from the large velocity-widths (~ 140 km s^{-1}) found in the Brackett-γ line in the sources M17-IRS1 and GL 490 (Simon et al. 1981).

a) Free-Free Emission from an Ionized Outflowing Wind

In an outflowing wind the electron density generally decreases monotonically with increasing distance from the central star or protostar. Consequently, the optical depth measured along lines of sight that pass close to the star will be larger than those that are farther away. The case of constant velocity outflows was considered independently by Wright and Barlow (1975) and Panagia and Felli (1975); they showed that except at the highest frequencies (where even the line of sight closest to the star is optically thin), the free-free emission has a spectral index close to +0.6. As both groups pointed out, this value corresponds well to that observed for P Cygni; although spectral indices covering a wide range of values have been observed in various young objects. These differences in spectral index may arise because of time variability in the outflow (Wright and Allen 1978), because of acceleration or deceleration of the flow (Felli and Panagia 1981) or because of the existence of an outer boundary to the ionized region (Simon et al. 1983). This outer boundary, which could be due either to the insufficiency of ionizing photons or to a dust-front or other physical neutral boundary, generally has the effect of raising the spectral index from around 0.6 toward the value 2.0. It should be stressed that because of the self-absorption in the central parts of these outflowing winds, the radio emission, unlike the infrared line emission, is generally dominated by flux from the <u>outer</u> parts of the region.

b) Recombination Rate in an Outflowing Wind

If a star is losing mass at a rate $\dot{M}$ at constant velocity v, the electron density N_e at a radius r in a fully ionized wind of pure hydrogen is given by:

$$N_e = \frac{\dot{M}}{4\pi M_H r^2 v},$$

where M_H is the mass of the hydrogen atom.

If, for the time being, we treat the wind as a variable-density isothermal H II region, we may calculate the total recombination rate between the stellar surface R_* and the outer boundary of the ionized region R_o:

$$\text{Total Recombination Rate} = \alpha(T) \int_{R_*}^{R_o} 4\pi N_e^2 r^2 dr$$

$$= \frac{\alpha \dot{M}^2}{4\pi M_H^2 v^2} \left[\frac{1}{R_*} - \frac{1}{R_o} \right] ,$$

where $\alpha(T)$ is the recombination coefficient to levels other than the ground state (see, for example, Spitzer 1978, p. 107).

This last equation has some interesting consequences; first it implies that at least half of all the recombinations occur between R_* and $2R_*$. Second, as discussed by Felli and Panagia (1981), it indicates that the nature of the emitting region is critically dependent on the rate of production of ionizing photons by the central star (N_ϕ) in the sense that if

$$N_\phi > \frac{\alpha \dot{M}^2}{4\pi M_H^2 v^2 R_*}$$

then the ionized region extends out to infinity. If N_ϕ is less than half this initial critical value, then the ionized region extends to less than one stellar radius above the stellar surface, and the ionized region might properly be considered to be part of the stellar atmosphere. Only for a very narrow range of N_ϕ does the ionized region bear any resemblance to a classic Strömgren sphere. The relationship between N_ϕ and $\dot{M}$ for main sequence stars is shown in Figure 2, which is based on Panagia's (1973) compilation of OB star properties. Sources are found on either side of the critical region.

c) Physical Characteristics in the Outflow Region

Before considering the formation of emission lines in these outflows, it is important to stress how greatly the physical conditions in these regions differ from those in "conventional" H II regions. As discussed above, most of the activity in these regions occurs within one stellar radius of the surface. Since an OB star has a typical radius of order $10R_\odot$ the characteristic diameter of an outflow region is of order only 3×10^{12} cm, some four orders of magnitude less than that of the ultracompact H II region W3(OH) (Dreher and Welch 1981). An immediate consequence of this small diameter is that the ionized region will be extremely dust-free, since a grain at a distance of $1R_*$ from a 20,000 K star would immediately be heated to far above its melting temperature.

An estimate of the typical electron density in an outflowing region

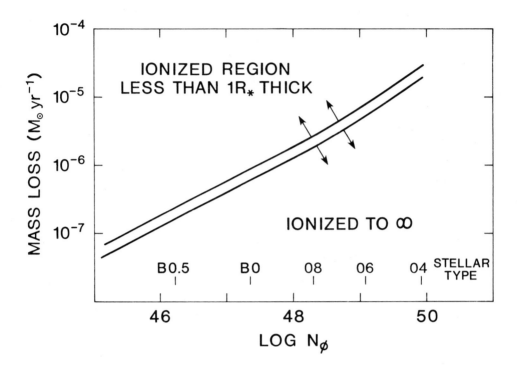

Fig. 2. Relationship between mass loss rate and ionization rate for early type stars.

requires the assumption of a mass loss rate and a wind velocity. For $\dot{M} = 10^{-6} M_\odot$ y^{-1} and v = 100 km s^{-1}, the electron density at 1R$_*$ above the surface is 1.5×10^{11} cm^{-3}. This density is about six orders of magnitude greater than that in W3(OH) and, as discussed below, is so high as to collisionally deexcite the upper levels of all the ionic forbidden lines that determine the thermal balance of a conventional H II region.

d) Optical Depth of IR Hydrogen Lines

For several infrared sources with strong Brackett lines it is easy to show that if the emission arises from a constant velocity outflow region then it must be optically thick. For GL 490, for example, a compact infrared source at a distance of 0.9 kpc (Harvey et al. 1979), Simon et al. (1983) measured a Brackett-α flux of 3.6×10^{-15} Wm^{-2} and a velocity of FWHM of 150 km s^{-1}. If this emission were optically thin it would emanate from a region of characteristic size 3×10^{12} cm and would by application of Planck's law, have a brightness temperature of order 6×10^4 K. Since this high temperature is incompatible with our anticipation that the electron temperature is not grossly different from 10^4 K, we must conclude that the Brackett-α line is optically thick out to a radius of several stellar diameters.

Simon et al. (1981, 1983) and Krolik and Smith (1981) have addressed the theory of optically thick hydrogen line formation in an outflowing wind; both groups draw on Castor's (1970) treatment of the Sobolev approximation. The most important assumptions entering into these treatments are

i) the velocity gradient along any line of sight is large enough that a newly emitted photon is reabsorbed either very close to its point of origin or not at all.

ii) the emission is under LTE at constant temperature.

iii) the radial velocity is either constant or follows a power law.

For the case of a constant velocity flow the following results emerge from the theory:

i) <u>Line Shapes</u>. The lines range from rectangular in the limit of low optical depth to parabolic in the limit of large optical depth.

ii) <u>Line Flux</u>. The flux density of an optically thick hydrogen line is reduced below its optically thin value by a factor of approximately $\tau^{-1/3}$. For a mass outflow of 10^{-6} $M_\odot/yr^{-1}$ at 100 km s^{-1}, the optical depth at the center of the Brackett-α line is about 3×10^4.

The flux density of an optically thick line varies as

$$S \propto \dot{M}^{4/3} v^{-2}$$

iii) <u>Line Ratios</u>. The Brackett-α line has a higher optical depth than the Brackett-γ line, so that the line ratio varies as

$$2.8 > \frac{B\alpha}{B\gamma} > 1.2$$
Thin   Thick

It is this variation with optical depth that makes it hard to use Brackett lines for measuring the extinction toward "protostellar" sources.

e) Specific Sources

The results of this theory have been applied to a number of sources that have measured Brackett lines. The sources GL 490, M17-IRS1, and S106 can be accounted for by a mass losses in the range 10^{-7} to 10^{-5} $M_\odot$ yr^{-1} at constant velocities of around 150 km s^{-1} (Krolik and Smith 1981, Simon et al. 1983), while a study of four infrared hydrogen lines in P Cygni by Felli et al. (1982) gives a mass loss rate of 1.5×10^{-5} $M_\odot$ yr^{-1} in a wind accelerating from 20 to 40 km s^{-1}.

The BN source in Orion does not appear to fit this picture. The close similarity of the Brackett-α and Brackett-γ line shapes has led Scoville et al. (1983) to conclude that the emission is optically thin with a mass loss of order 10^{-7} $M_\odot$ yr^{-1} and a decelerating flow. Its radio spectrum resembles that of an optically thick compact H II region (Moran et al. 1983).

f) Problems and Future Work

Although the treatment of thick lines outlined above has greatly improved our understanding of hydrogen lines in compact infrared sources, there are currently several major weaknesses in the theory that are related to the assumptions that enter into the model. Probably the most serious is the assumption of a constant, or power law, velocity flow. The escape velocity for main-sequence O star is about 1,000 km s^{-1}, so that changes in velocity of a few hundred km s^{-1} are almost inevitable as the gas leaves the surface of a star. The shapes and strengths of the infrared lines are very sensitive to the conditions in the acceleration or deceleration region. Also, the Sobolev approximation may not be valid in the case of slow winds where the expansion speed is not much greater than the sound speed (~10 km s^{-1}).

Another major simplification of the present theory concerns the assumption of LTE. A full theory would involve a proper examination of the ionization and thermal balance in the outflowing wind as a function of position. Large optical depths in the Balmer and higher lines will significantly change the level populations as compared to those of H II regions, while the absence of cooling by ionic forbidden lines renders dangerous the assumption of isothermal electron temperature.

An even more serious effect of high optical depths is pointed out by Simon et al. (1983). They show that in objects such as GL 490, the optical depth of Balmer continuum radiation may be substantial, and that as a consequence, hydrogen may ionized out of the n = 2 level as well as the n = 1 level. They go on to show that this effect may lead to a substantial increase in the ionization (and hence recombination) rate in sources with outflowing winds, and that this effect can account for the "excess line emission" discussed by Thompson (1982) and others. The fact that this excess line emission appears to be a property of objects with the luminosity of B stars rather than O stars is neatly explained by the fact that B stars have much larger ratios of Balmer to Lyman continua than do O stars.

REFERENCES

Becklin, E. E., Matthews, K., Neugebauer, G., Willner, S. P.: 1978, Astrophys J. 220, pp. 831-35.
Brocklehurst, M.: 1971, Monthly Notices Roy. Astron. Soc. 153, pp. 471-90.

Burgess, A.: 1958, Monthly Notices Roy. Astron. Soc. 118, pp. 477-95.
Castor, J. L.: 1970, Monthly Notices Roy. Astron. Soc. 149, pp. 111-27.
Clarke, W. H.: 1965, "Recombination Line Intensities in an Attenuated Hydrogen Plasma," University of California, Los Angeles (Ph.D. Thesis).
Dreher, J. W. and Welch, W. J.: 1981, Astrophys J. 245, pp. 857-65.
Felli, M., Oliva, E., Natta, A., Stanga, R., and Beckwith, S.: 1982, Poster presented at this meeting.
Felli, M. and Panagia, N.: 1981, Astron. Astrophys. 102, pp. 424-30.
Genzel, R., Reid, M. J., Moran, J. M., and Downes, D.: 1981, Astrophys. J. 244, pp. 884-902.
Giles, K.: 1977, Monthly Notices Roy. Astron. Soc. 180, pp. 57P-59P.
Grasdalen, G. L.: 1976, Astrophys. J. (Letters) 205, pp. L83-L85.
Harris, S.: 1975, Monthly Notices Roy. Astron. Soc. 170, pp. 139-53.
Harvey, P. M., Campbell, M. F., Hoffmann, W. F., Thronson, H. A., and Gatley, I.: 1979, Astrophys. J. 229, pp. 990-93.
Hilgeman, T. W.: 1970, "Infrared Observations of H II Regions," California Institute of Technology (Ph.D. Thesis).
Krolik, J. H. and Smith, H. A.: 1981, Astrophys. J. 249, pp. 628-36.
Lester, D. F., Dinerstein, H. L., Witteborn, F. C., Bregman, J. D., Cohen, M., and Rank D. M.: 1983, "Hydrogen Recombination Line Measurements for Two Compact H II Regions: Implications for Extinction and the Nature of the Underlying Sources" (preprint).
Menzel, D. H.: 1937, Astrophys. J. 85, pp. 330-39.
Moran, J. M., Garay, G., Reid, M. J., Genzel, R., Ho, P. T. P.: 1983, in A. E. Glassgold and P. J. Huggins (eds.), "Henry Draper Memorial Symposium," Annals N. Y. Acad. Science (in press).
Neugebauer, G., Morton, D., Oke, J. B., Becklin, E. E., Daltabuit, E., Matthews, K., Persson, S. E., Smith A. M., Soifer, B. T., Torres-Peimbert, S., and Wynn-Williams, C. G.: 1980, Astrophys. J. 238, pp. 502-509.
Osterbrock, D.E.: 1974, "Astrophysics of Gaseous Nebulae," Freeman, San Francisco.
Panagia, N.: 1973, Astron. J. 78, p. 929-34.
Panagia, N. and Felli, M.: 1975, Astron. Astrophys. 39, pp. 1-5.
Persson, S. E. and Frogel, J. A.: 1974, Astrophys. J. 188, pp. 523-27.
Penston, M. V., Allen, D. A., and Hyland, A. R.: 1971 Astrophys. J. (Letters) 170, p. L33.
Scoville, N. Z., Hall, D. N. B., Kleinmann, S. G., and Ridgway, S. T.: 1983, Astrophys. J. (in press).
Seaton, M. J.: 1960, Monthly Notices Roy. Astron. Soc. 119, pp. 90-97.
Simon, M., Righini-Cohen, G., Fischer, J., and Cassar, L.: 1981, Astrophys. J. 251, pp. 552-56.
Simon, M., Felli, M., Cassar, L., Fischer, J., and Massi, M.: 1983, Astrophys. J. (in press).
Spitzer, L.: 1978, "Physical Processes in the Interstellar Medium," Wiley, New York.
Thompson, R. I.: 1982, Astrophys. J. 257, pp. 171-78.
Wright, A. E. and Allen, D. A.: 1978, Monthly Notices Roy. Astron. Soc. 184, pp. 893-902.

Wright, A. E. and Barlow, M. J.: 1975, Monthly Notices Roy. Astron. Soc. 170, pp. 41-51.
Wynn-Williams, C. G.: 1982, Ann. Rev. Astron. Astrophys. 20, pp. 587-618.
Wynn-Williams, C. G. and Becklin, E. E.: 1974, Publ. Astron. Soc. Pacific 86, pp. 5-25.
Wynn-Williams, C. G., Becklin, E. E., Matthews, K., Neugebauer, G., and Werner, M. W.: 1977, Monthly Notices Roy. Astron. Soc. 179, pp. 255-64.
Wynn-Williams, C. G. Becklin, E. E., Matthews, K., Neugebauer, G.: 1978, Monthly Notices Roy. Astron. Soc. 183, pp. 237-44.

THE SIGNIFICANCE OF FAR-INFRARED SPECTRA OF THE INTERSTELLAR MEDIUM

Martin Harwit

Cornell University

This paper deals with the analysis of interstellar clouds through infrared and submillimeter observations. The determination of temperature, density, state of ionization and abundance of different chemical constituents is discussed for clouds exhibiting a range of optical depths. There is a preferred choice of spectral features useful for examining different types of gaseous complexes. The prospects for improving our understanding of chemical reaction schemes in interstellar clouds are outlined. Prime advances may be expected in improved data on the abundance of atoms; diatomic hydrides; other small, hydrogen-containing molecules; and their ions.

I. INFRARED EMISSION FROM THE INTERSTELLAR MEDIUM

For many years infrared and submillimeter studies of the interstellar medium concerned themselves solely with radiation by dust. Dust absorbs starlight and emits infrared radiation at wavelengths dictated by the temperature of the grains. The bulk of Galactic infrared emission originates in this fashion; though occasionally, in well-shielded clouds, the grains absorb little starlight and are heated primarily by collisions with ambient molecules.

When interstellar clouds are observed at a spectral resolution, $R \gtrsim 100$, a number of strong emission lines begin to appear, and occasionally one finds absorption lines as well (Ward et al. 1975, Storey et al. 1981). While these lines have only become accessible over the past seven years, they have already shown themselves to be valuable in providing us with the temperature and density of portions of the interstellar medium which we were unable to study before.

Of particular importance are observations of two kinds of regions.

1. Phases of the interstellar medium which previously were inaccessible -- clouds in the temperature range $30 \lesssim T_{gas} \lesssim 3000$ K: Such clouds are too cool to produce optical excitation lines, but sufficiently warm so that the lowest radio-emitting levels are largely populated

in proportion to their statistical weights, and only a weak dependence on temperature.

This class of sources comprises several quite different kinds of domains:
    a) gas at the interface between fully ionized and molecular clouds, including the ionization front, C II regions, and quite possibly a number of other phases remaining to be identified;
    b) shocked neutral regions in which high-lying levels of diatomic molecules can be collisionally excited;
    c) emission from the diffuse, low-density interstellar medium.

2) Gas sufficiently hot to emit optical lines, but obscured by dust to an extent that makes a determination of optical line intensities unreliable: Infrared radiation can pass through such clouds virtually unattenuated, and can provide information on local conditions.

This type of source again comprises several quite distinct domains:
    a) H II regions obscured by dust;
    b) planetary nebulae containing dust;
    c) obscured central portions of Seyfert galaxies and, one imagines, also quasars.

While we talk about shocks and ionized fronts, about molecular clouds and fully ionized gas, or about galaxies and planetary nebulae, the physical insights we derive from infrared and submillimeter observations are always of one and the same kind: Given high-quality data, we can obtain an average local temperature and density, for the atomic or molecular species studied, a column density for the chosen field of view, and, in high-resolution observations, the line-of-sight velocity distribution. How these characteristics are derived from raw data is discussed next.

II. COLLISIONS AND RADIATION

The far-infrared line emission we observe in interstellar clouds involves little more than collisional excitation and radiative de-excitation of particles, often with slight variations. We can describe these variations as added complexities by considering the following sequence of cases:

Case a. Tenuous Cloud, Low-Optical Depth; Two-Level Atom, Ion or Molecule:

In the simplest situation, an atom, ion, or molecule of species X experiences a collision with another atom, ion or electron and becomes excited from a state X_i to a higher energy state X_j. The collision is followed by radiative de-excitation $X_j \to X_i$, and the emitted radiation eventually enters our telescope and is registered by a detector.

i) The rate of radiation is directly proportional to:
n_j the number density of particles in state X_j
A_{ji} the Einstein coefficient for radiative de-excitation $X_j \to X_i$,
V the volume observed.

ii) The rate of radiation also is directly proportional to:
n_i the number density of particles in state X_i,
n_c the number density of colliding particles producing the excitation,
v_c the relative velocity at collision,
σ_{ij} the collisional cross section for excitation from $\underline{i}$ to $\underline{j}$,
V the volume observed.

If the gas is tenuous, the number of particles in the upper state X_j is limited by the relatively few exciting collisions, and we can simply write

$$n_j A_{ji} V = n_i n_c v_c \sigma_{ij} V \qquad (1)$$

provided that the line of sight through the clouds is sufficiently short to prevent significant absorption of radiation within the cloud.

Case b. Dense Cloud, Low-Optical Depth:

When the density of colliding particles n_c in the cloud is increased, n_j, the density of particles in state X_j, also increases until the thermal equilibrium ratio

$$n_j/n_i = [g_j/g_i] \exp[-E_{ij}/kT] \qquad (2)$$

is reached. Here g_i and g_j are the respective statistical weights of states X_i and X_j; E_{ij} is the transition energy, T is the temperature and k, the Boltzmann constant. The density for which the collisional de-excitation rate equals the radiative de-excitation rate is called the critical density, n_{crit}. Once the critical density is exceeded, the emission becomes increasingly independent of n_c. The emission rate is now given only by the left-hand side of equation (1). Combined with equation (2) the emission observed becomes

$$\text{Emission} = V A_{ji} n_i [g_j/g_i] \exp[-E_{ij}/kT] \text{ for } n_c \gg n_{crit} \qquad (3)$$

We note that
i) At the high temperature found in planetary nebulae and H II regions $E_{ij} \ll kT$ for all ground state fine-structure transitions and n_j/n_i approaches g_j/g_i. This is seen happening for O^{++} in Fig. 1; all the different temperature curves converge.

ii) The observed emission now is directly proportional to n_i, but independent of n_c. This contrasts to the low-density case where

iii) For $n_c \ll n_{crit}$ the observed emission is proportional to

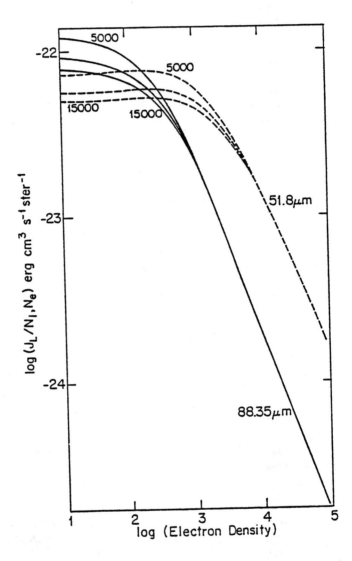

Figure 1 - Emission rate of O^{++} in its 88 and 52µ lines (after Simpson 1975, Melnick 1981).

$n_j n_c$ and therefore also to $n_i n_c$.

iv) As a homogeneously mixed gas is isothermally compressed, the emission rate at low densities first increases as the density squared. However, when the critical density is exceeded, the emission begins to rise more gradually -- asymptotically in proportion to the first power of the density.

Case c. Dense Cloud, High-Optical Depth:

As the depth of the observed gas cloud increases, self-absorption of radiation within the cloud becomes important. The absorption cross section is given by the Einstein coefficient, B_{ij}, related to the spontaneous emission by

$$B_{ij} = [g_j/g_i] A_{ji} \lambda_{ji}^2/8\pi \qquad (4)$$

λ_{ji} is the transition wavelength. The bandwidth over which the emission occurs is usually dictated by the line-of-sight Doppler velocity v within the cloud

$$\delta\nu]_{Doppler} = \nu_{ji}[v/c] = v/\lambda_{ji} \qquad (5)$$

so that the absorption cross section for radiation, within the Doppler broadened line is

$$\sigma_{ij}]_{Doppler} \sim A_{ji} [g_j/g_i] \lambda_{ji}^3 / 8\pi v \qquad (6)$$

The column density for unit optical depth is just the reciprocal of the cross section

$$N \ [\tau=1]_{Doppler} = [\sigma_{ij}]_{Doppler}]^{-1} \sim 8\pi v [g_i/g_j]/[A_{ji}\lambda^3] \qquad (7)$$

When the column density increases beyond this value, the radiation may be reabsorbed and re-emitted a number of times in succession.

We can think of four simple cases

i) If the density is low and the column density does not greatly exceed the value given in equation (7), then the radiation simply is absorbed and re-emitted several times in a row. Eventually it finds its way out of the cloud in a random walk, and there is no loss in observed emission. Equation (1) holds.

ii) If the density is low, and the column density very high, then successive reabsorption and re-emission occurs until emission far out in the wing of the line becomes probable. Radiation emitted in the line wings is not readily reabsorbed and can also escape from the cloud. Equation (1) still holds.

iii) If the density approaches the critical density or exceeds it, the radiation may not escape at all if collisional de-excitation takes place after repeated reabsorption and re-emission. The energy E_{ji} then is fed back into thermal energy of the gas, rather than being radiated away. The effective optical depth of a cloud with column density $N_{ij} \gg N_{ij}(\tau=1)$ becomes

$$\tau_{eff} \sim \left[\left(\frac{N_{ij}}{N_{ij}(\tau=1)}\right)\right]^2 \varepsilon, \quad \varepsilon = \frac{n_c}{n_c + n_{crit}} \tag{8}$$

Here ε is the probability for collisional de-excitation per absorption, and the square of the ratio of column densities gives the number of successive absorptions undergone before a photon reaches the edge of the cloud, in a random walk. The fraction of the radiation that escapes diminishes exponentially as the square of the column density.

$$I = I[0] \exp[-\tau_{eff}] \tag{9}$$

where $I[0]$ represents the intensity observed if there were no collisional de-excitation.

iv) If the density and optical depth become very high, equilibrium is established, and radiation escaping from the surface of the cloud, within the line width, has the intensity of blackbody radiation at the temperature of the gas in the cloud. The temperature derived from the line intensity always provides a lower limit to the actual temperature.

Case d. Multi-Level Atomic System:

When an atom, ion or molecule has several different states of excitation, individual levels can be populated along different collisional or radiative paths. To calculate expected line intensities we then need to know

• collisional cross sections for transitions between all possible levels X_k and X_j,

• Einstein coefficients A_{kj} for transitions between all possible levels k and j.

Case e. Several Different Species of Exciting Particles:

To calculate the expected emission, we need to know the collisional excitation cross sections for transitions between all possible levels, X_k, X_j, for each species of colliding particle, as well as all the respective Einstein coefficients.

Case f. Density Inhomogeneity Within the Field of View:

When the region observed is inhomogeneous, expressions such as equations (1) or (3) must be integrated over the available volume. In principle, we would need to know local densities throughout the volume in order to precisely calculate the flux escaping the region. We seldom have that much information available in practice.

III. WHAT CAN THE OBSERVATIONS TELL US?

Let us assume first that collisional cross sections and radiative transition probabilites are known for all the constituents in a given cloud, then the physical conditions within the cloud can be estimated in a succession of steps. Two particular cases exhibit the kind of argument that can be pursued.

a. Observations of O^{++} Transitions in an H II Region

i) For most H II regions we have data on the total column density of hydrogen, indirectly obtained from the emission measures that generally are available. The emission measure, however, is the line-of-sight integral of density squared, and we therefore have to make a guess about the depth of the observed region, and assume homogeneity along the line of sight if we wish to derive a density.

ii) Having done that, we can estimate the column density of oxygen, for example, by assuming cosmic abundance $\sim 7 \times 10^{-4}$ and a certain fractional population of oxygen in the doubly ionized state. We need to know this number only approximately in order to tell whether the optical depth in the O^{++} emission is likely to be low or high. For the clouds observed to date the optical depth has been low and self-absorption in the emission line has generally been neglected.

iii) In ionized hydrogen regions, protons and electrons are equally abundant, but the bulk of the excitation of O^{++} is due to collisions with electrons.

iv) The density of many HII regions lies in the range $n_e = 10^2$ to 10^4 cm^{-3}. This is the range in which the critical densities lie, for the two available transitions, the $^3P_1$ to $^3P_0$ transition at 88μ and the $^3P_2$ to $^3P_1$ transition at 52μ. Direct radiative transition from the $^3P_2$ to the ground $^3P_0$ state is too improbable to contribute significantly. Figure 1 shows that the ratio of intensities oberved at 88 and 52μ can be used as a measure of the density, and is quite insensitive to local temperature.

v) Temperature data, in any case, are available from optical and especially from radio observations for most of the HII regions studied. We are therefore able to derive an ambient electron density $\langle n_e \rangle$ for the doubly ionized oxygen region. If this differs from the root-mean square density, $\langle n_e^2 \rangle^{1/2}$, estimated from radio emission measures, we directly obtain a clumping factor that might be expressed as

$$\text{Clumping factor} \rightarrow \langle n_e \rangle / \langle n_e^2 \rangle^{1/2} \tag{10}$$

Melnick et al. (1979) obtained such a measure for several H II regions and found clumping factors ranging from ~1.5 to ~5. The value for M42 was ~3, substantially lower than estimated by Osterbrock and Flather (1959) by optical means.

vi) Finally, the absolute brightness of the observed region can be used in conjunction with the O^{++} line intensity ratio, to solve for the abundance of O^{++} ions along the line of sight.

vii) When a variety of different spectral lines are available for one and the same region, increasingly detailed information about the H II region can be derived. Thus, the pair of S^{++} lines at 18 and 33μ or the N^{++} transition at 57μ can be used to obtain supplementary data (Moorwood et al. 1980, Herter et al. 1982). Unfortunately, the different sets of observations often lead to rather different results, indicating that the various species of ions are not co-extensive and that the observations obtained, therefore, do not refer to one and the same physical region. H II regions are rather complicated domains consisting of irregular nested, onion-like layers, each layer containing a different set of ions. At increasing distances from the ionizing stars, the most energetic photons tend to become depleted and successively lower states of ionization are encountered in increasingly distant layers. At increasing distances from a star, O^{++} gives way to O^+, and S^{++} will partially overlap with both the O^{++} and the O^+ region. S^+, in turn, partially overlaps the O^+ region, but also extends out into the neutral hydrogen domain. Similar results are found in higher ionization states in planetary nebulae -- [OIV] 25.87μ and [NeV] 24.28μ (Shure et al. 1983). All this is readily seen from the sequence of ionization potentials $S°[10.4\ eV]$, $H°[13.6\ eV]$, $O°[13.6\ eV]$, $S^+[23.3\ eV]$, $O^+[35.1\ eV]$.

Optical data can be particularly useful for obtaining temperature, while infrared data provide corresponding densities, as Dinerstein's useful diagramatic approach has shown (cf. Fig. 2, Dinerstein 1982).

Stasinska and Baluteau have discussed such questions in a paper contributed to this meeting, while Roger Emery will be reporting on observations of ionized regions in greater detail in his own report. I will therefore not pursue this topic further.

b. Border Regions Separating Ionized from Molecular Gas

Observations have been made in the $C^+(157μ)$ fine-structure emission line emanating from the vicinity of HII regions. The radiation appears to come from two types of domains:

i) The carbon radio-recombination line region which forms the interface between H II and H_2 regimes. Calculations show that if this

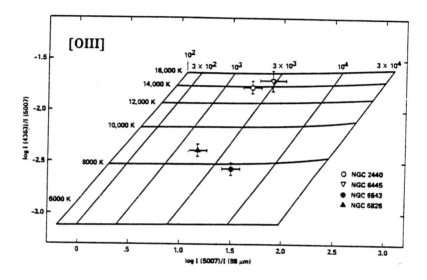

Figure 2 - Optical and far-infrared line emission from planetary nebulae (after Dinerstein 1982). The 4363Å/5007Å ratio is highly temperature-sensitive. The 52μ/88μ ratio would be highly sensitive to density. Since no 52μ data were available, the somewhat temperature-dependent 5007Å/88μ ratio was used as a density indicator. That temperature dependence accounts for the slanted constant-density lines.

gaseous phase is at a temperature of the order of 100 K in NGC 2024, and 225 K in Orion, roughly the observed ratio of intensities for the $C^+(157\mu)$ and $O^\circ(63\mu)$ emission are obtained (Russell et al. 1980). Neutral oxygen has an inverted level structure with a ground state $^3P_2$, a first excited fine-structure state $^3P_1$ and an upper excited state $^3P_0$. The wavelength of the transition between the two lower states is 63μ; the wavelength of the transition between upper states is 145.5μ. Recent observations of the Trapezium region in M42 show that the intensity ratio of the 63 and 145μ lines is of the order of 10 (Stacey et al. 1982). This neutral atomic oxygen emission cannot come from the ionized gas, since oxygen is ionized wherever hydrogen is -- both ionization potentials are 13.6 eV. The emission cannot come from the ionization front, either, since the front is too thin and the column density too small to produce the high observed flux. If the radiation were to emanate from the hot, shocked region known to exist in the Kleinmann-Low Nebula, the expected CO emission would be much higher than actually observed (see below). The warm neutral C^+ region, however, can emit this radiation provided the 63μ emission is self-absorbed by roughly a factor of 2. This produces both the correct range of absolute fluxes, for the C^+ and the two 0° lines, as well as the observed ratio of 145 and 63μ intensities. Figure 3 shows how different line intensities vary as a function of temperature in low-optical-depth clouds, at densities far below n_{crit}. Figure 4 shows how the intensity of the 145 and 63μ lines varies as a function of temperature as well as density of different ambient particles.

ii) Studies of wide regions around M17 and NGC 2024 show that there is an added, tenuous, more extended C^+ cloud extending out to several H II region radii (Kurtz et al. 1982). There also is a suggestion of considerable emission in the 157μ line, from the Galactic plane, probably from the borders of the aggregated molecular clouds. Of the order of 1% of the luminosity of the Galaxy may be emitted in this one line, if a first set of observations proves to be typical (Stacey et al. 1982a).

c. Molecular Regions at High Temperatures:

The Kleinmann-Low region of the Orion complex contains a hot molecular hydrogen cloud believed to be shock-heated. Watson et al. (1980) first studied the far-infrared CO emission from this region. The transitions observed are rotational transitions $J \to J-1$. A variety of transitions from $J = 34$ downward have been seen. McKee et al. (1982) developed models that showed the distribution of emission among the different transitions as a function of temperature and local density. On this basis, one can show that the observed line strengths fit a curve expected for gas at a density of order $n_{H_2} = 10^6$ cm^{-3} and $T = 1000$ K (Fig.5.) The intensity in the lowest transitions is appreciably higher, suggesting a wider distribution of lower temperature gas, as well.

Some of the transitions around $J = 15$ to 20 are particularly insensitive to the local density, once a hydrogen threshold density of order

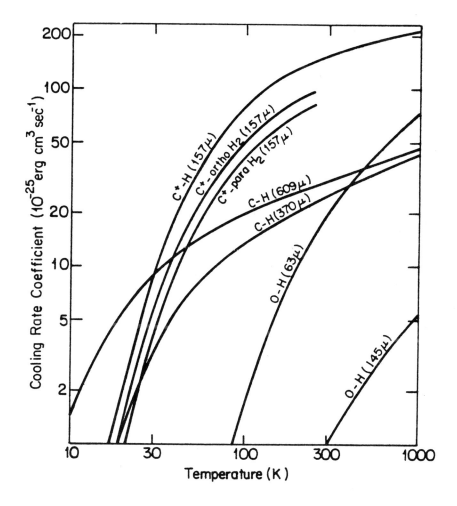

Figure 3 - Emission rate for different interstellar constituents for collisions with H, or ortho- or para-H_2 (data taken from Flower 1977, et al. and Launay & Roueff 1977).

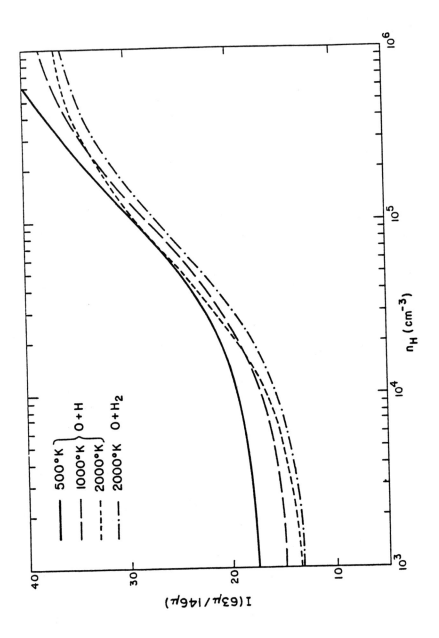

Figure 4 - Ratio of the 63.2 and 145.5μm line intensities as a function of the number density of atomic and molecular hydrogen (Melnick 1982).

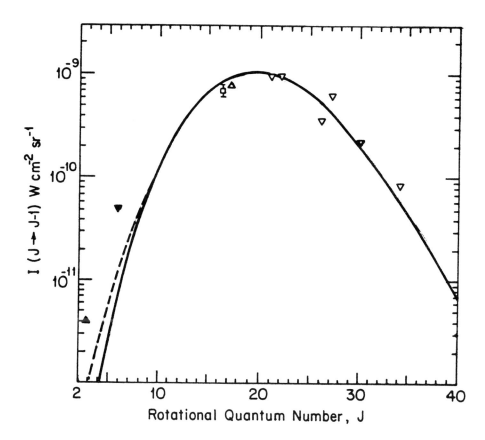

Figure 5 - CO emission observed from the Kleinmann-Low Nebula in Orion. Data are taken from observations of several groups (after Stacey et al. 1983).

$n_{H_2} \sim 10^6$ cm^{-3} is attained, and also are insensitive to temperature once a minimum temperature of order 750 K has been reached. Stacey et al. (1981) pointed out that this circumstance could be used to estimate the total mass of CO present in the cloud, $\sim 8 \times 10^{30}$ g. If one guesses a quarter of the carbon to be present in the form of CO, and carbon to have its cosmic abundance, $\sim 1.7 \times 10^{-4}$, one finds the mass of molecular hydrogen heated above 750 K to be ~ 1.5 M$_0$. This mass is an order of magnitude higher than the mass of shocked hydrogen deduced from near-infrared vibrational transitions of H_2 (Shull and Beckwith 1982). But the H_2 observations refer to a hotter component at a temperature $\sim$ 2000 K, and the amount of gas at these higher temperatures may well be small.

Between the various H_2 vibrational transitions and the CO data, we are able to construct at least a rough temperature/mass profile for the shocked domain, even if we do not yet have an accompanying set of data showing the spatial distribution of the hot CO.

Recently, the Berkeley group announced the detection of a far-infrared transition of ammonia, NH_3 (Genzel 1982), but since Dan Watson (this volume) will be talking in detail about molecular domains, I will not discuss these results further, except to mention molecules in the context of chemical reaction schemes.

IV. CHEMICAL REACTION SCHEMES

The Kleinmann-Low nebula has been especially well-studied in a variety of far-infrared lines. In particular, there are data on OH emission (Storey et al. 1981), atomic oxygen emission (Furniss et al. 1982), H_2O emission (Phillips et al. 1978) and 3.28µ solid water-ice emission (Sellgren 1981). Since oxygen is far more abundant normally than is carbon, we would expect strong emission or absorption in at least one of these lines, unless all of the oxygen was in the form of O_2, (cf. Fig. 6).

The observations suggest (Stacey et al. 1981) that oxygen, in fact, is most likely to be in the form of O_2 -- but only by default. None of the other observed features is sufficiently strong to represent an oxygen-containing species with an atomic oxygen column density exceeding that of CO by the requisite factor of 3 to 10.

Models of shocked hydrogen by Iglesias and Silk (1978), Hollenbach and McKee (1980), Draine and Roberge (1982) and Chernoff et al. (1982) differ on the fate of oxygen in shocks. There is some divergence of opinion on whether O_2 or H_2O should be the dominant final constituent. The observational results side with O_2, since H_2O is not found to be strong, either in gaseous emission or in the near-infrared water-ice contribution. Direct millimeter observations of O_2 would be useful in clarifying the situation, but telluric opacity is a difficulty in such observations.

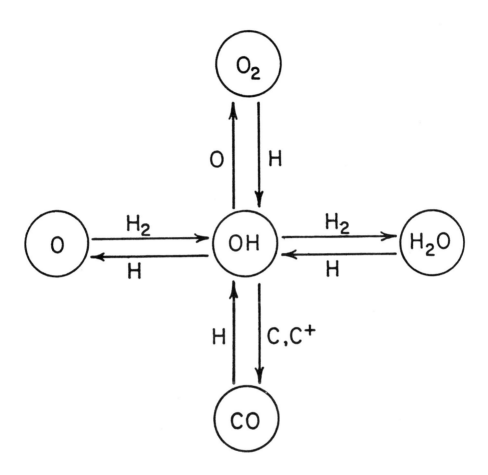

Figure 6 - Oxygen gas chemistry.

Competing gas-phase reactions involving carbon may become similarly constrained within the next few years. Figure 7 shows two different possibilities for gas-phase reactions involving hydrogen and carbon. We can anticipate being able to observe emission lines from all the constituents shown in the two diagrams within the not-too-distant future. Quantitative data on these and other reactions involving atoms, their ions, as well as diatomic hydrides of atoms and ions should become available fairly soon, since many diatomic hydrides have their low-lying rotational transitions in the far infrared.

V. GALACTIC ABUNDANCE GRADIENTS

Galactic abundance gradients are being studied in HII regions. The low opacity of the interstellar medium, in the far-infrared makes fine-structure transitions particularly promising features for observing these ratios (Herter et al. 1982). Complexities of the HII region structure so far have precluded definitive results, but a larger sample of observations could provide the desired information.

VI. ENERGY BUDGETS OF GAS CLOUDS

Far-infrared observations have given us an opportunity to directly measure the energy budget available to interstellar clouds. We can now compare the luminosities of the energizing stars in a cloud complex, to the amount of starlight that goes directly into ionizing and heating the gas, and to the amount of radiation that is appropriated to the acceleration of clouds -- to gas dynamics.

We cannot observe this allotment of energy directly, but we do see the dissipation of the cloud motions, through emitted line radiation. If we then set the rate of dissipation equal to the rate of energy supply, we obtain an estimate of the total energy supplied.

Table 1 shows the radiated energy budget of the M42 region in Orion. The luminosity of the stars is approximated by the total far-infrared emission from surrounding dust. Optical lines and infrared lines from ionized oxygen, doubly ionized sulphur, and from other ionized species reflect the amount of energy going into ionization and into heating the H II region. Far-infrared neutral oxygen and C^+ lines tell us the amount of energy dissipated in the border regions between H II and molecular domains. H_2 near-infrared emission and far-infrared CO emission, seen in transitions between high-lying rotational states, tell us of the energy dissipated in shocks.

We are now able, at last, to see the way the energy is budgeted in M42. Similar studies should be undertaken for other sources as well.

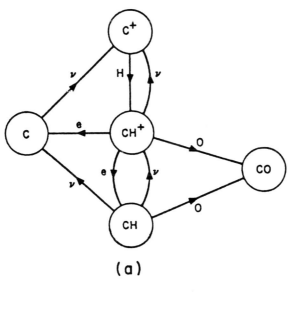

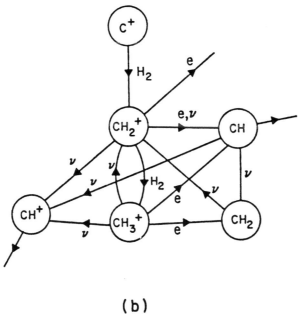

Figure 7 - CH, CH^+ and CO chemical schemes (a) in the absence of H_2, (b) with H_2 present (after Dalgarno cf. general reference 1).

TABLE 1

Orion Cooling

Ionized Region and Fronts:

1. Free-Free Emission $\sim 10^{-14}$ Watt cm^{-2}

2. Optical Lines: H-α $\sim 6 \times 10^{-15}$ Watt cm^{-2}
   All Others (3200-7300Å) $\sim 3 \times 10^{-14}$ Watt cm^{-2} } $\sim 4 \times 10^{-14}$ Watt cm^{-2}

3. Fine-Structure Lines in 18μ
   [SIII] $\sim 2.4 \times 10^{-15}$ Watt cm^{-2}

   Mid- and Far-IR 51.8 + 88μ
   [O III] $\sim 8 \times 10^{-15}$ Watt cm^{-2}

   63μ [O I] $\sim 8 \times 10^{-15}$ Watt cm^{-2} } $> 2 \times 10^{-14}$ Watt cm^{-2}

   etc.

Shocks:

4. H$_2$ Total Emission from Shocks $\sim 3 \times 10^{-15}$ Watt cm^{-2}
   (partly extinguished)

   CO in Shocks (all lines) $\sim 2 \times 10^{-15}$ Watt cm^{-2}

CII-Halo:

5. [CII] Region 157μ 7×10^{-16} Watt cm^{-2}

Dust:

6. Trapezium Dust Continuum 9-38μ 7×10^{-14} Watt cm^{-2}

   Total dust continuum 20-100μ 7×10^{-12} Watt cm^{-2}

VII. THE EXISTING BODY OF DATA ON COLLISION CROSS SECTIONS, EINSTEIN COEFFICIENTS AND TRANSITION ENERGIES

A considerable compilation of calculated collisional cross sections has been accumulating over the past few years. The observer interested in interpreting his data, however, runs into several difficulties:

a. Excitation cross sections for collisions with protons have not been re-examined since initial calculations were published long ago by Bahcall and Wolf (1968). There is some question about how reliable these results were, since no comparison with other theoretical work is available. Recalculation with more modern techniques would be reassuring.

b. Few calculations exist on excitation cross sections for collisions with molecular hydrogen. These are needed for any detailed analysis of the intensity of neutral carbon emission, carbon monoxide emission, and the interpretation of line ratios of $C°$ and various CO lines seen in molecular clouds.

c. Theorists understandably cannot publish all the details of their calculations. However, the observer frequently is given nothing more than the total cooling rates for atoms and molecules summed over all possible spectral lines. If he is not to have to go through a whole calculation all over again, he needs to be given a breakdown of the total cooling rate, into its different line-components. The observer often is able to detect no more than a single unobscured line -- not the entire family of available emission features.

d. Einstein coefficients have become available for a wide variety of transitions in the past few years. The calculations, however, appear to be supported by relatively few laboratory measurements. To be sure, it is difficult to obtain Einstein coefficients for metastable species that are hard to manufacture in the laboratory. But without at least some experimental checks, there could be appreciable errors in the calculated transition probabilities. We need to be sure of these probabilities if we are to derive correct physical conclusions from observations.

e. Wavelengths for many of the fine-structure transitions are poorly known. Predicted line positions often are in error by a micron or more. Improved laboratory data or calculations would greatly aid the observer, since a small wavelength displacement can place a spectral line directly at the position of a strong telluric absorption feature. Doppler motion of the Earth then leads to seasonal line modulation, difficult to take into account.

ACKNOWLEDGEMENTS

Support for the author's infrared observations has come through

NASA Grants NGR 33-010-146 and NSG-2347. These are acknowledged with pleasure.

GENERAL REFERENCES

1. A useful general guide to atomic processes and chemical reaction schemes is: A. Dalgarno, F. Masnou-Seeuws and R. W. P. McWhirter, 1975, <u>Atomic and Molecular Processes in Astrophysics</u>, Swiss Society of Astronomy and Astrophysics, Fifth Advanced Course, Saas-Fee, Geneva Observatory CH-1290 Sauverny/Switzerland.

2. A complete list of fine-structure transitions in the ground term, for the twenty most abundant elements, and ionization states I through VII, appears in the recently published article by J. Schmid-Burgk, 1982, Landolt-Börnstein, New Series, Volume VI/2, Section 7.5, Physics of the Interstellar Gas, subsection 2.1, Springer.

3. A previous review paper that lists equipment flown on NASA's Kuiper Airborne Observatory and describes some of the chief results is due to D. M. Watson, 1981, Queen Mary College, London, Conference on Submillimeter Wave Astronomy.

4. Two early theoretical papers of considerable historical interest are S. R. Pottasch, 1968, "The Infrared Lines and the Temperature and Ionization of the Interstellar Medium," Bull. Astron. Inst., Netherlands 19, p. 469 and V. Petrosian "Infrared Line Emission from H II Regions," Ap. J., 159, p. 833.

SPECIFIC REFERENCES

0. Bahcall, J.N. and Wolf, R.A.: 1968, Ap. J. 152, p. 701.

1. Chernoff, D. F., Hollenbach, D. J., McKee, C. F.: 1982, Ap. J. Lett. 259, pp. L97-L101.

2. Dinerstein, H. L.: 1982, NASA Technical Memorandum 84279.

3. Draine, B. T. and Roberge, W. G.: 1982, Ap. J. Lett. 259, pp. L91-L96.

4. Flower, D. R., Launay, J.-M. and Roueff, E.: 1977, <u>Les Spectres des Molecules Simples au Laboratoire et en Astrophysique</u>, Universite de Liege, p. 137.

5. Furniss, I., Jennings, R. E., King, K. J., Lightfoot, J. F., Emery, R. J., Fitton, B. and Naylor, D. A.: 1982, M.N.R.A.S., to be published.

6. Genzel, R.: 1982, in <u>The Scientific Importance of Submillimetre Observations</u>, European Space Agency, SP-189, p. 141.

7. Herter, T., Helfer, H. L., Pipher, J. L., Briotta, Jr., D. A., Forrest, W. J., Houck, J. R., Rudy, R. J., and Willner, S. P.: 1982, Ap. J. 262, pp. 153-163.

8. Herter, T., Briotta, Jr., D. A. Gull, G. E., Shure, M. A. and Houck, J. R.: 1982, Ap. J. 262, pp. 164-170.

9. Hollenbach, D. J. and McKee, C. F.: 1980, Ap. J. Lett. 241, pp. L47-L50.

10. Iglesias, E., Silk, J.: 1978, Ap. J. 226, pp. 851-857.

11. Kurtz, N., Smyers, S., Russell, R., Harwit, M. and Melnick, G., Melnick, G.: 1982, Ap. J., to be published.

12. Launay J.-M. and Roueff, E.: 1977, Astron. Astrophys. 56, pp. 289-292.

13. McKee, C. F., Storey, J. W. V., Watson, D. M. and Green, S.: 1982, Ap. J. 259, pp. 647-656.

14. Melnick, G., Gull, G. E. and Harwit, M.: 1979, Ap. J. Lett. 227, pp. L35-L38.

15. Melnick, G.: 1981, Cornell University Ph.D. Thesis.

16. Melnick, G.: 1982, private communication.

17. Moorwood, A. F. M., Baluteau, J.-P., Anderegg, M., Coron, N., Biraud, Y. and Fitton, B.: 1980, Ap. J. 238, pp. 565-576.

18. Osterbrock, D. E. and Flather, E.: 1959, Ap. J. 129, pp. 26.

19. Phillips, T. G., Scoville, N. Z., Kwan, J., Huggins, P. J. and Wannier, P. G.: 1978, Ap. J. Lett. 222, pp. L59-L62.

20. Russell, R. W., Melnick, G., Gull, G. E. and Harwit, M.: 1980, Ap. J. Lett. 240, pp. L99-L103.

21. Sellgren, K.: 1981, Ap. J. 245, pp. 138-147.

22. Stacey, G. J., Kurtz, N. T., Smyers, S. D., Harwit, M., Russell, R. W. and Melnick, G.: 1981, Ap. J. 257, pp. L37-L40.

23. Stacey, G., S. Smyers, N. T., Kurtz and Harwit, M.: 1982a, Ap. J. Lett., submitted.

24. Stacey, G., Smyers, S., Kurtz, N. and Harwit, M.: 1982, Ap. J. Lett., to be published.

25. Shull, J. M. and Beckwith, S., 1982, "Interstellar Molecular

Hydrogen," <u>Annual Reviews of Astronomy and Astrophysics</u>, p. 20.

26. Shure, M. A., Herter, T., Houck, J. R., Briotta, Jr., D. A., Forrest, W. J., Gull, G. E., and McCarthy, J. R.: 1983, to be published.

27. Simpson, J. P.: 1975, Astr. Ap. 39, 43.

28. Stacey, G. J., Kurtz, N. T., Smyers, S. D., and Harwit, M.: 1983, MNRAS 202, pp. 25P-29P.

29. Storey, J. W. V., Watson, D. M. and Townes, C. H.: 1981, Ap. J. Lett. 244, pp. L27-L30.

30. Ward, D., Dennison, B., Gull, G. E. and Harwit, M.: 1975, Ap. J. Lett. 202, pp. L31-L32.

31. Watson, D. M., Storey, J. W. V., Townes, C. H., Haller, E. E. and Hansen, W. L.: 1980, Ap. J. 239, L129.

ASTROPHYSICAL INTERPRETATION OF MOLECULAR SPECTRA

Nicholas Z. Scoville

Five College Radio Astronomy Observatory, U. Mass, Amherst

I. INTRODUCTION

Though infrared photometric studies over the last two decades have revealed numerous "objects" embedded in the dense clouds out of which stars are presumed to condense, there remains no single source one may point to and confidently say "Indeed, that is a protostar, no need to search further" (cf. Wynn-Williams 1982). The problem is that despite the fact that the photometry gives an excellent assessment of the energy generated within, it tells painfully little of the object's structure--that is the density, temperature, or dynamics as a function of radius. These latter "details" are clearly needed to define the evolutionary state of the source.

As sensitive, high resolution spectrometers are developed throughout the infrared we can anticipate great progress in understanding not only the young-stellar objects but also the active galaxy nuclei so luminous in the far-infrared. In the infrared the variety of atomic and molecular spectroscopic transitions is capable of probing conditions ranging from hot circumstellar HII regions, molecular envelopes, and shock fronts at > 2000 K down to cold, low density interstellar gas at < 10 K. The ability to measure both physical conditions and kinematics (not possible of course by photometry) aids in the separation of the physical regimes and in the building of a coherent dynamic/evolutionary model.

In the following sections I briefly review the characteristics of some of the observed molecular transitions (§II) and theoretical considerations important for understanding their excitation (§III). The observational data are described fully in the reviews by Hall and Watson in this volume.

II. OBSERVED MOLECULAR TRANSITIONS

To illustrate the types of molecular transitions occurring in the infrared, energy levels of H_2 and CO are shown in Figures 1-2. Here the rotational energies are not drawn to proper scale relative to the vibrational energies. For CO the energy difference for $J = 1 \rightarrow 0$ corresponds to 5.5 K, while for H_2 it is 170 K (due to the much smaller moment of inertia). The lowest rotational transition of CO ($J = 1 \rightarrow 0$) occurs at $\lambda = 2.6$ mm while that of H_2 ($J = 2 \rightarrow 0$) is at $\lambda = 28$ µm. Vibrational energies ($v = 1 \rightarrow 0$) for both molecules are vastly greater, 3000 and 6000 K respectively; their fundamental vibrational transitions are in the near infrared at $\lambda = 2-5$ µm.

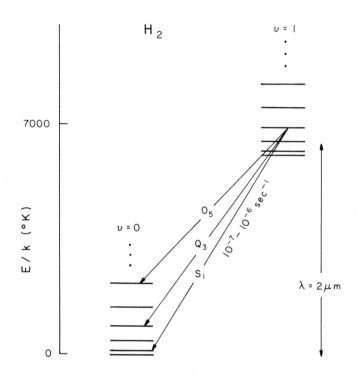

FIGURE 1: Energy levels and radiative decay rates are shown for H_2. The three types of electric quadrupole emission of H_2 are shown. (Rotational levels are not to scale.)

Also shown in the figures are the spontaneous decay rates characteristic of the vibrational transitions. For CO (and most molecules other than H_2) the radiative transitions between vibrational states obey the dipole selection rule, $\Delta J = \pm 1$, on the rotational quantum number. There are no strong selection rules in undergoing changes of vibrational quantum number (v) but typically the matrix elements for higher overtone transitions are reduced by a factor ~ 100 for each unit of Δv since the strengths of the vibrational transitions depend on higher order anharmonic terms in the molecular potential. The decay rates for CO $v = 1 \rightarrow 0$ and $v = 2 \rightarrow 0$ (1 sec^{-1} and 30 sec^{-1}; summing the R&P transitions--cf. Kirby-Docken and Liu 1978) are typical of infrared fundamental and overtone transitions. On the other hand the H_2 decay rates are exceedingly low, $10^{-7} - 10^{-6}$ sec^{-1} (Turner et al. 1977). This is due to the fact that the nuclei in H_2 are identical fermions and the only permitted transitions are weak electric quadrupole transitions ($\Delta J_{u\ell}$ = -2, 0, 2 denoted S, Q, and O; see Field, Somerville, and Dressler 1966).

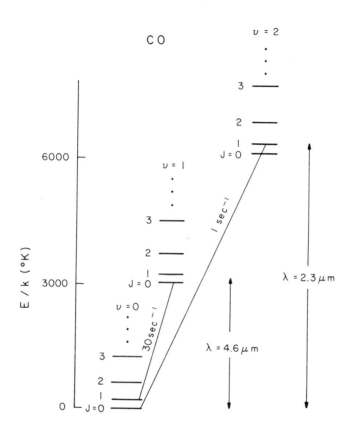

FIGURE 2: Energy levels and radiative decay rates for CO are shown. (Rotational energies are not drawn to scale.)

This difference, a factor of 10^8 in the radiative rates for CO and H_2, has a profound bearing on our interpretation of the CO and H_2 emission lines in the near infrared. Both molecules require a high temperature environment (T > 1000 K) for collisional excitation of these states. On the other hand, since radiative decay out of CO v = 1 states is a factor of 10^8 more rapid, the collision rate and therefore the density must be a factor of $\sim 10^8$ higher to maintain population in v = 1 for CO. (In fact the collision rate required to excite H_2 in v = 1 is not much greater than that required to maintain thermalization in the millimeter CO lines.) In the next section it is also shown that the radiative excitation is vastly different for the two molecules.

III. CONSIDERATIONS FOR EXCITATION OF MOLECULES

There exist two basic methods by which the physical conditions in a source may be deduced from spectroscopic line measurements. In the first the strength of an absorption line formed in front of a continuum source is used to measure a gas column density which can then give a volume density estimate if the path length is known by other means. Alternatively the strengths of several observed transitions may yield estimates for the collisional or radiative excitation rates. Here we outline simple theoretical considerations for this last mode of analysis--reviewing the excitation of a two level system pumped by collisions and by radiation, first in the case of optically thin, then optically thick transitions. More complete numerical treatments for vibrational excitation of CO and H_2 are given by Scoville, Krotkov, and Wang (1980); Chernoff, Hollenbach, and McKee (1983); and Draine, Roberge, and Dalgarno (1983). The far-infrared emission of CO is modeled by McKee et al. (1982). The collisions might be provided by a hot atomic or molecular hydrogen gas; the radiation field might be provided by a central infrared source. In this analysis we neglect pumping by absorption of UV photons in the electronic transitions of the molecules.

a) Optically Thin Case

In order to establish significant population in the upper state of two molecular levels there must be a collisional or radiative pump rate $P_{\ell u}$ comparable to the radiative decay rate $R_{u\ell}$. Thus,

$$P_{\ell u} > R_{u\ell} \qquad (1)$$

In general for cases when the decay transitions are optically thin, $R_{u\ell}$ is approximately given by the product of the largest spontaneous decay coefficient $A_{u\ell}$ and the upper state population n_u.

If the pump is collisional (for example collisions with neutral H), then the net upward pump rate is

$$P_{\ell u} - P_{u\ell} = n_H(n_\ell C_{\ell u} - n_u C_{u\ell}) \qquad (2)$$

where $C_{\ell u}$ and $C_{u\ell}$ are the excitation and deexcitation rate coefficients $\langle\sigma v\rangle$ averaged over a Maxwellian velocity distribution at kinetic temperature T_K. From the condition of detailed balance,

$$C_{\ell u} = C_{u\ell}(g_u/g_{\ell\mu})e^{-(E_u - E_\ell)/kT_K}.$$

In statistical equilibrium of the two levels, the <u>net</u> pump rate must balance the <u>net</u> decay rate and we arrive at the familiar conditions:

$$T_K > (E_u - E_\ell)/k \tag{3a}$$

and

$$n_H > A_{u\ell}/\langle \sigma v \rangle_{u\ell} \tag{3b}$$

which must be satisfied if there is to be significant upper state population in the case $\tau < 1$.

Alternatively, if the pump is radiative then the net external pump rate is

$$P_{\ell u} - P_{u\ell} = U_{\ell u}(n_\ell B_{\ell u} - n_u B_{u\ell}) \tag{4}$$

where $B_{u\ell}$ is the stimulated emission coefficient and $U_{\ell u}$ is the local radiation energy density from an external source such as a star or dust--not from the molecules themselves. If we characterize this external source with a dilution factor W, color temperature T_c, and opacity τ_c, then

$$U_{\ell u} = \frac{8\pi h \nu^3}{c^3} \frac{W(1 - e^{-\tau_c})}{e^{h\nu/kT_c} - 1}. \tag{5}$$

Combining Equations (1) and (4) with the Einstein relations $[B_{u\ell} = (g_\ell/g_u)B_{\ell u} = (c^3/8\pi h\nu^3)A_{u\ell}]$, we find the radiative pump condition

$$\frac{W(1 - e^{-\tau_c})}{e^{h\nu/kT_c} - 1} > 1 \tag{6}$$

implying that <u>the radiation field seen by the molecules must be almost that of a black body with temperature $> h\nu/k$</u> since the numerator in (6) is always $\leqslant 1$.

b) Optically Thick Case

The rather severe requirements for collisional pumps may be considerably relaxed in a dense region where the <u>effective</u> radiative decay rates are greatly reduced from the pure spontaneous rates due to trapping of line photons. When the radiative transitions become opaque to their own emission, a given spontaneous decay photon cannot readily escape the region. Instead, this photon may be reabsorbed and reemitted many times before it either escapes entirely or is thermalized via collisional deexcitation following an absorption. A common procedure used in treating the radiative decay when transitions become opaque has been to employ a net radiative decay rate $R_{u\ell} - R_{\ell u}$ which, for the case of no background radiation, is simply given as the spontaneous decay rate $A_{u\ell}$ multiplied by a photon escape probability β:

$$R_{u\ell} - R_{\ell u} = n_u A_{u\ell} \beta , \qquad (7)$$

The escape probability depends on the optical depth τ of the transition (Castor 1970 and Lucy 1971) and for a spherical region with constant velocity gradient dV/dr in the radial direction (i.e., $v \propto r$)

$$\begin{aligned} \beta &= \frac{1}{\tau}(1 - e^{-\tau}) \\ &\simeq 1 \qquad\qquad \text{for } \tau \ll 1 \\ &\simeq \frac{1}{\tau} \qquad\qquad \text{for } \tau \gg 1. \end{aligned} \qquad (8)$$

with

$$\tau = \frac{\lambda^3 g_u}{8\pi \, dV/dr} A_{u\ell} \left\{ \frac{n_\ell}{g_\ell} - \frac{n_u}{g_u} \right\} . \qquad (9)$$

This formalism has been previously developed and applied to the analysis of millimeter rotational lines by Scoville and Solomon (1974) and Goldreich and Kwan (1974). In cases where the requisite large velocity gradients do not exist, the escape probability is considerably more complex to evaluate since it must depend on conditions across the entire cloud but generally the qualitative dependence on τ is similar (e.g., Adams 1972).

When Equation (7) is used for the net decay rate, the conditions for significant upper state population are drastically changed from the optically thin case. The collisional pump requirement (3b) now becomes

$$n_H > \frac{8\pi}{\lambda^3} \frac{dV}{dr} / \{n_\ell \langle \sigma v \rangle_{u\ell}\} \qquad (10a)$$

If [M/H] is the molecular abundance ratio and f_e is the fraction of the molecules occupying the lower state than (10a) becomes

$$n_H > \left[\frac{8\pi}{\lambda^3} \frac{dV}{dr} / \{ [M/H] \, f_e \, \langle\sigma v\rangle_{n\ell} \} \right]^{1/2} \quad (10b)$$

In the optically thick limit the critical density has become independent of the radiative decay rate; a high molecular abundance is now the requesite for appreciable population in the upper state. *This behavior arises because the more rapid spontaneous decay occuring when the line strength is large is exactly compensated by the greater optical depth and consequently greater photon trapping in the transition.*

Finally for the interesting case of radiative pumping in optically thick molecular transitions the pump requirement is found to be identical to that in the thin limit [Equation (6)]. This surprising result is strictly true only when the pump radiation is supplied by a source such as a star which is external to the molecular gas and when the escape probability is independent of direction such as a radial flow with $V \propto r$. The reason for the identical excitation requirement (independent of τ) is that when $\tau > 1$ the net decay rate is decreased by the escape probability β but at the same time the probability of external pumping photons reaching the molecules is reduced by the same factor.

c) Numerical Results

In Figure 3 the results of numerical solution of the full rate equations for CO are presented. For these computations the first eight vibrational states were included and collisional excitation up to $\Delta v = 3$ transitions were included. Two cases with low and "normal" CO/H abundance were treated. (The artificially low abundance forces all transitions to be optically thin for the assumed velocity gradient $dV/dr = 10$ km s^{-1} AU^{-1}.) In both instances the CO rotational temperature was 3500 K implying a fractional population f_e (Equation 10b) of 1% in a typical state at $J = 50$.

The adopted rate coefficient for excitation of CO was $C_{01} = 10^{-10}$ cm^3 sec^{-1}; one therefore expects according to the two level treatment that a density $n_H \simeq 10^{11}$ cm^{-3} will be required to appreciably populate $v = 1$ in the optically thin case (cf. Equation 3b). This result is in fact upheld in the numerical calculations (dashed curve in Figure 3). For the more realistic CO abundance the transitions should become thick at $n_H \simeq 10^9$ cm^{-3} and one must use Equation 10b predicting that the critical density is reduced to $n_H \simeq 10^{10}$ cm^{-3}. This effect is verified in Figure 3 by comparison of the solid curves (thick transitions) with the dashed curves (for the thin case).

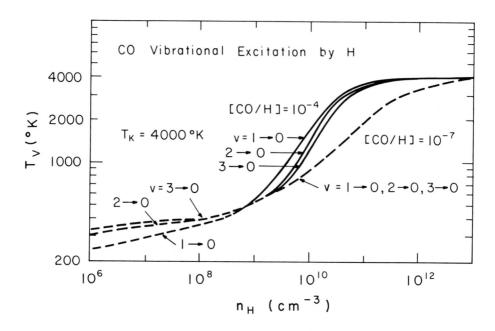

FIGURE 3: The excitation temperatures for the first three vibrational transitions of CO are shown at $T_K = 4000$ as a function of n_H. For a normal CO/H ratio of $\sim 10^{-4}$ the $v = 0 - 1$ lines become thick at $n_H \sim 10^9$ cm^{-3} (assuming $T_{ROT} = 3500$ K and a velocity gradient of 10 km s^{-1} AU^{-1}). Note that in the optically thick case the critical density required for excitation is reduced a factor of 10 due to photon trapping.

IV. OBSERVED SPECTROSCOPIC TRACERS

The formal excitation requirements presented in the previous section can now be used to place qualitatively the various physical regimes probed in the infrared. In Table 1 the selected transitions (by no means an exhaustive list) are grouped by regime--low energy (T < 200 K), medium energy (T ≃ 200-1000 K), and high energy (T > 1000 K). In all cases the optical depth estimates given for the observations in Orion are extremely approximate--they are meant only to be indicative of the excitation considerations.

For most of these transitions the primary mode of excitation has been considered to be collisional rather than radiative. For three of the transitions noted by "R" in the table, some of the excitation could conceivably arise from radiative pumping or scattering of continuum photons from a central source. If this is not true and collisions provide the excitation of all transitions, then the range of density sampled in each regime is more than a factor of 10^5 except in the cool gas! This is an impressive dynamic range one is equipped to measure.

TABLE 1: Selected Spectroscopic Probes

	λ	τ	Critical n_{H_2} (cm^{-3})	Obs. Ref.
LOW ENERGY (T < 200 K)				
Low J CO pure-rotational emission	~1mm	>10(CO)	~300	a
		<1($^{13}CO, C^{18}O$)	~3000	a
Low J CO rotation-vibration absorption	4.6μm	~100(CO)	~300	b
		~1(^{13}CO)	~3000	b
	2.3μm	~1(CO)	~300	b
MID ENERGY (T ≃ 200-1000 K)				
H_2 S(2)	12μm	<<1	~10^3	c
High J CO pure-rotational emission	75-157μm	<1	~10^6	d
OH rotational emission	120μm	>1	~10^8(R)	e
Mid J CO rotation-vibration emission	4.6μm	~1	~10^{10}(R)	b
HIGH ENERGY (T > 1000 K)				
H_2 rotation-vibration emission	2-4μm	~10^{-6}	10^5	f
CO overtone bandhead emission	2.3μm	~1	10^{10}(R)	b

This research is supported in part by NSF grant #AST-82-12252. This is contribution #542 of the Five College Radio Astronomy Observatory.

NOTES:

"R" indicates transitions possibly pumped by radiation or alternatively the observed line photons are in fact continuum photons from a central source scattered in an assymetric gas envelope.

a - see Goldsmith in this volume.

b - Scoville et al. (1983); Hall in this volume.

c - Beck et al. (1979).

d - Storey et al.(1981); Watson in this volume.

e - Storey, Watson, and Townes (1981); Watson in this volume.

f- Beckwith, Persson, and Neugebauer (1979), Scoville et al. (1982); Hall in this volume.

REFERENCES:

Adams, T.F. 1972, Ap.J., 174, 439.
Beck, S.C., Lacy, J.H., and Geballe, T.R. 1979, Ap.J.(Letters), 234, L213.
Beckwith, S., Persson, S.E., and Neugebauer, G. 1979, Ap.J., 227, 436.
Castor, J.I. 1970, M.N.R.A.S., 149, 111.
Chernoff, D.F., Hollenbach, D.J., and McKee, C.F. 1983 (in preparation).
Draine, B.T., Roberge, W.G. and Dalgarno, A. 1983, Ap. J. 264, 485.
Field, G.B., Somerville, W.B., and Dressler, K. 1966, Ann.Rev. of Astron.Astrophys., 4, 207.
Goldreich, P. and Kwan, J. 1974, Ap.J., 189, 441.
Kirby-Docken, K. and Liu, B. 1978, Ap. J. Suppl. 36, 359.
Lucy, L.B. 1971, Ap.J., 163, 95.
McKee, C.F., Storey, J.W.V., Watson, D.M., and Green, S. 1982, Ap.J., 259, 647.
Scoville, N.Z., Hall, D.N.B., Kleinmann, S.G., and Ridgway, S.T. 1982, Ap.J., 253, 136.
Scoville, N.Z., Hall, D.N.B., Kleinmann, S.G., and Ridgway, S.T. 1983, Ap.J. (in press).
Scoville, N.Z., Krotkov, R., and Wang, D. 1980, Ap.J., 240, 929.
Scoville, N.Z. and Solomon, P.M. 1974, Ap.J.(Letters), 187, L67.
Storey, J.W.V., Watson, D.M., and Townes, C.H. 1981, Ap.J.(Letters), 224, L27.
Turner, J., Kirby-Docken, K., and Dalgarno, A. 1977, Ap.J.Suppl., 35, 381.
Wynn-Williams, C.G. 1982, Ann.Rev. of Astron.Astrophys., 20, 587.

EXCITATION CONDITIONS IN MOLECULAR CLOUDS

G Winnewisser

H Ungerechts

I. Physikalisches Institut der Universität Köln

I. Introduction

Interstellar radio spectroscopy over the last 12 years has shown that molecular and atomic transitions can be used as a powerful tool for probing the physical conditions and kinematics of the interstellar gas. By using different molecular transitions we have learned to investigate in quite some detail the molecular clouds, from their tenuous outer surroundings to their dense interiors. In effect molecular transitions are now being used to investigate interstellar gas ranging in temperature from 10K in cold dark clouds to 3000K in molecular shock fronts and circumstellar shells. Astronomical objects which can be investigated in molecular lines include quiescent dark clouds, young protostellar objects and molecular envelopes of evolved stars. Since molecular spectroscopy both in the radio and infrared region allows the properties of this wide range of objects to be derived with a fair amount of certainty, one may hope to understand their morphology and evolutionary status.

Molecular transitions are found in abundance throughout the entire centimeter, millimeter, submillimeter and far-infrared region of the electromagnetic spectrum owing to the fact that the energy separations of the rotational and ro-vibrational levels lie generally in the range between 1K and 5000K. The intensity of the transitions increases with frequency giving intensity preference to millimeter, submillimeter and far- infrared transitions. Owing to intrinsic properties associated with each molecular transition, different molecular lines trace different physical conditions within interstellar clouds.

The purpose of this paper is to present a short review on line formation mechanisms both for absorption and emission lines (II) and to give a summary of our present knowledge of molecular cloud energetics (III). The latter is important in the context of molecular excitation, as many molecular transitions can be excited only in the vicinity of powerful energy sources. Finally we shall refer to some recent observational data pertaining to the different molecular transitions discussed (IV).

II. Line Formation

Interstellar molecular lines have produced spectacular examples of non- equilibrium excitation. The most extreme cases of non-equilibrium phenomena are provided by interstellar formaldehyde H_2CO, which is cooled below the 2.7K background radiation and by maser emission of OH, H_2O, SiO and CH_3OH which show extremely high brightness temperatures. As any understanding of the non- equilibrium phenomena requires elaborate calculations of collisional cross sections and of radiative transfer in the molecular lines, we will discuss in detail only the most simple case of equilibrium excitation in order to present the most basic processes.

The population of molecular energy levels in interstellar clouds generally is determined by an equilibrium of the radiative and collisional transition rates. In table I we have summarized the relevant formulae for the simple and hypothetical case where only two energy levels are considered. Downward transitions are due to spontaneous emission of radiation (rate coefficient A_{ul}), induced emission ($I_\nu B_{ul}$) and collisions (C_{ul}), upward transitions are by absorption of radiation ($I_\nu B_{lu}$) and collisions (C_{lu}). A and B are the usual Einstein coefficients, the collision rates are the product of the density of the collision partners (mostly H_2 molecules) and the

EXCITATION CONDITIONS IN MOLECULAR CLOUDS

Table 1

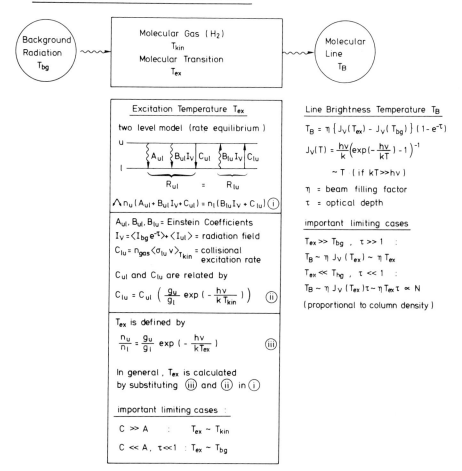

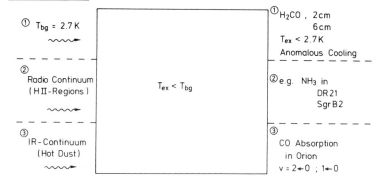

collisional cross sections averaged over a thermal velocity distribution characterized by the gas kinetic temperature T_{kin}. I_ν is the average intensity of the radiation field, which is the sum of the background radiation from outside the cloud decreased by the optical depth factor $\exp(-\tau)$, the line radiation generated by all other molecules, and any continuum radiation originating within the cloud (e.g. from warm dust).

It is convenient to express the ratio of the populations of the upper and lower level of a transition in terms of the excitation temperature T_{ex}, which is defined by $n_u/n_l = g_u/g_l \exp(-h\nu/kT_{ex})$. At very low densities collisional processes vanish in the equilibrium equation and T_{ex} will be equal to the brightness temperature of the continuum radiation T_{bg}. For typical molecular line frequencies, T_{bg} is at least 2.7K, the temperature of the isotropic cosmic background radiation.

On the other hand, at sufficiently high densities collisions dominate and the excitation temperature equals the gas kinetic temperature. T_{kin} itself is determined by the equilibrium of the energy loss from the cloud and the energy input provided by various astrophysical processes, which we are going to refer to in section III.

In the intermediate density range, the excitation temperature of a particular transition may be far from thermal equilibrium depending on the detailed interaction of the various collisional and radiative processes. Indeed T_{ex} may be lower than T_{bg} (e.g. in the H_2CO molecule) or higher than T_{kin} by many orders of magnitude (e.g. in the maser transition of the H_2O molecule).

If the cloud can be represented by a homogeneous plane-parallel slab, the brightness temperature of the emitted line is given by the formulae given in the right column of table I. The optical depth τ in turn is determined by the molecular line strength (essentially the Einstein A_{ul}), the excitation temperature, the line profile and the molecular column density. The beam filling factor η gives the ratio of the solid angle subtended by the radiation source to the telescope's beam solid angle. Clearly, the line will be in absorption (i.e. $T_B < 0.0$) if $T_{ex} < T_{bg}$ and in emission ($T_B > 0.0$) if $T_{ex} > T_{bg}$.

Absorption Lines

Three different types of sources may provide the continuum background for interstellar molecular absorption lines, as shown in the lower part of table I.

(a) If the population of two energy levels is inverted by collisions with H_2 molecules, the excitation temperature of the transition may be lower than the brightness temperature of the cosmic background radiation (2.7K). In this case an absorption line will occur without the presence of a localized continuum source. Indeed, over a large range of temperatures and densities the H_2CO-molecule is "colder" than the isotropic background and hence its 2cm and 6cm lines are seen in absorption.

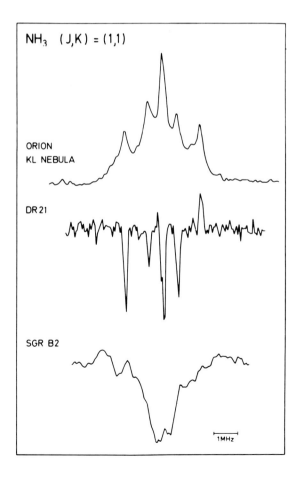

Figure 1: Spectra of the $NH_3(1,1)$ line in Orion, SgrB2 and DR21. All these spectra were observed with the 100m telescope of the MPIfR at Effelsberg.

(b) HII- regions which are excited by the radiation from massive young stars tend to be associated with large complexes of molecular clouds, out of which these stars formed. If a part of the molecular gas happens to be located in front of the hot ionized gas, molecular lines will be seen in absorption. As the brightness temperature of the continuum radiation from HII- regions is significant only below a certain "cut-off" frequency, this type of absorption line is most important in the centimeter wavelength region, e.g. for the inversion transitions of ammonia (NH_3) at ~1.3cm in the molecular clouds associated with SgrB2 and DR21 (see figure 1). The ammonia (1,1) transition seen in absorption towards DR21 is not excited in thermal equilibrium as is evidenced by the strong intensity anomalies of the quadrupole hyperfine structure components which were first discussed by Matsakis et al.(1977).

(c) In regions where massive stars are presently forming either the protostars themselves or associated hot dust may provide powerful sources of thermal continuum radiation in the IR and FIR. Again, if there is cooler interstellar matter in front of these continuum sources, infrared molecular absorption lines may be observed. This configuration is basically the explanation for the v=2<--0 and 1<--0 vibrational bands of carbon monoxide CO seen in absorption towards the BN- object in Orion (figure 6).

Emission Lines

In all cases where $T_{ex} > T_{bg}$ emission occurs. For this to happen molecules have to be pumped into the upper level of the transition by some process. In most cases this occurs by collisions with H_2 molecules (and He atoms). However in more localized surroundings, e.g. regions with strong embedded radiation sources (and hot dust) radiative pumping may also be of some importance, especially in the high excitation levels. As no special geometrical configuration is necessary for emission lines to occur, they are well suited for investigations of the structure, dynamics and evolution of molecular clouds. The physical properties of the clouds such as density, temperature, column density and velocity determine which lines are emitted, their intensity and the shape of profile. These properties allow us to discriminate between the various cloud types and the different regions within a particular cloud by observing the appropriate lines. Some characteristic parameters derived from NH_3 observations in five molecular clouds are well suited to demonstrate this point and indicate their range, as summarized in table II. Today molecular emission lines are known from interstellar gas having temperatures from ~10K to >3000K. Although there are no generally accepted criteria for a classification of molecular clouds, it seems there are at least three general types of sources characterized by different temperature:

(1) **Cold clouds**

In their interior these clouds have temperatures typically close to 10K and densities $>10^4 cm^{-3}$. They are surrounded by somewhat warmer and less dense shells. A well known example is the complex of dark clouds in the Taurus constellation, where in sources such as TMC1 a surprisingly complex chemistry was found including the long carbon chain molecules. These clouds are preferentially studied in low excitation lines of CO, NH_3, H_2CO and the cyanopolyynes $(H-(C-C)_n-CN)$ (with n = 0...5). NH_3 and cyanopolyyne lines from such clouds are the most narrow lines known in interstellar space with widths close to 0.2km/s (see figure 3).

(2) **Warm clouds**

In molecular clouds near (within ~1.0pc) luminous stars or IR-sources higher temperatures are found, T~15 - 50K. E.g. in S140/L1204 the gas temperature as calculated from NH_3 and CO observations shows a clear maximum around an embedded group of IR-sources, which are believed to be young stars. Farther away from these sources the temperature drops to values characteristic for the cold clouds. Another example in this category is the molecular cloud surrounding the bipolar nebula S106.

(3) **Hot centers of molecular clouds, high velocity flows and molecular shockfronts**

Closer to embedded luminous stars (typically within ~0.1pc) gas temperatures are even higher, ~100- 600K. In the region of the KL- nebula and the IR- source IRc2 in Orion, NH_3 inversion lines indicate temperatures of ~200K (Genzel et al., 1982). In these hot regions molecular transitions in the IR are excited as well. In the Orion region, high excitation pure rotational transitions of CO are emitted from regions with temperatures >700K (see e.g. Watson et al., 1980, Stacey et al., 1982). These transitions

Table II

Physical conditions in molecular clouds (derived from NH_3 observations)

Cloud	Distance (pc)	Size (pc)	$T_R(NH_3)$ (K)	$T_{ex}(CO)$ (K)	linewidth (km/s)	lg $n(H_2)$ (cm^{-3})	lg $N(NH_3)$ (cm^{-2})
L183	100	0.1*0.2	9	9	0.2	4.5	15.1
TMC1	140	0.1*0.6	10	10	0.3	4.5	15.5
S106	600	0.5*1.2	18	30	1.5	5	15.0
S140/L1204	910	1.2*1.2	26	33	2.0	5	15.0
Ori-KL/IRc2	480	<0.1	200		10.0	8-9	18.7

Notes : L183 and TMC1 are high density cores of cold dark clouds (Ungerechts et al., 1982). S106 is a molecular cloud surrounding a bipolar nebula (Stutzki et al., 1982). S140/L1204 is a warm region of a dark cloud with an embedded cluster of young stars having a total luminosity $2*10^4$ times that of the sun. Ori-KL/IRc2 means the hottest part of the Orion cloud seen in maps of the NH_3 lines obtained with the VLA near the KL- nebula and IRc2 (for this particular interpretation see the paper by Genzel et al., 1982). T_R is the rotational temperature of the ammonia molecule (determined by the ratio of the populations of the metastable rotational levels) and $T_{ex}(CO)$ is the excitation temperature of the $J=1\longrightarrow 0$ line of carbon monoxide.

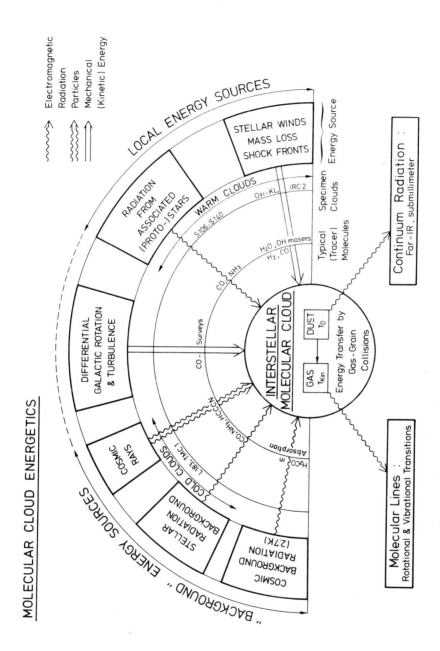

Figure 2: Schematic representation of the energetics of molecular clouds, see text, section III.

as well as the pure rotational quadrupole transitions of H_2 and the ro-vibrational lines of both H_2 and CO are characteristic of hot interstellar molecular gas which requires extreme excitation conditions (see e.g. Davis et al., 1982, Scoville, 1980). Characteristic temperatures of this hot gas are ~1500K for the H_2 transitions and >3000K for the overtone bandhead emission of CO. The necessary temperatures and densities may be provided in hot circumstellar shells, high velocity flows originating from protostars and in molecular shock fronts. Generally associated with the outflow of matter are intense time variable H_2O maser emission regions.

Self-Absorption in Molecular Lines

The "background" against which molecular absorption is seen may also be provided by radiation in the line itself. For this to happen there must be either a cold absorbing cloud in front of a warmer cloud emitting the line or a temperature gradient along the line of sight in one cloud whereby the lower temperature region of the cloud faces the observer. Generally this cloud configuration leads to a narrow absorption dip superimposed on a broad emission line.

III. Molecular Cloud Energetics

We have tried to summarize the knowledge about the most important processes providing energy to molecular clouds in a schematic way in figure 2. Isolated cold clouds are heated by absorption of cosmic rays in the gas and of the general stellar background radiation by the dust. These two energy sources are sufficient to heat the gas to temperatures of ~8 to 10K, which is presently the lowest value observed for dark cloud cores. In large cloud complexes, a significant amount of energy may be provided by dissipation of kinetic energy which is stored in the differential galactic rotation and in the turbulent motion within the cloud complexes. More localized energy sources are provided in regions of active star formation. The stars and protostars may heat the dust in the clouds directly by their radiation, or cause high velocity flows (stellar winds, mass loss or radiatively driven flows), thus providing kinetic energy to shock fronts and to the turbulence in the clouds.

Apart from the immediate neighbourhood of young stars, the main flow of energy through the interstellar clouds does not proceed via the gas, but the dust, which absorbs radiation at UV, visual and IR wavelengths and emits continuum radiation in the IR, FIR and submillimeter regions of the spectrum. At sufficiently high densities ($>10^{4.5}$ cm^{-3}), however, the gas will be thermally coupled to the dust by gas- grain collisions.

IV. Some examples of recent observations

$NH_3(J,K)=(1,1)$ line in TMC1

A spectrum of the inversion transition of NH_3 in the $(J,K)=(1,1)$ rotational state which we observed in TMC1 is shown in figure 3. The magnetic hyperfine structure is partly resolved. This is an example for a line from a quiescent cold dark cloud where the temperature is ~10K and the linewidth is almost purely thermal. The relative intensities of the components have the "normal" LTE- values, i.e. all HfS transitions are characterized by the same value of the excitation temperature, which must be close to the kinetic temperature of the gas.

$NH_3(J,K)=(1,1)$ line in S106

In figure 4 we present a spectrum of the same line in the molecular cloud associated with the bipolar nebula S106. The linewidth is much larger than in TMC1 and the magnetic hyperfine structure is not resolved. The intensity ratio of the quadrupole satellites is completely different from what would be expected under LTE conditions. To explain these anomalies, probably excitation conditions similar to those in the absorbing layer in front of DR21 are required.

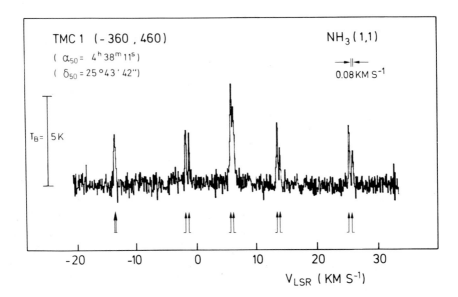

Figure 3: Spectrum of the $NH_3(1,1)$ line in TMC1. It was obtained with a maser frontend and a 1024 channel autocorrelator on the 100m telescope. The positions of the strongest magnetic hyperfine structure components are indicated by the arrows.

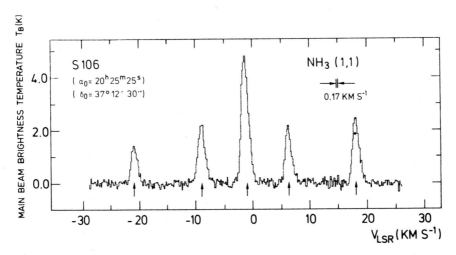

Figure 4: Spectrum of the $NH_3(1,1)$ line in S106. It was obtained with a maser frontend and a 384 channel autocorrelator on the 100m telescope. The intensities of the outer HfS satellites expected under LTE conditions are indicated by two full circles.

H_2 infrared spectrum in Orion

Among all regions in the galaxy where massive stars have recently formed, the Orion cloud is closest to the sun and hence the most thoroughly investigated specimen. IR- emission from H_2 molecules was observed in this source in rotational- vibrational as well as purely rotational transitions. The emission is extended over more than 1.5 arc min. In figure 5 we reproduce a spectrum of the v=1-->0 transition at the H_2 line emission peak from Davis et al.(1982). At this position the best fit to the observed line intensities is obtained with an H_2 excitation temperature of ~1500K. The total power radiated in the H_2 vibrational lines is ~200-400 solar luminosities. Current models indicate that the emission lines arise from a shock front which is formed where an energetic outflow of matter encounters the ambient gas. The source of this flow is associated with the cluster of IR sources and probably dominated by IRc2.

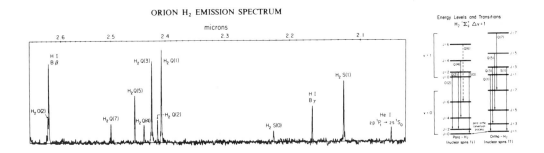

Figure 5: Infrared spectrum of H_2 in Orion (Davis et al., 1982). The measurements were made with a Fourier transform spectrometer on the 91cm telescope of the NASA Kuiper airborne observatory. In the right part the energy levels and transitions involved in the observed spectrum are shown. Some lines of atomic hydrogen and helium happen to fall in this spectral range as well.

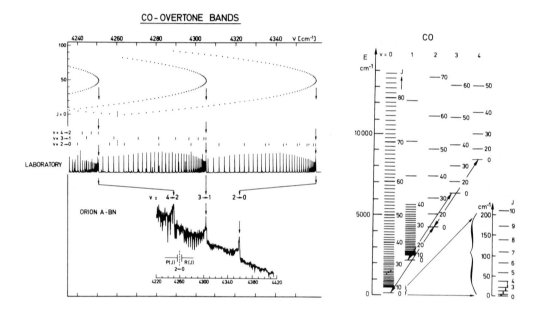

Figure 6: Infrared spectrum of CO towards the BN- object in Orion (Scoville, 1980) compared to a calculated Fortrat- diagram and the appropriate laboratory measurements (Mantz and Maillard, 1974). In the theoretical spectrum, only every fifth J- line is shown. In the BN spectrum, the bandheads of three overtone vibrational transitions are seen in emission and the central part of the 2-0 band is seen in absorption. The origin of the 2-0 band is indicated by the broken vertical lines. In the energy level diagram the three bands are indicated by arrows, along with the pure rotational transitions. The lowest CO rotational levels of v=0 are drawn on an expanded scale.

CO infrared spectrum towards the BN- object in Orion

In figure 6 we reproduce a CO- spectrum in the wavelength range 2.26-2.37μm observed towards the BN- object in the Orion cloud. Absorption in the v=2<--0 band is seen superimposed on emission in three overtone bandheads (v=4-->2, 3-->1 and 2-->0) as well as IR continuum emission from hot dust. Since the highest level involved in the emission (v=4, J~50) is at an energy equivalent to ~19000K above the ground, collisional excitation is only possible by gas at temperatures above ~3000K and densities $>10^{10}$ cm^{-3}. These requirements may be met in molecular shocks or in the very hot and dense circumstellar shells associated with the young stars at distances of a few AU. The gas causing the absorption lines must be much colder and farther out from the stars.

References

Davis,D.S., Larson,H.P. and Smith,H.A., 1982, Astrophys.J. 259, 166

Genzel,R., Downes,D., Ho,P.T.P. and Bieging,J., 1982, Astrophys.J. 259, L103

Mantz,A. and Maillard,J.P., 1974, J.Mol.Spectrosc. 53, 466

Matsakis,D.N., Brandshaft,D., Chui,M.F., Cheung,A.C., Yngvesson,K.S., Cardiasmenos,A.G., Shanley,J.F., and Ho,P.T.P., 1977, Astrophys.J. 214, L67

Scoville,N.Z., 1980, IAU Symp. 96, 187

Stacey,G.J., Kurtz,N.T., Smyers,S.D., Harwit,M., Russell,R.W. and Melnick,G., 1982, Astrophys.J. 257, L37

Stutzki,J., Ungerechts,H. and Winnewisser,G., 1982, Astron.Astrophys. 111, 201

Ungerechts,H., Walmsley,C.M. and Winnewisser,G., 1982, Astron.Astrophys. 111, 339

Watson,D.M., Storey,J.W.V., Townes,C.H., Haller,E.E. and Hansen,W.L., 1980, Astrophys.J. 239, L129

SECTION III: GALACTIC SOURCES

FAR-INFRARED SPECTROSCOPY OF NEUTRAL INTERSTELLAR CLOUDS

Dan M. Watson*

Department of Physics, University of California, Berkeley

* Present address: Downs Laboratory of Physics, 320-47, California Institute of Technology, Pasadena, California 91125, USA.

Abstract

A summary is presented of airborne observations of the far-infrared fine structure lines of neutral atomic oxygen and singly-ionized carbon, and of the far-infrared rotational lines of CO, OH, NH_3 and HD, together with a brief description of the analysis and interpretation of the spectra. The "state of the art" in instrument performance and the prospects for improved sensitivity and resolution are also surveyed.

1. Introduction

Many of the processes which are encountered by observers of interstellar matter are illustrated by the spectrum shown in Figure 1, which is based upon observations of a 1' diameter region centered on the Kleinmann-Low nebula in Orion. The line of sight intersects an H II region (M42), a rather warm molecular cloud (OMC-1), a dissociated interface region which separates the two, and an "active" molecular region (Orion-KL) which displays the symptoms of very recent star formation. The spectral features which are due to the various different sources are, apart from total brightness, probably fairly typical of sources of their respective types. Until recently, the data between wavelengths of 1 mm and about 20 μm were unavailable due to the opacity of the atmosphere in this part of the spectrum and to the lack of suitable spectroscopic instruments. This range is particularly interesting, though, since, as can be seen in Figure 1, it covers most of the emitted energy and the brightest spectral lines. The lines include some of the best available probes of abundances and physical conditions within the various emitting regions. One of our pleasant tasks at this Symposium is to report the significant advances in the observational capabilities with regard to far-infrared and submillimeter spectroscopy which has been brought about by the availability of airborne telescopes, notably the NASA Kuiper Airborne Observatory, and the continuing progress in instrumentation.

The purpose of this is to review the results of far-infrared spectroscopic observations of neutral interstellar clouds, and to discuss the present and short-term-future status of the instrumentation with which these observations are carried out. It will thus overlap with those in the papers by Drs. Emery, Harwit and Scoville in this volume, and with recent reviews by Phillips and Keene (1982) and Watson (1982a). The first two sections cover the present observations of neutral atomic and molecular regions and contain simple reviews of the physics of the

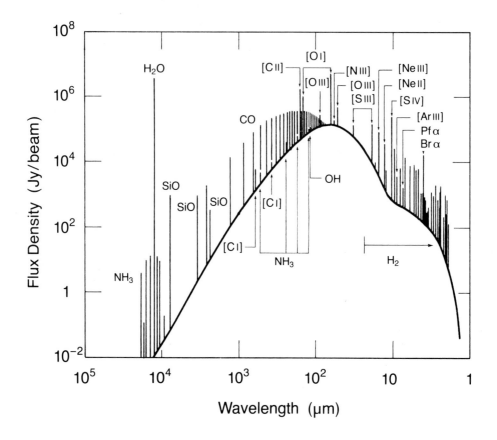

Figure 1: The spectrum of Orion-KL. Omitted for clarity are numerous molecular rotational lines in the millimeter-wavelength region which are present with antenna temperatures as high as about 20 K, the free-free continuum emission of M42, and all of the hydrogen recombination lines but the "alphas."

formation of the lines and the analysis of their strengths. For each regime, a discussion of the spectroscopic observations of the best-studied object, the Orion nebula and molecular cloud complex (the observations of which are shown in compressed form in Figure 1), is included as an example. Since the observations of other objects are still fairly limited on the scale of most of the fields of astronomy, we are not able to present here a comprehensive view of physical and chemical conditions in the various different types of neutral interstellar clouds as derived from far-infrared lines. It is to be emphasized, however, that the small size of the body of data reflects mostly the limitation on observing time on suitable telescopes, since the present instruments have performance adequate for many interesting projects. Accordingly, we mention in various places the sensitivity of present spectrometers in terms of ability to detect certain common objects. (For reasons of the author's familiarity, the sensitivity figures apply to the tandem

Fabry-Perot spectrometer described by Storey, Watson and Townes [1980] and Watson [1982b], hereafter referred to as the UCB Fabry-Perot. This instrument is used on the NASA Kuiper Airborne Observatory). The paper concludes with a discussion of the prospects for improved sensitivity through instrument development and the eventual availability of orbital platforms for far-infrared and submillimeter telescopes.

2. Neutral Atomic Regions

Among the more prominent features in the spectrum shown in Figure 1 are several atomic fine structure lines which may be presumed to be formed in warm, neutral material associated with the molecular cloud and the dissociation region. These lines are produced by magnetic dipole transitions within multiplets of given spin and orbital angular momentum whose states of different total angular momentum are split by the spin-orbit interaction. Since the transitions are simply dipole transitions, the spontaneous transition probabilities (A-coefficients) can be computed with high accuracy. As is usually the case in forbidden transitions, the lines are collisionally excited. In completely neutral clouds, the important collision partners are H and H_2, which appear to excite the fine structure levels with comparable efficiency (cf. Flower and Launay 1977a,b). At higher ionization levels ($H^+/H \gtrsim 10^{-3}$ for [C II], $\gtrsim 0.1$ for [O I]), collisions with electrons and protons dominate the fine structure excitation. In either case, the fine structure lines are sufficiently easily excited that they are the main coolants of neutral atomic interstellar gas (cf.Dalgarno and McCray 1972). Table 1 is a list of the far-infrared and submillimeter fine structure lines for some of the more abundant interstellar species which exist in neutral atomic regions. For the rest of this section, we focus our attention on the lines of [C II] and [O I], the observations of [C I] having been reviewed recently by Phillips and Huggins (1981) and Phillips and Keene (1982).

Because of the similarity of the ionization potentials of hydrogen and oxygen, the O exists primarily as O^0 in H I regions. The similarity of the ionization potential of carbon to the dissociation energy of CO suggests that in these regions, C^+ would be the predominant form of carbon. Thus, to first order, we expect the primary sources of far-infrared [C II] and [O I] lines to be coextensive.

Much of the interest in the far-infrared [C II] and [O I] lines arises from their role as the primary coolant in diffuse ($n_H = 1-100$ cm^{-3}) clouds; the [C II] 157.7 μm line does nearly all of the cooling in typical H I regions (Dalgarno and McCray 1972). Detection of these lines in diffuse clouds would also yield information on element abundances and depletion, and free electron abundances (since C^+ would also be the main source of electrons). At higher densities, such as one finds adjacent to or within molecular clouds in places where the molecules have been dissociated by ultraviolet light or shocks, the [O I] 63.2 μm line is usually the dominant coolant. In the latter regions, the [O I] and [C II] lines can serve as probes of density and temperature as well as cooling rate, as we describe below. Regions of this latter type may in fact provide the bulk of the total [O I] and [C II] emission of normal spiral galaxies.

Analysis of [C II] and [O I] Intensities

The atomic hydrogen column density required to produce unit optical depth in the [O I] or [C II] lines is $0.3-3\times10^{21}$ cm^{-2} for gas with cosmic abundances of oxygen and carbon. This range corresponds to a range of visual extinction of $A_V = 1.5-15$ (cf. Savage and Mathis 1979). Normally, one would expect the gas that produces the fine structure line emission to be less than an ultraviolet light penetration depth thick (else the gas would be in molecular form). In many cases this

Table 1
Selected Fine Structure Lines Arising in Neutral Regions

Transition	Species	λ (μm)	A (s^{-1})	Excitation Potential (eV)	Ionization Potential (eV)	Detected?
$2p: {}^2P_{3/2} \to {}^2P_{1/2}$	C$^+$	157.73	2.36×10^{-6}	11.260	24.383	yes
$3p: {}^2P_{3/2} \to {}^2P_{1/2}$	Si$^+$	34.8	2.13×10^{-4}	8.151	16.345	no
$2p^2: {}^3P_1 \to {}^3P_0$	C^0	609.133	7.93×10^{-8}	...	11.260	yes
${}^3P_2 \to {}^3P_1$		370.414	2.68×10^{-7}			no
$3p^2: {}^3P_2 \to {}^3P_1$	Si^0	129.68	8.25×10^{-6}	...	8.151	no
${}^3P_2 \to {}^3P_1$		68.474	4.20×10^{-5}			no
$2p^4: {}^3P_1 \to {}^3P_2$	O^0	63.1837	8.95×10^{-5}	...	13.618	yes
${}^3P_0 \to {}^3P_1$		145.526	1.70×10^{-5}			yes
$3p^4: {}^3P_1 \to {}^3P_2$	S^0	25.246	1.40×10^{-3}	...	10.360	no
${}^3P_0 \to {}^3P_1$		56.322	3.02×10^{-4}			no

Notes to Table 1: The potentials and A-coefficients come from Moore (1971) and Wiese, Smith and Glennon (1966), respectively. The wavelengths are from Bashkin and Stoner (1978, Si$^+$, S^0), Saykally and Evenson (1980, C^0), Evenson (1982, personal communication, O^0 ${}^3P_1 \to {}^3P_2$, Si^0), Davies et al. (1978, O^0 ${}^3P_0 \to {}^3P_1$), and Crawford et al. (1983, C$^+$).

would correspond to $A_V \lesssim 1$ (cf. de Jong, Dalgarno and Boland 1980). An extreme case would be that of a young H II region, for which the dissociation region is formed by an ultraviolet radiation field from a relatively nearby, hot star, which would be thousands of times stronger than the interstellar radiation field. Even in this case, the CO-dissociating/C^0-ionizing radiation would still not penetrate very far (only to about $A_V \sim 10$), so we anticipate that these lines will most often be optically thin, or at least not very optically thick. For the purposes of illustration, we present here the simple formulae for the optically thin case. The line intensity I (power per unit area and solid angle) can be written as follows:

$$I = \frac{h\nu}{4\pi} A_{kj} \int f_k n \, dl , \qquad (1)$$

where A_{kj} is the A-coefficient for the transition from the upper state k to the lower state j, n is the density of the emitting species, f_k is the fraction of the population which occupies the upper state, and the integration is taken along the line of sight l. The fractional populations f_k are determined for given densities and temperatures by supposing that the populations have reached a steady state under collisional excitation:

$$f_j \left[\sum_k n_H \gamma_{jk} + \sum_{k<j} A_{jk} \right] = \sum_k f_k n_H \gamma_{kj} + \sum_{k>j} f_k A_{kj} , \qquad (2)$$

where γ_{jk} is the collisional rate coefficient for the transition $j \to k$, n_H is the atomic hydrogen density, and $\sum_k f_k = 1$. The dependence on temperature is implicit in the γ_{jk}'s. Since higher multiplets are too high in energy to be significantly populated in the density-temperature regimes we consider here, we ignore all but the ground term fine structure levels, and Equations 2 reduce to a system of two or three equations, the solution of which is trivial. Figure 2 shows the solution of Equations 2 for the fine structure levels of O^0 and C^+ in a useful range of temperature and density.

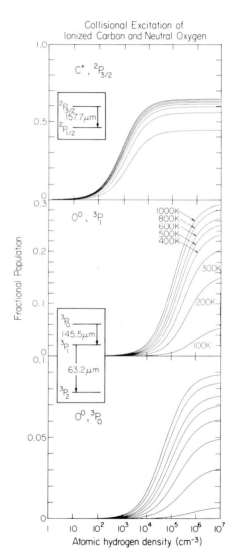

Figure 2: *Fractional populations f_k for the fine structure levels of ionized carbon and neutral oxygen, calculated for a range of density and temperature under the assumption of optically thin fine structure lines. The collisional rate coefficients of Launay and Roueff (1977a,b) were used in the calculation.*

If the hydrogen density is low, so that collisional de-excitation of the upper level k is negligible (ie. "subthermal" excitation), f_k is given simply by

$$f_k = \frac{n_H}{A_{kj}} \sum_{j \geq k} \gamma_{0j} \qquad (3)$$

(where γ_{0j} is the collisional rate coefficient to level j from the ground state), so that the line intensity is proportional to the atomic emission measure, $\int n_H n \, dl$. For high densities, f_k is given by its thermal equilibrium value:

$$f_k = \frac{g_k e^{-E_k/kT}}{\sum_j g_j e^{-E_j/kT}} \qquad (4)$$

(where g_j is the degeneracy of level j), and in this limit the line intensity is proportional to the atomic column density, $\int n \, dl$. The reference densities by which "low" and "high" are determined are $n_H \sim 10^3$ cm^{-3} for [C II] 157.7 μm and $n_H \sim 10^5$ cm^{-3} for [O I] 63.2 μm, as can be seen in Figure 2. It is often a good approximation to take the levels of C$^+$ to be in thermal equilibrium, and the levels of O^0 to be in the opposite (subthermal) limit. Line ratios involving these two species, for instance [O I] 63.2 μm/ [C II] 157.7 μm, should therefore be a good hydrogen density indicator, since the C/O abundance ratio should be fairly constant. The [O I] 63.2 μm/ [O I] 145.5 μm ratio, on the other hand, is quite insensitive to density - for a fixed temperature, this ratio varies by only about a factor of two over the density range 1–10^7 cm^{-3}. These line intensity ratios are plotted in Figure 3. Note that for most of the temperature-density combinations which one might expect to apply to H II/ H$_2$ interfaces, molecular cloud edges, and shock-dissociated regions within molecular clouds, the two families of curves meet at reasonably large angles, which indicates that observations of [O I] and [C II] lines can permit a useful determination of temperature and density in these types of objects.

[O I] and [C II] Observations of Galactic H II/ H$_2$ Complexes

A log of the present [O I] and [C II] observations is presented in Table 2. In Table 3 we note the C$^+$ column densities derived from the Galactic observations under the assumption that the line is optically thin and thermalized at high temperature. The average line optical depth can only be computed for Sgr A since this is so far the only velocity-resolved observation, and the result turns out to be rather small. In the other sources, the linewidths could be similar to those of the C$^+$ radio recombination lines ($\sim$ 5 km s^{-1}), which would imply line optical depths on the order of 1. As noted above, the latter result is on the order of what one might expect for the dissociation regions of very young H II regions like M42. However, it has been noted that in the slightly more highly evolved H II region complex M17, the region of [C II] 157.7 μm emission is much larger than the C$^+$ radio recombination line region (Russell et al. 1981); in this and similar cases there is no reason to expect the [C II] profile to resemble the recombination line profile. Observations at much higher spectral resolution than is presently available are thus necessary to determine the line optical depths.

In addition to the [C II] observations listed in Table 2, Stacey et al. (1983b) have attempted to measure the scale height of galactic [C II] 157.7 μm emission by chopping on and off the moon with a 7'×7' beam as the moon occulted the galactic plane at $l_{II} = 8°$. They observed a half-width of 1/6°, which would imply a much narrower distribution than the far-infrared continuum emission measured in a similar beam at this longitude, and much narrower than the typical

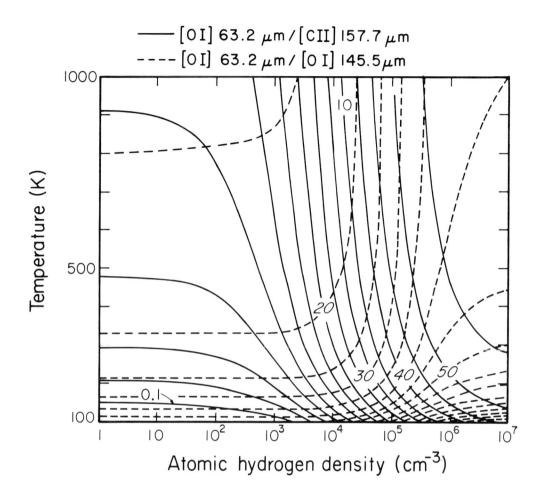

Figure 3: The line intensity ratios [O I] 63.2 μm/[C II] 157.7 μm (solid contours, upright numerals) and [O I] 63.2 μm/[O I] 145.5 μm (broken contours, slanted numerals), plotted as functions of atomic hydrogen density and temperature. The contour intervals are logarithmic for [O I] 63.2 μm/[C II] 157.7 μm and linear for [O I] 63.2 μm/[O I] 145.5 μm. The results of the same calculation which produced Figure 2 were used here, along with the assumption that the C^+/O^0 relative abundance ratio is equal to the cosmic C/O abundance ratio of 0.5.

distributions found for CO $J = 1\rightarrow 0$, H I 21 cm and H II regions, all of which extend roughly ±1° from the galactic plane.

From the C^+ column density values given in Table 3 for the Galactic H II region complexes, the mass of photodissociated atomic gas associated with each complex may be computed (which, because of the assumption that the [C II] line is both optically thin and thermalized, should be taken as a lower limit to the actual

Table 2
Log of Far-Infrared [O I] and [C II] Observations

Object	[O I] 63.2 μm Reference	[O I] 145.5 μm Reference	[C II] 157.7 μm Reference
M42	Melnick, Gull and Harwit 1979; Storey, Watson and Townes 1979; Naylor et al. 1982; Werner et al. 1982; Furniss et al. 1983; Werner et al. 1983	Stacey et al. 1983a	Russell et al. 1980; Crawford et al. 1983
NGC 2024	...	...	Russell et al. 1980, 1981; Kurtz et al. 1982
M82	Watson et al. 1983a,b	...	Watson et al. 1983b
NGC 6357	Moorwood et al. 1980	...	...
Sgr A	Lester et al. 1981; Genzel et al. 1983	...	Crawford et al. 1983
M17	Melnick, Gull and Harwit 1979; Storey, Watson and Townes 1979	...	Russell et al. 1981
W49	...	...	Crawford et al. 1983
DR21	Storey, Watson and Townes 1979; Werner et al. 1983	...	...
NGC 7027	Melnick et al. 1982; Werner et al. 1983	...	...
NGC 7538	Werner et al. 1983	...	...

mass). It is interesting to note that in each case the deduced mass of the dissociation region is larger by a substantial factor (2-10) than the mass of the corresponding ionized region. The total mass of the H II regions in our Galaxy is estimated to be about 10^7 $M_\odot$ (see, for instance, Osterbrock 1974). If there is a comparable or greater mass than this in dissociation regions, as seems likely, the total galactic [C II] emission from these regions would be at least as great as that from the 10^9 $M_\odot$ of diffuse matter which also exists in our Galaxy.

Table 3
[C II] Intensities and C$^+$ Column Densities

Object	I (erg s^{-1} cm^{-2} sr^{-1})	$N(C^+)$ (cm^{-2})	τ_L
M42	5.2×10^{-3}	3.3×10^{18}	6/Δv[km s^{-1}]
W49	2.8×10^{-3}	1.7×10^{18}	4/Δv[km s^{-1}]
M17	2.4×10^{-3}	1.5×10^{18}	4/Δv[km s^{-1}]
NGC 2024	2.6×10^{-3}	1.6×10^{18}	4/Δv[km s^{-1}]
Sgr A	3.0×10^{-3}	1.9×10^{18}	0.02

The [O I] and [C II] Lines in Orion

In the [O I] 63.2 μm line, M42 has been mapped completely with a 90" beam (Furniss et al. 1983), and partially with smaller beams (eg. 44"; Werner et al. 1982). There are additional observations of all three [O I] and [C II] lines with beams large enough to cover essentially the whole nebula (Melnick, Gull and Harwit 1979; Russell et al. 1980; Stacey et al. 1983a), and with approximately 1' beams centered on the Trapezium (Stacey et al. 1983a; Crawford et al. 1983). Very little of the [C II] emission, and essentially none of the [O I] emission, can be provided by the ionized gas in M42 (Russell et al. 1980). The far-infrared lines can be presumed to arise in the C$^+$-recombination-line-emitting dissociation region (Stacey et al. 1983a; see also Jaffe and Pankonin 1978). The measured intensity ratios are [O I] 63.2 μm/ [C II] 157.7 μm = 8 and [O I] 63.2 μm/ [O I] 145.5 μm = 16 in the large beams and [O I] 63.2 μm/ [C II] 157.7 μm = 11 and [O I] 63.2 μm/ [O I] 145.5 μm = 11 in the Trapezium. The implied density/temperature combination for the large beam, which represents an average over the whole nebula, is n_H = 10^4 cm^{-3} and T = 10^3 K (see Figure 3). In the small beam, the gas is slightly denser and significantly warmer, as expected if the Trapezium stars are the heating source.

In models of the structure of dissociation regions formed at the edges of molecular clouds by the interstellar radiation field, approximately half of the hydrogen turns out to be in the form of H$_2$ (Federman, Glassgold and Kwan 1979). If H II/ H$_2$ dissociation regions have the same molecular hydrogen content and if temperatures as high as 10^3 K were present in such a dissociation region, one might on this basis expect strong, widely-distributed H$_2$ rotation-vibration emission. The lack of evidence for such emission in Orion led to the suggestion by Stacey et al. (1983a) that the temperature is actually subtantially lower than 10^3 K. In the model proposed by these authors, the temperature is 220 K and the relatively small [O I] 63.2 μm/[O I] 145.5 μm ratio is preserved by making the [O I] 63.2 μm line optically thick, with a resulting atomic hydrogen column density of 6×10^{21} cm^{-2} in the dissociation region.

Present Sensitivity and Observations of [O I] and [C II] Lines

There are many objects, galactic and extragalactic (see, for example, Figure 4), which can be detected in [O I] and [C II] with present instruments. For example, let us consider a "standard" observation consisting of a two-hour integration covering eight resolution elements, require 10-1 signal-to-noise, and compute from the sensitivity the smallest mass of gas lying within the beam which would be detected. For the [O I] 63.2 μm line, which is usually subthermal, the detectable mass is

$$ M = \frac{8 \times 10^5 \, M_\odot}{n_H [\text{cm}^{-3}] T[K]} \left[\frac{D}{1 \text{kpc}} \right]^2 . $$

In subthermal [C II] 157.7 μm emission at $T = 100$ K, which would apply to diffuse cloud material, the current sensitivity is

$$ M = \frac{270 \, M_\odot}{n_H [\text{cm}^{-3}]} \left[\frac{D}{1 \text{kpc}} \right]^2 . $$

For [C II] 157.7 μm which is thermalized at high temperature, we have

$$ M = 9 \times 10^{-2} \, M_\odot \left[\frac{D}{1 \text{kpc}} \right]^2 . $$

Some interesting corollaries of these expressions are:

(1) Orion would be detectable in [O I] 63.2 μm or [C II] 157.7 μm no matter where it were placed in the Galaxy;

(2) Our galaxy, for which we estimate a mass in the form of H II/ H_2 "dissociation regions" of $\gtrsim 10^7 \, M_\odot$ and in the form of diffuse H I regions of $10^9 \, M_\odot$, would be detectable in [C II] 157.7 μm from a distance of 20 Mpc (The bulk of this mass would be covered if a beam size of about 2' were used);

(3) With the parameters as given by Morton (1975), the neutral cloud in front of ζ Oph, one of the best-studied H I regions, is detectable in [C II] 157.7 μm.

3. Molecular Clouds

Some far-infrared molecular lines are also prominent in the spectrum shown in Figure 1: lines of CO, OH and NH_3 which arise in hot molecular matter associated with Orion-KL, the region of most recent star formation in OMC-1. These are examples of the many rotational (and rotation-inversion) transitions of light molecules that lie in the far-infrared and submillimeter spectral regions. In the case of the rotational emission lines shown in Figure 1, the transitions are collisionally excited. In general, collisional excitation of the rotational levels prevails over radiative excitation except for regions very close to warm, optically thick far-infrared continuum sources (cf. the paper on this topic by Dr. Scoville in this volume), although radiative trapping can have a strong influence on the energy level populations. Since $h\nu/k$ is comparable to usual molecular cloud temperatures, submillimeter and far-infrared molecular emission lines are important coolants in molecular clouds, as well as serving as probes of abundance,

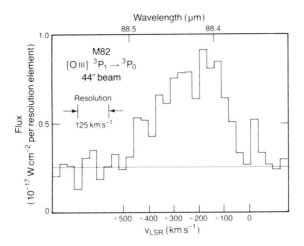

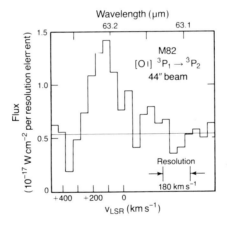

Figure 4. Above: The [O III] 88.4 μm line from the nucleus of the galaxy M82 (Watson et al. 1983a). Below: [O I] 63.2 μm emission in the same beam position as above. Note that the profiles of the [O I] and [O III] lines are quite different, with the [O I] line showing a smaller velocity dispersion and a blueshift of about 90 km s^{-1} relative to the [O III] centroid. The peculiar distribution of warm neutral atomic gas implied by this difference in profiles is discussed by Watson et al. (1983a).

temperature and density. For the purposes of interstellar chemistry, observations in this part of the spectrum are crucial, since the most important radicals and molecular ions have their lowest-lying transitions here. Since the derivation of physical parameters and abundances from molecular line strengths is not commonly treated for the excitation regimes which apply to far-infrared molecular lines, we discuss this in the following section.

The A-coefficients of far-infrared rotational lines, which are proportional to $\mu^2 \nu^3$, are typically much larger than those at lower (radio) frequencies. This can lead to strong self-absorption in the lower-lying transitions and subthermal populations in the upper states for molecules with typical dipole moments (~1 D). Two of the molecules which are discussed below, OH (μ = 1.67 D) and NH_3 (μ = 1.47 D), fall in this category. There are some molecules, however, which have very small (or zero) dipole moment, and would usually be optically thin and thermalized. This applies to HD, which has a rotational dipole moment of only 5.9×10^{-4} D (and, of course, H_2, which has no dipole moment). In between these two extremes is CO, with a dipole moment of 0.112 D, which in the present measurements turns out to have optically thin far-infrared lines but states with a range of departure from thermal equilibrium. For all fairly abundant molecules except the ones with exceedingly small dipole moments, the line absorption coefficient κ_{kj} for the transition $k \to j$, averaged over the equivalent linewidth $\Delta \nu$, which is given by

$$\kappa_{kj} = \frac{1}{\Delta \nu} \frac{h\nu}{c}(n_j B_{jk} - n_k B_{kj})$$
$$= \frac{8\pi^3}{3hc} \frac{\nu}{\Delta \nu} \frac{g_k}{g_j} \mu^2_{kj} n_j \left(1 - \frac{n_k g_j}{n_j g_k}\right) , \quad (5)$$

is much larger than that of the dust, given by

$$\kappa_D = 10^{-24} \left[\frac{n_{H_2}}{1 \text{ cm}^{-3}}\right]\left[\frac{125 \, \mu m}{\lambda}\right] \quad (6)$$

(Savage and Mathis 1979; Whitcomb et al. 1981). Under this assumption ($\kappa_L \gg \kappa_D$) and in the Sobolev escape probability approximation (see de Jong, Chu and Dalgarno [1975] and references therein), the excess of line intensity above the continuum, per unit frequency, is easily shown to be

$$\Delta I_\nu = S_L(1 - \beta_L) - B_\nu(T_D)(1 - e^{-\tau_D}) , \quad (7)$$

where the subscripts L and D refer to line and continuum respectively, β is the line escape probability, and S_L is the source function of the line, which is given by

$$S_L = \frac{2h\nu^3}{c^3}\left[\frac{n_j g_k}{n_k g_j} - 1\right]^{-1} , \quad (8)$$

In the opposite case ($\kappa_L \ll \kappa_D$), which applies to HD and H_2, we get instead

$$\Delta I_\nu = \tau_L B_\nu(T) e^{-\tau_D} , \quad (9)$$

where it has been assumed that the dust and gas are in equilibrium. (This latter formula is similar to that derived for radio recombination lines and free-free continuum in thermal equilibrium).

Consider a molecule which has a typical dipole moment (~ 1 D) and collisional excitation cross-section (~ 10^{-15} cm^2), and an abundance high enough along a certain line of sight to give the transition with wavelength λ an optical depth of 100. The molecular hydrogen density necessary for the collisional excitation rate to be equal to the A-coefficient for this transition is $n_{H_2} \sim A/\tau \langle\sigma v\rangle \sim 10^7$ cm^{-3}($100\mu m/\lambda$)^3 - that is, even strongly trapped far-infrared lines require relatively large densities for significant excitation. Thus it is often the case that the upper states of far-infrared and submillimeter molecular transitions are characterized by small, and usually subthermal, populations $(n_k g_j / n_j g_k \ll e^{-h\nu/kT} \ll 1)$. In quiescent clouds, where the gas and dust temperatures are similar and lower than $h\nu/k$ for far-infrared transitions, this usually makes the line source function enough lower than the Planck function that $S_L(1-\beta_L) < B_\nu(T_D)(1-e^{-\tau_D})$, and the line appears in absorption. This situation reverses rapidly for wavelengths increasing toward the submillimeter, because $h\nu/k$ decreases enough to become comparable to common molecular cloud temperatures (raising n_k, which can nevertheless remain subthermal) and τ_D also decreases greatly. Thus submillimeter molecular lines from quiescent matter may appear in emission (and, incidentally, act as major coolants of the gas). In "non-quiescent" molecular clouds such as regions influenced by "protostellar" mass outflow and shocks, very hot molecular gas can be produced in the presence of much colder dust (which is not very well coupled thermally to the gas and cools off much more quickly than the gas does), and $S_L(1-\beta_L) > B_\nu(T_D)(1-e^{-\tau_D})$ obtains even for far-infrared lines.

In the following items, we note the relations useful in the analysis of the far-infrared lines under the various limits that apply to the present observations.

$\kappa_L \gg \kappa_D$, optically thin

In this case, Equations 1 and 2 can be applied. Line fluxes are then proportional to the column density in the upper state of the transition for emission lines, or the lower state for absorption lines, and the derivation of density, temperature, and total column density proceeds straightforwardly. This limit applies to the far-infrared CO lines in Orion, which are discussed below. The collisional excitation cross-sections for CO-H_2 have been computed to an accuracy of about 30% (Green and Chapman 1983). McKee et al. (1982) have computed CO line emission coefficients in the optically thin limit using these cross-sections; these results are presented in Figure 5.

$\kappa_L \gg \kappa_D$, optically thick

In this limit, and under the escape probability approximation, Equations 1 and 2 may still be used, with the A-coefficients simply multiplied by the escape probabilities. Thus in the optically thick, strongly subthermal case for emission lines, the rate at which line photons leave the cloud is proportional to the rate of collisional excitations, since the photons may be absorbed and re-emitted many times on their way out of the cloud, but collisional de-excitations are not frequent enough to destroy a significant number of them. Thus the line intensity is proportional to the molecular emission measure, just as it is in the optically thin case (cf. Equation 3), although the repeated scattering of the line photons causes the constant of proportionality to be geometry-dependent. For example, in the case of a plane-parallel cloud with a large velocity gradient, for which the line escape probability is

$$\beta_L = \frac{1-e^{-3\tau_L}}{3\tau_L}$$

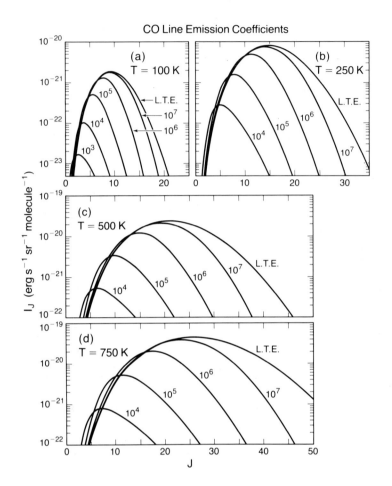

Figure 5: Line emission coefficient $I_J = \dfrac{h\nu A_{J \to J-1}}{4\pi} \dfrac{N_J}{N_{CO}}$ for CO $J \to J-1$ rotational transitions as a function of J for several values of temperature and molecular hydrogen density. The I_J come from computer solutions to Equations 2 for the lowest 61 rotational levels of CO. From McKee et al. (1982).

(Scoville and Solomon 1974), the intensity of a ground-state transition is

$$I = \frac{3h\nu}{4\pi} \sum_k \gamma_{0k} n_{H_2} N_{molecule} \qquad (10)$$

(Watson et al. 1983c). Thus the molecular column density may be estimated, provided an estimate of n_{H_2} is made (for instance, from measurements of other, optically thin lines). This applies to shocked OH emission, as is discussed below.

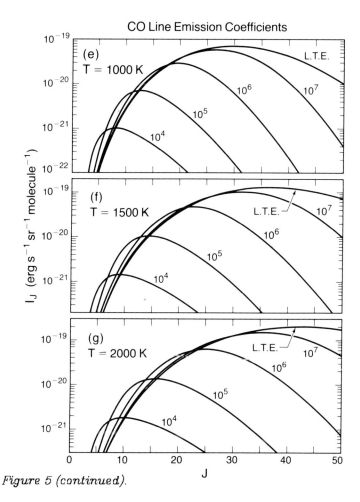

Figure 5 (continued).

Of course, there exist compact regions of high density (like the region around IRc2 in Orion-KL - see below) where the combined effects of collisional excitation and radiative trapping are sufficient to thermalize far-infrared transitions. In this case the line source function is the Planck function, the line is in emission, and its intensity can be used to determine the temperature of the surface of the emitting region.

$\kappa_L \ll \kappa_D$, **optically thin**

In dense, quiescent material where the dust and gas are in thermal equilibrium, the transport of line radiation in this limit is determined by dust absorption (cf. Equation 9). This has the effect of decreasing the line intensity at large column densities. For example, the following expression for the $J \to J - 1$ transition of HD may be derived:

$$I_J = (6.3\times10^3 \text{ W cm}^{-2} \text{ sr}^{-1} \text{ K}) \frac{\chi_{HD} J^4 \tau_D e^{-(64.2K)J(J+1)/T}}{T}. \qquad (11)$$

Many objects are characterized by gas and dust with different temperatures, with the dust usually existing at the lower temperature. Then the effect of dust on the line could be negligible and the line would merely obey the rules for optically thin thermalized lines. This is the case, for example, in the near- and mid-infrared lines of H_2 from shocked regions.

Far-Infrared Observations of Quiescent Molecular Material

OH Absorption Observations

The fundamental rotational lines of OH at 119.2 and 119.4 μm have been detected in absorption against the continuum sources of Sgr B2, Sgr A and W49 (Storey, Watson and Townes 1981, and unpublished observations with the UCB tandem Fabry-Perot spectrometer on the KAO). An example of these measurements is shown in Figure 6. The lines appear to be optically thick in all cases so only lower limits can be derived for the OH column densities. The lines in Sgr B2 and Sgr A are centered at v_{LSR} = −80 km s^{-1} and −120 km s^{-1} respectively, and thus would appear to be dominated by absorption from the foreground component of the expanding molecular ring at R_G = 180 pc (Scoville 1972).

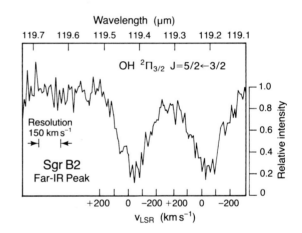

Figure 6: OH absorption lines at 119.2 and 119.4 μm in a 40 " beam centered on the far-infrared continuum peak of Sgr B2. Accounting for the finite resolution, the opacity in the lines is essentially 100%. (Observation made with the UCB tandem Fabry-Perot spectrometer on the NASA Kuiper Airborne Observatory).

HD

The $J = 1 \rightarrow 0$ transition of HD at 112.08 μm has been unsuccessfully searched for in several molecular clouds using the UCB tandem Fabry-Perot spectrometer on the KAO. The best upper limit (6×10^{-19} W cm^{-2}, in a 1' beam; Watson et al. 1983c) comes from an observation of Orion, and gives an upper limit to the deuterium abundance of D/H = 1.5×10^{-5} (using Equation 12 with T_D = 65 K and τ_D = 0.3; Werner 1982). This places no unexpected constraints on the interstellar deuterium abundance, which is thought to be near 10^{-5} (York and Rogerson 1976). The measurement, however, was limited by confusion due to weak telluric absorption, rather than instrumental sensitivity. Significantly higher spectral resolution than what is presently available (50 km s^{-1}) is required to improve on these measurements, but once this is obtained, it will probably be possible to detect the line

FAR-INFRARED SPECTROSCOPY OF NEUTRAL INTERSTELLAR CLOUDS

in several molecular clouds.

Far-Infrared Molecular Lines in Orion

So far, only two objects have been detected in far-infrared molecular emission lines: Orion-KL and Sgr A. In the latter object, only the CO $J = 16 \rightarrow 15$ line at 162.8 μm has been detected (with the UCB Fabry-Perot on the KAO, at a flux of 9×10^{-19} W cm^{-2} in a 44" beam centered 40" SW of the galactic center), and further observations are required before the source parameters may be derived. Near-infrared H$_2$ line measurements have been made in this object, however, and are discussed by Dr. Gatley in this volume.

The Orion-KL observations are summarized in Table 4. The results are discussed in detail in the more recent references (cf. Watson 1982b, Watson et al. 1983c) and are briefly described in the items below.

Table 4
Far-Infrared Molecular Lines from Orion-KL

Line	λ (μm)	Intensity† (10^{-2} erg s^{-1} cm^{-2} sr^{-1})	Reference
CO $J = 16 \rightarrow 15$	162.81	0.59	Stacey et al. 1983c
CO $J = 17 \rightarrow 16$	153.31	0.71	Stacey et al. 1982
NH$_3$ $K = 3$, $J = 4 \rightarrow 3$, $a \rightarrow s$	124.65	0.16	Townes et al. 1983
CO $J = 21 \rightarrow 20$	124.19	1.0	Watson et al. 1980, Storey et al. 1981
OH $^2\Pi_{3/2}$ $J = 5/2 \rightarrow 3/2$	119.44 / 119.23	0.17 / ~0.2	Storey, Watson and Townes 1981, Watson et al. 1983c
CO $J = 22 \rightarrow 21$	118.58	1.1	Watson et al. 1980, 1983c
HD $J = 1 \rightarrow 0$	112.08	<0.01	Watson et al. 1983c
CO $J = 26 \rightarrow 25$	100.46	0.38	Watson et al. 1983c
CO $J = 27 \rightarrow 26$	96.77	0.64	Storey et al. 1981
CO $J = 30 \rightarrow 29$	87.19	0.24	Storey et al. 1981
CO $J = 34 \rightarrow 33$	77.07	0.086	Watson et al. 1983c

† Averaged over the beams, which were 45-60" in diameter.

CO

It can easily be shown that the far-infrared CO lines from Orion-KL are optically thin. Thus Equation 1 can be used to give the upper-state column densities, and the density and temperature of the emitting gas may be estimated by comparing the shape of the column density function to the solutions of Equation 2 (for instance, those shown in Figure 5) for a range of different temperatures and densities. A single-component model results in estimates of $n_{H_2} \sim 10^6$ cm^{-3} and $T \sim 750$ K (see Figure 7). These parameters may be applied to the analysis of the H_2 $v = 0$ $S(2)$ line at 12.8 μm (Beck et al. 1982), most of which must arise from the same region that produces the CO emission. From the CO $J = 22 \rightarrow 21/H_2$ $v = 0$ $S(2)$ ratio, with the relatively small density and temperature corrections applied via the above estimates, a CO/H_2 relative abundance ratio of 1.2×10^{-4} is derived. Finally, the use of this relative abundance with one of the lower-lying far-infrared CO lines (which are nearly thermalized and fairly temperature-independent) yields a mass of 1.5 $M_\odot$ for the shocked gas. This gas is evidently related to, but is cooler than and further "downstream" from, the smaller amount of 2000 K gas which dominates the H_2 $v = 1 \rightarrow 0$ (Beckwith 1982, and references therein). Since it is understood that there must be a temperature gradient behind the shock, the above estimates for temperature and density should not be taken to be very precise. However, the CO/H_2 relative abundance, which comes from the flux ratio of two optically thin lines arising from nearly-thermalized states of similar energy, and the derived total mass, are better estimates.

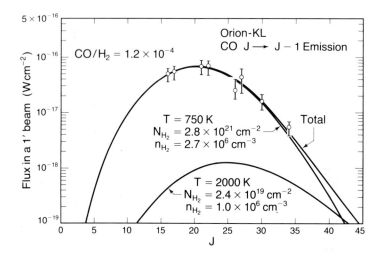

Figure 7: *Observed CO line fluxes from Orion-KL, scaled to a 1' beam (Watson et al. 1983c).*

Although the above density and temperature estimates are not very precise, they are good enough to rule out the possibility of a non-dissociative, non-magnetic (NDNM) shock being responsible for the H_2 and CO emission in Orion-KL. Such a shock would have a density jump from the pre-shock value of n_0 to approximately $6n_0$, with an accompanying temperature jump from T_0 to T_{max}. The postshock gas would be approximately at constant pressure, so that $n = 6n_0 T_{max}/T$. The maximum temperature and pre-shock density n_0 are constrained by the near-infrared H_2 line measurements to be

$$T_{max} \sim 2000 \text{ K},$$

$$n_0 = 4\pi I(H_2)/m_{H_2}v_s^3 \gtrsim 10^6 \text{ cm}^{-3},$$

since the shock speed v_s must be $\lesssim 15$ km s^{-1} to keep the molecular hydrogen from being dissociated in this type of shock shock (Kwan 1976). By the time it had cooled to 750 K, the gas would have a density of about 10^8 cm^{-3}, and all of the presently-observed far-infrared CO lines would be thermalized at high temperatures, in conflict with the observations.

Because of this difficulty with NDNM shock models, theoretical attention has turned to hydromagnetic shocks (Draine 1980), which can be made to fit the observations very well (Chernoff, Hollenbach and McKee 1982; Draine and Roberge 1982). These models are discussed in this volume by McKee, Chernoff and Hollenbach. As pointed out in the latter paper, far-infrared CO observations are very useful in the study of molecular shocks and, therefore, in the relation of the apparently very common phenomenon of protostellar mass outflow activity to star formation.

OH

The OH emission lines at 119.2 and 119.4 μm are seen to have about the same extent in angle and velocity as the far-infrared CO lines (Watson et al. 1983c). Thus we apply the above values of density and temperature and the observed OH flux to Equation 10, whence we obtain

$$N(\text{OH}) \sim 10^{15} \text{ cm}^{-2}$$

$$\tau_L \sim 4$$

$$\text{OH}/H_2 \sim 5 \times 10^{-7}$$

(cf. Watson et al. 1983c). In the calculation, the collisional excitation rate coefficients of Dewangan and Flower (1982) were used. The result represents a very large overabundance of OH, about a factor of 10 higher than what is inferred for the quiescent parts of the Orion molecular cloud (Zuckerman and Turner 1975). Such an overabundance is the expected outcome of the elevated temperature in a post-shock molecular environment, in which endothermic or high-threshold reactions between neutral molecules can occur (Hollenbach and McKee 1979,1980)

NH_3

Two rotation-inversion lines of NH_3 have recently been detected toward the core of the Orion molecular cloud: $J = 4 \rightarrow 3$, $K = 3$, $a \rightarrow s$ at 124.6 μm (Townes et al. 1983) and $J = 1 \rightarrow 0$, $K = 0$, $s \rightarrow a$ at 523.7 μm (Keene, Blake and Phillips 1983). The $J = 4 \rightarrow 3$ line may be ascribed to the very dense, warm gas within a few arcseconds of the compact infrared source IRc2 (this gas is usually called the "hot core", and is described in detail by Genzel et al. 1982). The optical depth of this line is very large and the levels involved are populated according to thermal equilibrium; the line flux is consistent with a surface temperature of about 200 K for the emitting region. Perhaps the most significant result of the $J = 4 \rightarrow 3$ observation is that since the line is seen in *emission*, radiative excitation of the non-metastable levels of NH_3 by the far-infrared dust emission is ruled out. Collisional excitation of this line requires a molecular hydrogen density of about 10^7 cm^{-3}.

Most of the $J = 1 \to 0$ emission (Keene, Blake and Phillips 1983) arises in warm material belonging to the quiescent cloud which lies just outside of the region of shocked and/or flowing gas (ie. in the "spike" component). The linewidth and derived size of the $J = 1 \to 0$ source are both larger than the corresponding parameters for the 22 GHz inversion lines of the "spike" component, by factors which are consistent with the much larger optical depth of the $J = 1 \to 0$ line.

Present Sensitivity in Far-Infrared Molecular Line Observations

Figures 8 and 9 are typical of the Orion-KL observations. Another way of expressing the present instrumental sensitivity is that at the wavelength of the $J = 16 \to 15$ line of CO (162.8 μm), a two hour integration covering eight resolution elements results in a 10σ limit which is a factor of about 70 lower than the intensity of the same line in Orion-KL. Judging from the observed $H_2 \, v = 1 \to 0 \, S(1)$ fluxes, there may be at least 10 or so shocked regions which can be detected with the present sensitivity (Watson 1982b).

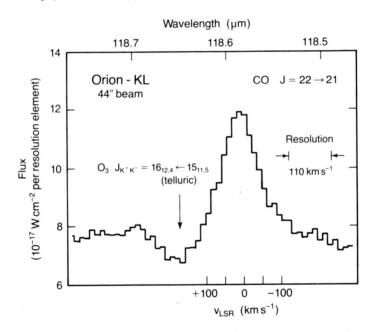

Figure 8: The CO J = 22→21 line in Orion-KL (Watson et al. 1983c). The total integration time for this spectrum was 3 minutes, or 3 s per point.

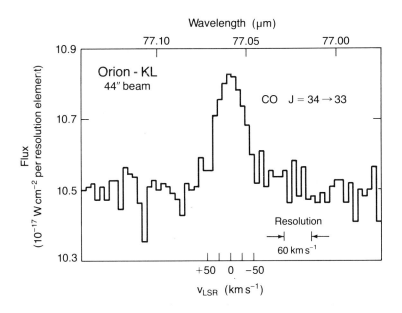

Figure 9: Detection of the CO J = 34→33 line in Orion-KL (Watson et al. 1983c). The total integration time was 42 minutes (40 s per point).

4. Instruments and Prospects for Improved Sensitivity

The Fabry-Perot spectrometer mentioned above is similar to many of the presently-working far-infrared instruments in that it employs extrinsic germanium photoconductive detectors and incoherent detection, narrow-band, cryogenically-cooled filtering, and is limited in sensitivity only by photon noise from the thermal background of the telescope environment. (As is discussed by several authors, narrow-band spectrometers have a sensitivity advantage over broad-band multiplexing spectrometers in the far-infrared). It works in the wavelength range 40-200 μm. The highest resolution which has been used is 50 km s^{-1}, and the noise equivalent power is typically $1-2\times10^{-14}$ W Hz$^{-1/2}$. The spatial resolution is usually chosen to be about 45", a beam not very much larger than the diffraction-limited beam size on the 91.4 cm KAO telescope. As described above, the present sensitivity is sufficient for many of the interesting projects in the far-infrared spectrum. Nevertheless, there is still quite a bit of room for improvement. With more carefully optimized filtering, there is a modest factor to be gained at the current resolution with a single-detector system. Measures which would potentially provide larger sensitivity gains include the following:

Multiple detectors. Detector arrays can be used to provide a spatial and/or a spectral multiplex advantage, depending upon how they are used. One could contemplate the use of an imaging array behind a Fabry-Perot spectrometer, for example. An alternative is to use one array dimension along the dispersion of a diffraction grating to spectrally multiplex. One example of this latter method is the new cooled grating spectrometer (resolution $\sim$ 150 km s^{-1} in the far-infrared) nearing completion at NASA Ames Research Center, which is being constructed as a "facility" instrument for the KAO (E.F. Erickson, personal communication). This device will have a single beam and (eventually) 40 spectral channels. Another example which uses fewer detectors but lends itself more easily to higher spectral resolution is the tandem Fabry-Perot-grating spectrometer described by Poulter (1982), in which two groups of detectors are placed along the dispersion axis of a moderate-resolution grating, each in the appropriate position for a different line (eg. the two [O III] lines at 88.4 and 51.8 μm). The Fabry-Perot is adjusted so that each detector receives one high order, and the Fabry-Perot thus can be piezoelectrically scanned over the two lines simultaneously. This system is in use on the University College balloon-borne telescope.

Higher spectral resolution. Since the present spectrometers are background limited, the per-detector sensitivity to a spectral line improves with increasing resolution ($NEP \propto \Delta\nu^{-1/2}$), until the resolution is comparable to the linewidths. Since insufficient resolution is beginning to hamper the interpretation of the observations, as has been apparent in the above discussions, there is a great incentive to undertake this development. It is possible to build higher-resolution versions of the present instruments - for instance, by using an additional Fabry-Perot or grating in the UCB tandem Fabry-Perot. However, the use of heterodyne detection may produce more sensitive high-resolution receivers, since arbitrarily high resolution could be obtained in this way with high optical efficiency. Work is now under way to extend the heterodyne techniques of both the millimeter-wave (ie. Schottky diode mixers) and the mid-infrared regions (ie. photodetector mixers) into the far-infrared. Mid-infrared photomixer heterodyne receivers have been shown to approach the quantum noise limit, given by

$$T_N = \frac{h\nu}{k\eta(1-\varepsilon)}\left[1 + \frac{\eta\varepsilon}{e^{h\nu/kT}-1}\right]$$

(where η is the product of detector quantum efficiency and cold transmission losses, and ε is the effective emissivity of the telescope environment, including coupling losses; chopping has also been taken into account). Since detectors similar to the mid-infrared photomixers are available in the far-infrared, there is no reason why quantum-limited far-infrared receivers cannot be built. This would represent a very substantial improvement over present performance.

Spaceborne telescopes. Atmospheric absorption is by no means negligible for the present suborbital telescopes. Absorption lines of water, oxygen and ozone are often very troublesome and sometimes still render important experiments impossible. At KAO altitudes, for instance, water lines near 63 μm render the atmosphere opaque to [O I] 63.2 μm radiation in the velocity range of roughly +500-1500 km s^{-1}, a range which covers many of the most interesting relatively nearby galaxies (including the bulk of the Virgo cluster!). Since direct-detection spectrometers on warm telescopes are strongly background limited, a cryogenically-cooled telescope, such as SIRTF, would also give an enormous sensitivity gain. Thus great progress will be made in this field when telescopes in space become available.

Acknowledgements

The new results from the UCB Fabry-Perot spectrometer were obtained in collaboration with M.K. Crawford, H.L. Dinerstein, R. Genzel, D.F. Lester, J.W.V. Storey, C.H. Townes and M.W. Werner. I am grateful my collaborators and to the far-infrared spectroscopy groups at Cornell University and ESTEC for permitting the inclusion of their unpublished data in this paper. I also thank E.N. Grossman, J.B. Keene and T.G. Phillips for critical readings of the manuscript. Far-infrared spectroscopy at Berkeley is supported in part through NASA grant NGR 05-003-511.

References

Bashkin, S. and Stoner, J.B. Jr. 1978, *Atomic Energy Levels and Grotrian Diagrams* v.2 (Amsterdam: North-Holland).
Beck, S.C., Bloemhof, E.E., Serabyn, E., Townes, C.H., Tokunaga, A.T., Lacy, J.H. and Smith, H.A. 1982, *Ap. J. (Letters)* **253**, L83.
Beckwith, S. 1982, *Ann. N.Y. Acad. Sci.* **395**, 118.
Chernoff, D.F., Hollenbach, D.J. and McKee, C.F. 1982, *Ap. J. (Letters)* **259**, L97.
Crawford, M.K., Watson, D.M., Genzel, R. and Townes, C.H. 1983, in preparation.
Dalgarno, A. and McCray, R.A. 1972, *Ann. Rev. Astron. Ap.* **10**, 375.
Davies, P.B., Handy, B.J., Murray Lloyd, E.K. and Smith, D.R. 1978, *J. Chem. Phys.* **68**, 1135.
de Jong, T., Chu, S.-I. and Dalgarno, A. 1975, *Ap. J.* **199**, 69.
de Jong, T., Dalgarno, A. and Boland, W. 1980, *Astron. Ap.* **91**, 68.
Dewangan, D.P. and Flower, D.R. 1982, *M.N.R.A.S.* **199**, 457.
Draine, B.T. 1980, *Ap. J.* **241**, 1021.
Draine, B.T. and Roberge, W.R. 1982, *Ap. J. (Letters)* **259**, L91.
Federman, S.R., Glassgold, A.E. and Kwan, J. 1979, *Ap. J.* **227**, 466.
Flower, D.R. and Launay, J.-M. 1977a, *J. Phys. B.* **10**, L229.
Flower, D.R. and Launay, J.-M. 1977b, *J. Phys. B.* **10**, 3673.
Furniss, I., Jennings, R.E., King, K.J., Lightfoot, J.F., Emery, R.J., Fitton, B. and Naylor, D.A. 1983, *M.N.R.A.S.*, to be published.
Genzel, R., Downes, D., Ho, P.T.P. and Bieging, J.H. 1982, *Ap. J. (Letters)* **259**, L103.
Genzel, R., Watson, D.M., Townes, C.H., Dinerstein, H.L., Hollenbach, D.J., Lester, D.F., Werner, M.W. and Storey, J.W.V. 1983, *Ap. J.*, submitted.
Green, S. and Chapman, S. 1983, *Chem. Phys. Lett.*, submitted.
Hollenbach, D.J. and McKee, C.F. 1979, *Ap. J. (Supplement)* **41**, 555.
Hollenbach, D.J. and McKee, C.F. 1980, *Ap. J. (Letters)* **241**, L47.
Jaffe, D.T. and Pankonin, V. 1978, *Ap. J.* **226**, 869.
Keene, J., Blake, G. and Phillips, T.G. 1983, *Ap. J. (Letters)*, to be published.
Kurtz, N.T., Smyers, S.D., Russell, R.W., Harwit, M. and Melnick, G. 1983, *Ap. J.* **264**, 538.
Kwan, J. 1976, *Ap. J.* **216**, 713.
Launay, J.-M. and Roueff, E. 1977a, *J. Phys. B.* **10**, 879.
Launay, J.-M. and Roueff, E. 1977b, *Astron. Ap.* **56**, 289.
Lester, D.F., Werner, M.W., Storey, J.W.V., Watson, D.M. and Townes, C.H. 1981, *Ap. J. (Letters)* **248**, L109.
McKee, C.F., Storey, J.W.V., Watson, D.M. and Green, S. 1982, *Ap. J.* **259**, 647.
Melnick, G., Gull, G.E. and Harwit, M. 1979, *Ap. J. (Letters)* **227**, L29.

Melnick, G., Russell, R.W., Gull, G.E. and Harwit, M. 1981, *Ap. J.* **243**, 170.
Moore, C.E. 1971, *Atomic Energy Levels* (NSRDS-NBS35).
Moorwood, A.F.M., Salinari, P., Furniss, I., Jennings, R.E. and King, K.J. 1980, *Astron. Ap.* **90**, 304.
Morton, D.C. 1975, *Ap. J.* **197**, 85.
Naylor, D.A., Emery, R., Fitton, B., Furniss, I., Jennings, R.E. and King, K.J. 1982. In *Regions of Recent Star Formation*, ed. Roger, R.S. and Dewdney, P.E. (Dordrecht: Reidel), 73.
Osterbrock, D.E. 1974, *Astrophysics of Gaseous Nebulae* (San Francisco: Freeman).
Phillips, T.G. and Huggins, P.J. 1981, *Ap. J.* **251**, 533.
Phillips, T.G. and Keene, J. 1982. In *The Scientific Importance of Submillimeter Observations*, ed. de Graauw, T. and Guyenne, T.D., 45.
Poulter, G. 1982, Ph.D. Thesis, University College, London.
Russell, R.W., Melnick, G., Gull, G.E. and Harwit, M. 1980, *Ap. J. (Letters)* **240**, L99.
Russell, R.W., Melnick, G., Smyers, S.D., Kurtz, N.T., Gosnell, T.R., Harwit, M. and Werner, M.W. 1981, *Ap. J. (Letters)* **250**, L35.
Savage, B.D. and Mathis, J.S. 1979, *Ann. Rev. Astr. Ap.* **17**, 73.
Saykally, R.J. and Evenson, K.E. 1980, *Ap. J. (Letters)* **238**, L107.
Scoville, N.Z. 1972, *Ap. J. (Letters)* **175**, L127.
Scoville, N.Z. and Solomon, P.M. 1974, *Ap. J. (Letters)* **187**, L67.
Stacey, G.J., Smyers, S.D., Kurtz, N.T., Harwit, M., Russell, R.W. and Melnick, G. 1982, *Ap. J. (Letters)* **257**, L37.
Stacey, G.J., Smyers, S.D., Kurtz, N.T. and Harwit, M. 1983a, *Ap. J. (Letters)* **265**, L7.
Stacey, G.J., Smyers, S.D., Kurtz, N.T. and Harwit, M. 1983b, *Ap. J. (Letters)*, submitted.
Stacey, G.J., Smyers, S.D., Kurtz, N.T. and Harwit, M. 1983c, *M.N.R.A.S.* **202**, 25P.
Storey, J.W.V., Watson, D.M. and Townes, C.H. 1979, *Ap. J.* **233**, 103.
Storey, J.W.V., Watson, D.M. and Townes, C.H. 1980, *Int. J. IR and mm Waves* **1**, 15.
Storey, J.W.V., Watson, D.M. and Townes, C.H. 1981, *Ap. J. (Letters)* **244**, L27.
Storey, J.W.V., Watson, D.M., Townes, C.H., Haller, E.E. and Hansen, W.L. 1981, *Ap. J.* **247**, 136.
Townes, C.H., Genzel, R., Watson, D.M. and Storey, J.W.V. 1983, *Ap. J. (Letters)*, submitted.
Watson, D.M., Storey, J.W.V., Townes, C.H., Haller, E.E. and Hansen, W.L. 1980, *Ap. J. (Letters)* **239**, L129.
Watson, D.M. 1982a. In *Submillimeter Wave Astronomy*, ed. J.E. Beckman and J.P. Phillips (Cambridge: Cambridge University Press), 91.
Watson, D.M. 1982b, Ph.D. Thesis, University of California, Berkeley.
Watson, D.M., Genzel, R., Townes, C.H., Werner, M.W. and Storey, J.W.V. 1983a, *Ap. J. (Letters)*, submitted.
Watson, D.M., Crawford, M.K., Genzel, R. and Townes, C.H. 1983b, *Ap. J. (Letters)*, to be submitted.

Watson, D.M., Genzel, R., Townes, C.H. and Storey, J.W.V. 1983c, *Ap. J.*, submitted.

Werner, M.W. 1982, *Ann. N.Y. Acad. Sci.* **395**, 79.

Werner, M.W., Dinerstein, H.L., Hollenbach, D.J., Lester, D.F., Genzel, R. and Watson, D.M. 1982, *B.A.A.S.* **14**, 616.

Werner, M.W. *et al.*, 1983, in preparation.

Whitcomb, S.E., Gatley, I., Hildebrand, R.H., Keene, J., Sellgren, K. and Werner, M.W. 1981, *Ap. J.* **246**, 416.

Wiese, W.L., Smith, M.W. and Glennon, B.M. 1966, *Atomic Transition Probabilities* (NSRDS-NBS4).

York, D.G. and Rogerson, J.B. Jr. 1976, *Ap. J.* **203**, 378.

Zuckerman, B. and Turner, B.E. 1975, *Ap. J.* **197**, 123.

MOLECULAR ASTRONOMY AT SUBMILLIMETER WAVELENGTHS

D. Buhl

NASA/Goddard Space Flight Center, Greenbelt, MD 20771

ABSTRACT

The difficulties of making astronomical observations in the submillimeter spectrum are described in detail. Receiver technology and atmospheric absorption provide the biggest challenge. Techniques of overcoming these problems are discussed and some initial astronomical results are illustrated.

I. INTRODUCTION

The region of the spectrum from 300 to 3000 GHz (100 to 1000 microns), which is referred to as the submillimeter, is one of the least explored territories in the electromagnetic spectrum. This is due to a combination of a lack of adequate instrumentation for detecting the radiation and an almost total absorption of the radiation by the earth's atmosphere. The problem of overcoming these two obstacles is one of the most difficult challenges in astronomy.

The most sensitive way to build a high resolution submillimeter receiver is to use the heterodyne technique which at mm wavelengths results in a receiver with a system temperature of a few hundred $^{\circ}$K. The first problem at submillimeter wavelengths is the lack of suitable local oscillators (LO). Mm tubes with harmonic multipliers have been used but their power output is generally inadequate for normal Schottky mixers. To compensate for this an InSb bolometer mixer has been developed (Phillips and Jefferts 1973) which requires very small amounts of LO power. However, this device has a very limited bandwidth and so must be tuned for each point in the spectrum. A second approach is to use a laser local oscillator which can produce several mW of power on a particular gas line. The problem then becomes one of finding a line sufficiently close

[1] Visiting Astronomer at the Infrared Telescope Facility which is operated by the University of Hawaii under contract from the National Aeronautics and Space Administration.

to the molecular line so that the resulting IF frequency can be handled by microwave amplifiers. Current lasers and mixers are capable of producing sensitive heterodyne receivers over the 300 to 3000 GHz band. System temperatures of 6000 $^\circ$K have been obtained (Fetterman et al. 1981), comparable to the early mm receivers. Further work is necessary to improve the sensitivity and tunability of the submillimeter laser receiver. In addition, other types of submillimeter local oscillators are being explored. Considerable progress is expected in these areas in the next few years.

The atmospheric transmission is nonexistent from sea level at submillimeter wavelengths and is only marginally useful in a few windows from high mountain sites. Most of the spectrum is only available from aircraft, balloons and satellites. Even from the best mountain infrared observing site in the world (Mauna Kea) we have been able to obtain atmospheric transmission exceeding 20% for only about 30% of our observing time in the middle of the 450 micron window. Most of the spectral region is obscured by terrestrial water vapor lines, hence the need for high altitude observations which are above the bulk of the atmospheric water layer.

The majority of the molecular lines detected in stars and gas clouds in our Galaxy arise from lines in the mm region of the spectrum. This is due to the large number of rotational transitions of molecules which lie in the mm region. As the energy of the transitions increase so does the line strength, suggesting that more intense lines will be seen at submillimeter wavelengths. Hence, from the point of view of molecular astronomy the submillimeter spectrum represents the most interesting potential harvest of new results in the history of molecular cloud studies.

II. SUBMILLIMETER RECEIVERS

A submillimeter heterodyne receiver consists of local oscillator tuned to a frequency near the line of interest (the frequency difference being the IF frequency), a diplexer to combine the signal and LO, a mixer to down convert the signal to an IF frequency, a low noise IF amplifier to provide gain and a spectrometer to disperse the signal into a spectrum. The specific example of a laser heterodyne receiver is shown in Figure 1. The LO in this case consists of a CO_2 laser operating at 10 microns pumping a submillimeter laser containing a molecular gas such as CH_3OH or $HCOOH$. Transitions between energy levels of the gas exhibit gain at submillimeter wavelengths and laser action is induced in an optical cavity. The output of the laser is fed into a diplexer which couples it to the signal from the telescope with minimum loss. A GaAs diode mounted in a quasioptical corner reflector (Fetterman et al. 1978) provides the mixing between the laser and signal beams. This type of mixer works

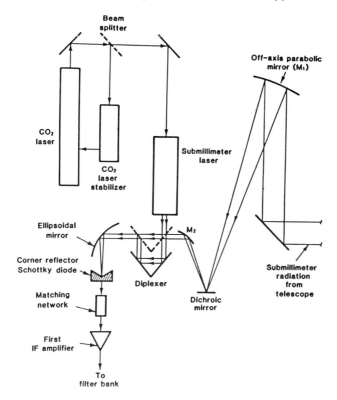

Fig. 1 Diagram of a submillimeter laser heterodyne receiver as configured for the Mauna Kea IRTF telescope by Goddard and Lincoln Laboratory.

efficiently to frequencies above 3000 GHz. It does, however, require several milliwatts of LO power. The signal has now been converted to an IF frequency of a few GHz where a GaAs FET amplifier provides the first gain in the receiver. For this reason it is important that this amplifier be very low noise. Amplifiers with noise temperatures in the 10 to 100°K range are possible. In mm wave receivers cooling both the mixer and preamp is important. Photon or shot noise is not significant at mm and submillimeter wavelengths.

Once the signal has been amplified to a suitable power level it is fed into a spectrometer so that a simultaneous display of the entire line spectrum can be obtained. A combination of a traditional RF filter bank providing low resolution broad bandwidth coverage and an acousto-optical (AO) spectrometer providing high resolution and a large number of channels is being used with the laser submillimeter receiver. An AO spectrometer appears to be ideally suited to the submillimeter wavelength region. It is also possible to miniaturize an AO system for flight experiments.

III. ATMOSPHERIC TRANSMISSION

Due to extremely strong lines of water vapor only very high mountain sites are suitable for submillimeter observing. An example

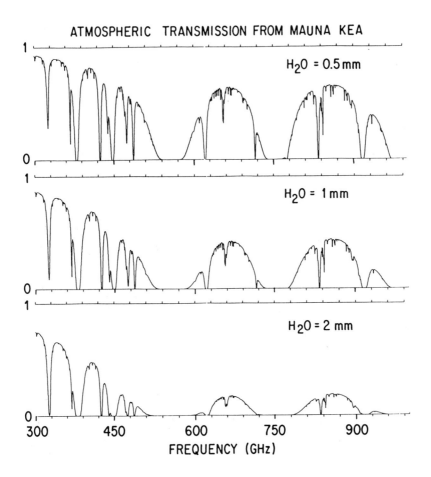

Fig. 2 Atmospheric transmission from 300 to 1000 GHz for various amounts of precipitable water vapor above the telescope. The spectrum is obtained from a program developed by Traub and Stier (1976).

of the transmission from Mauna Kea during optimum periods of low water vapor is shown in Figure 2. Transmission for frequencies higher than 1000 GHz is zero. Precipitable water vapor levels of 1mm can only be obtained at the best mountain sites, and only reaches

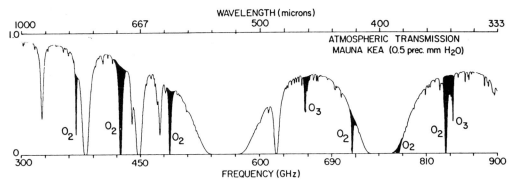

Fig. 3   Absorption lines due to O_2 and O_3 in the atmospheric spectrum. Since these lines arise from high up in the atmosphere they will be seen again in the KAO spectrum.

such low values occasionally. Not all of the absorption lines in the spectrum are due to water vapor. Figure 3 shows the spectrum with lines of O_2 and O_3 blacked in. These lines originate from higher up in the atmosphere than the water vapor layer and so are more difficult to eliminate. The broad absorption features at 550 GHz and 750 GHz are due to very strong water vapor lines.

To improve the atmospheric transmission requires going from Mauna Kea at 4200m to the NASA KAO aircraft at 12000m. At this altitude the precipitable water vapor drops to 10 microns. Even so lines of H_2O, O_2 and O_3 still appear in the spectrum (Figure 4).

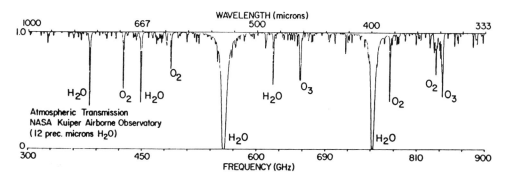

Fig. 4   Relatively clear transmission as seen from 12000 m in the KAO. Strong H_2O lines at 550 and 750 GHz are still prominent.

However, the line width is much narrower and experiments which were not possible from Mauna Kea can now be done from the KAO. Most of the submillimeter spectrum is now available for molecular astronomy

observations. Limitations on what can be done are imposed more by
the 91cm size of the KAO telescope than by atmospheric absorption.
Further out in the submillimeter spectrum at frequencies above 1000
GHz the KAO is the only feasible telescope. Even from 12000m
atmospheric lines are prominent (Figure 5). Careful selection of an
experiment requires detailed consideration of the atmospheric
absorption spectrum.

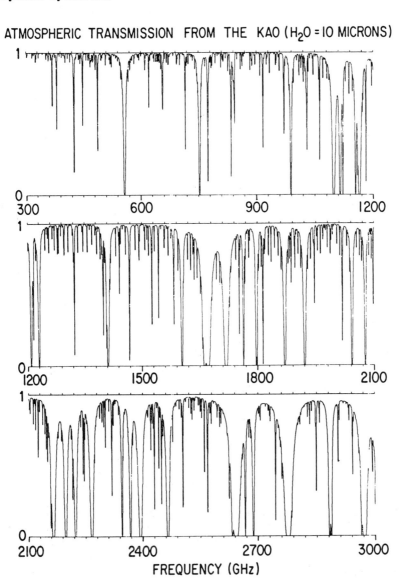

Fig. 5   Atmospheric transmission from 300 to 3000 GHz illustrating
the numerous lines still present from the KAO.

IV. GALACTIC MOLECULAR LINES

Rotational transitions of molecules start in the cm and mm wavelength range of the spectrum and proceed upwards in frequency into the submillimeter. Very light molecules such as OH have their first rotational transition in the submillimeter Watson (1982). Two important diatomic molecules (SiO and CO) are shown in diagrams which illustrate the energy levels of various transitions (Figure 6).

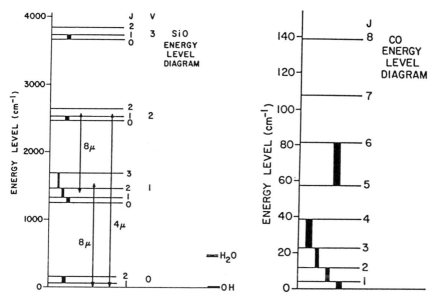

Fig. 6 Energy level diagrams showing the effects of both rotational and vibrational energy. Transitions shown with bars have been observed from astronomical sources. In addition several lines of CO up to J=30 have been detected (Storey et al. 1981).

Starting from the bottom, the first two transitions of CO are in the mm band whereas the first six transitions of SiO (not all are shown) are in the mm band since it is a heavier molecule. The SiO diagram also illustrates what happens when vibrational energy is added to rotational energy. The rotational transitions remain at approximately the same wavelength but vibration raises their level of excitation enormously. In this diagram the rotational splittings have been greatly exaggerated with respect to the vibrational levels: vibrational energy being a factor of 1000 larger than rotational energy. All the lines indicated have been detected in astronomical objects. In addition, CO lines at J=30, 27, 22 and 21 have been observed, indicating that the region from which they arise is quite hot (Watson et al. 1980 and Storey et al. 1981). Each molecule has a unique energy level diagram. The two illustrated here are relatively simple since they are diatomic and hence the structure is linear. Other molecules will present a more complex pattern.

Each molecular transition has a different groundstate energy level or threshold of excitation. Under these circumstances it is necessary to examine many transitions of a molecule to determine the local physical conditions inside a molecular cloud. This is illustrated in Figure 7 where several transitions of CO are shown in the BN/KL region of the Orion nebulae. There is a narrow 7 km/s

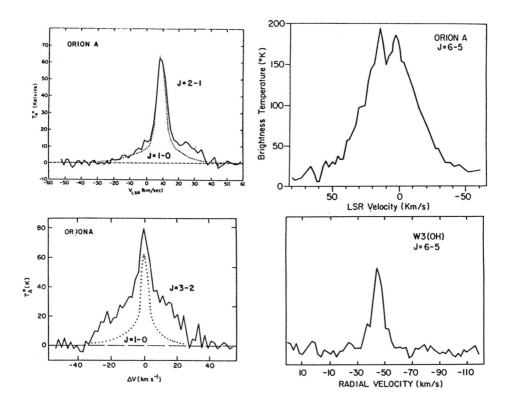

Fig. 7 Spectra of the J=1,2,3 and 6 lines of CO in Orion showing the dramatic increase in the broad source with J (Wannier and Phillips 1977, Phillips et al. 1977 and Koepf et al. 1982). The spectra of the J=6 line in W3 (OH) is shown for comparison (Chin et al. 1983).

wide feature which is associated with the cold part of the cloud. This feature appears in emission in the first three transitions studied and is possibly seen in absorption in the J=6 profile. The broad part of the CO emission (50 km/s wide) is associated with the hot part of the nebulae and when mapped appears to come from a region 1 min of arc in diameter. This extended feature becomes rapidly stronger with J and by J=6 has almost swallowed the narrow line. In W3 (OH) there is no obvious evidence of a similar broad feature. The rapid growth of the broad component implies temperatures >200 °K (Buhl et al. 1982 and Goldsmith et al. 1981). Detection of several lines up to J=30 by Watson et al. 1980 and Storey et al. 1981 suggests temperatures as high as 2000 °K. Most of these conclusions are tentative as many other transitions of CO remain to be explored.

V. CONCLUSION

The vast majority of the rotational transitions of molecules occur in the mm and submillimeter wave region of the spectrum (Figure 8). In addition, the strength of the lines increases at higher frequencies until a condition is reached where the energy levels are no longer populated. For CO, the levels appear to be populated as

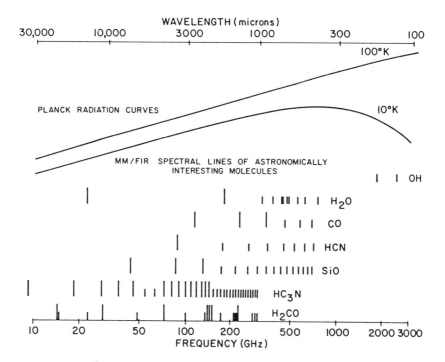

Fig. 8 The line spectra of several molecules in the mm and submillimeter bands. Bent molecules such as H_2O and H_2CO produce an irregular pattern of lines.

high as J=30 at the high frequency end of the submillimeter band. The situation for other molecules has yet to be determined.

The immediate goals for submillimeter heterodyne receivers are a number of important lines in the 300 to 900 GHz region (Figure 9). Some of these can be seen from Mauna Kea and some will require the KAO. NH_3, being a very light molecule, has its first rotational line at 572 GHz. This line is blanketed by the 550 GHz water line from Mauna Kea. The CO, HCO^+ and HCN lines originate at 100 GHz and are uniformly spaced. H_2O on the other hand is a bent molecule and has a very complex spectrum. Two lines at lower frequencies have been detected in Orion, one being the strongest molecular line ever observed. Which additional H_2O lines will be important is difficult to predict and terrestrial H_2O absorption will complicate the situation. These are illustrations of the simplest and most abundant of the molecular species. The search for more exotic molecules is an important component of any submillimeter astronomy program. This will be aided by several proposed

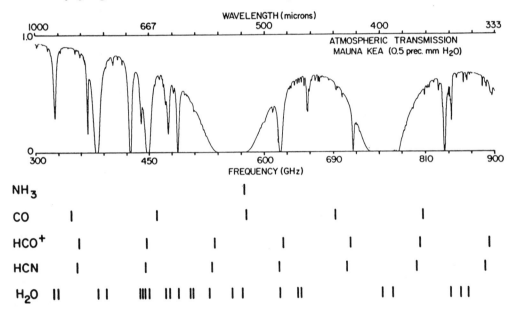

Fig. 9 Important molecules for submillimeter molecular exploration. These molecules produce strong spectra in the mm band and can be expected to be more intense in the submillimeter.

submillimeter space telescopes. Current plans call for an LDR telescope to be launched by NASA in the 1990's and a FIRST telescope to be launched by ESA. These facilities will eliminate the atmospheric problem and open up the entire submillimeter spectrum. At the same time new technology will be necessary for the instruments on these telescopes.

VI. REFERENCES

Buhl, D., Chin, G., Koepf, G. A., Peck, D. D., and Fetterman, H. R.: 1982, Submillimeter Wave Astronomy (Beckman and Phillips ed.), Cambridge University Press, 111.
Chin, G., Buhl, D., Petuchowski, S. Peck, D. D. and Fetterman, H. R.: 1983, in preparation.
Fetterman, H. R., Tannewald, P.E., Clifton, B. J., Parker, C. D., Fitzgerald, W. D. and Erickson, N. R.: 1978, Appl. Phys. Let., 33, 2, 151.
Fetterman, H. R., Koepf, G. A., Goldsmith, P. F., Clifton, B. J., Buhl, D., Erickson, N. R., Peck, D. D., McAvoy, N. and Tannenwald, P. E.: 1981, Science 211, 580.
Goldsmith, P. F., Erickson, N. R., Fetterman, H. R., Clifton, B. J., Peck, D. D., Tannenwald, P. E., Koepf, G. A., Buhl, D. and McAvoy, N.: 1981, Ap. J. 243, L79.
Koepf, G. A., Buhl, D., Chin, G., Peck, D. D., Fetterman, H. R., Clifton, B. J. and Tannenwald, P. E.: 1982, Ap. J. 260, 584.
Phillips, T. G., Huggins, P. J., Neugebauer, G. and Werner, M. W.: 1977, Ap. J., 217, L161.
Phillips, T. G. and Jefferts, K. B.: 1973, Rev. Sci. Inst. 44, 1009.
Storey, J. W. V., Watson, D. M., Townes, C. H., Haller, E. E. and Hansen, W. L. : 1981, Ap.J. 247, 136.
Traub, W. A. and Stier, M. T.: 1976, Appl. Opt., 15, 364.
Wannier, P. G. and Phillips, T. G.: 1977, Ap. J., 215, 796.
Watson, D. M., Storey, J. W. V., Townes, C. H., Haller, E. E. and Hansen, W. L. :1980, Ap. J., 239, L129.
Watson, D. M. :1982, Submillimeter Wave Astronomy (Beckman and Phillips ed.), Cambridge University Press, 91.

SUBMILLIMETER OBSERVATIONS OF MOLECULES AND THE STRUCTURE OF GIANT MOLECULAR CLOUDS

Paul F. Goldsmith

Five College Radio Astronomy Observatory, U. of Massachusetts

ABSTRACT

We discuss the contributions of submillimeter molecular line observations to our understanding of the structure of giant molecular clouds, with emphasis on determination of the density and temperature structure as well as the velocity field in these objects. The presence of large scale temperature gradients is expected when looking towards embedded heating sources; the frequency with which the expected effects are seen in molecular line data gives important clues to the nature of the velocity field of the surrounding material. Density determinations using multiple-transition studies of different molecules are found to demand a large range of densities within most elements of the highest angular resolution ($\sim 1'$) maps obtained, but evidence for smooth, large scale density gradients is surprisingly limited. Studies of this type, of different clouds in several transitions of CS and CO are discussed. In this work, observations at submillimeter wavelengths are particularly important, for when used in conjunction with longer wavelength data, the range of conditions which can be sensitively probed is significantly increased.

I. INTRODUCTION

Molecular clouds are recognized at present as a major (Ref. 1) and possibly the dominant (Ref. 2) mass component of the interstellar medium. Major uncertainties exist, however, concerning the internal structure of these objects in terms of their density and temperature structure, as well as of the velocity distribution of the constituent material. Submillimeter wavelengths constitute an important window into those clouds which are warm enough, $T \geq 15K$, to radiate appreciably in this spectral region; this includes almost all types of molecular clouds but it is readily apparent that submillimeter emission will be most intense from those regions significantly heated above the $\sim 10K$ temperature characteristic of quiescent dark clouds.

The submillimeter region is considered "officially" to cover wavelengths between 1000 μm and 100 μm corresponding to frequencies 300 GHz to 3000 GHz and temperatures 14.4 K to 144 K. In this discussion I have taken the liberty to extend the range of data discussed to wavelengths as long as 7 mm, but will concentrate on data in the 1 mm range. This is a result of a combination of factors: the difficulty of carrying out observations at wavelengths significantly shorter than 1 mm (as described in the contribution of D. Buhl), and the fact that at present the shorter-wavelength data are limited to relatively small regions of the molecular clouds of interest. Despite the difficulty of submillimeter observations, the quantity of data published is already quite extensive. Thus, this discussion of observational results is not complete. Rather, I have attempted to select certain submillimeter studies which illustrate the value of observations in this wavelength range, as well as mixing in a proportion of new data.

II. TEMPERATURE STRUCTURE OF MOLECULAR CLOUDS

Neighboring or embedded stars, protostellar objects, and HII regions are sources of heating for molecular clouds that can readily be identified with objects that can be studied at wavelengths other than the millimeter-submillimeter spectral region. However, numerous other heating sources - including dissipation of gravitational, turbulent, and magnetic energy as well as heating by cosmic rays - have been proposed to be significant in the thermal balance of molecular clouds (see Ref. 3 for discussion). The combination of these possible energy inputs (which are difficult to study directly by any method) makes the thermal structure of molecular clouds subject to considerable uncertainty.

Several different techniques are available for studying the temperature in molecular clouds. The relative intensities of different optically thin transitions of symmetric top molecules, particularly NH_3 and CH_3CN, have been used as temperature probes (Refs. 4 and 5). The utility of methyl cyanide is limited by its restricted spatial extent as well as complexities in its interpretation (Ref. 6). Observations of CH_3CN have indicated different temperatures for the narrow, spatially extended emission in Orion (T ~ 100K) and the small, hot core with T ≳ 200K (Ref. 7). Gas temperatures obtained using ammonia appear to follow those obtained from carbon monoxide (discussed below) and should serve as a useful comparison in certain situations (Ref. 8).

The most widely used probe of gas temperatures in molecular clouds is carbon monoxide (^{12}CO). For this technique to be valid we require that (1) collisional excitation (together with any radiation trapping present) bring the excitation temperature (T_{EX}) of the transition being observed close to the kinetic temperature (T_K) of the gas; and (2) the optical depth of the transition be large compared to unity. If these conditions are satisfied, the brightness temperature of the transition will be close to the gas kinetic temperature. The rotational transitions of ^{12}CO are excellent candidates for cloud thermometers. Since the spontaneous decay rate for the J = 1 → 0 transition is 7×10^{-8} s^{-1} and a characteristic rate coefficient for collisions with H_2 is 10^{-10} cm^3 s^{-1}, a density of only 10^3 cm^{-3} will produce significant excitation, even for optically thin lines. Although the details depend on the exact radiative transfer model, the effect of radiation trapping for optically thick lines is to reduce the effective spontaneous decay rate by a factor of the order of τ, the optical depth. Thus, since ^{12}CO has optical depths generally exceeding 10 and often close to 100 in dark and giant molecular clouds, this transition should be thermalized ($T_{EX} \rightarrow T_K$).

The major remaining question is how the emergent flux at a particular frequency relates to the temperature at any position within the cloud; this depends largely on the velocity field of the molecular material. For the case in which the scale size for velocity fluctuations is much smaller than the photon mean free path (microturbulence) one expects to measure the temperature close to the near surface of the cloud. On the other hand, if there are correlated mass motions on a scale large compared to a photon mean free path one can expect to see the temperature distribution at different locations in the cloud. In fact, the peaking of ^{12}CO temperatures at the positions of heating sources known to be deeply embedded within the clouds was one of the major reasons for considering models with large systematic velocity gradients (Ref. 9).

Submillimeter transitions of CO provide an important means of supplementing and verifying temperature information obtained from the lowest rotational transition, which has to date been the only one extensively observed. The transitions J → J - 1 for CO have frequencies ν = 115 GHz * J and equivalent temperatures $h\nu/k$ = 5.5K * J. The spontaneous decay rates are given by (Ref. 10)

$$A_{J \rightarrow J-1} = 2.2 \times 10^{-7} \, (s^{-1}) \, \frac{J^4}{2J + 1.} \quad (1)$$

The optical depth for different transitions depends on the degree of excitation and hence on the hydrogen density. In some cases all levels of significance may have their populations in thermal equilibrium and the optical depth varies as

$$\tau_{J \rightarrow J-1} \simeq J \sinh \left(\frac{hBJ}{kT}\right) \, \exp \left(- \frac{hB}{kT} J^2 \right) \quad (2)$$

where B is the rotational constant ($\sim$ 57.6 GHz; $\frac{hB}{k} \sim 2.8$ K).

Thus for reasonably warm gas the J = 2 → 1 transition has an optical depth 4 times larger than the J = 1 → 0 transition, but in very cold material its optical depth can be much less than that of the lower transition. When the level populations are not in thermal equilibrium, the situation is more complex; as an example we show in Figure 1 the dependence of the optical depth of the two lowest CO transitions on hydrogen density (Ref. 11). Over an important range of densities, $n(H_2) > 10^3$ cm^{-3}, the J = 2 → 1 transition is far more optically thick than J = 1 → 0.

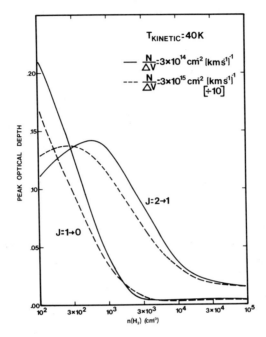

Fig. 1 - Dependence of optical depth of J = 1 → 0 and J = 2 → 1 transitions of CO on hydrogen density. The kinetic temperature and CO column density are fixed; two values of the latter have been used with the radiative transfer being treated by the LVG cloud model. From Ref. 11.

Maps of the molecular clouds NGC2068 and NGC2023 in the J = 3 → 2 transition of CO (λ = 870 µm) have been made by White, Phillips, and Watt (Ref. 12). The line intensity peaks at the location of the embedded IR sources thus confirming the picture that we are "seeing in" to the central regions of these clouds. The optical depth of the J = 3 → 2 transition in these warm (T ~ 30-40 K), dense ($n(H_2) \geq 3 \times 10^4$ cm^{-3}) clouds is expected (eq. 2) to be significantly (3-4 times) larger than that of the lower transitions. Nevertheless, the peak temperature - which should be close to the kinetic temperature - agrees with that measured in the J = 1 → 0 transition. Phillips et al. (Ref. 13) have made strip maps of several sources in the J = 2 → 1 transition of several CO isotopic species. The ^{12}CO peak intensities generally show close agreement with the J = 1 → 0 data (at least within the calibration uncertainties). Most of the sources observed by Koepf et al. (Ref. 14) have J = 6 → 5 ^{12}CO intensities fairly close to those in the lower frequency transitions, but signal to noise and calibration difficulties limit the accuracy with which comparisons can be made at present.

In Figure 2 maps of the Orion molecular cloud in the J = 2 → 1 and J = 1 → 0 transition of ^{12}CO are presented. The J = 1 → 0 data were obtained with the FCRAO 14m antenna and the J = 2 → 1 data with the MWO 5m antenna. Both maps have been convolved to 1'.5 angular resolution and the quantity plotted is the maximum antenna temperature corrected for atmospheric losses and antenna beam coupling. On the whole, the two maps follow each other closely, with peaks occuring at the KL nebula at the (0,0) position, and at the bright optical bar located approximately 3' south (Ref. 15). Note that most of the cold spots are seen in both maps, as are the regions of higher temperature. The two major heat sources are located quite differently with respect to the molecular gas: the IR source is buried deep within the cloud and the young stars (the Trapezium and $\theta^2 A$) are on the near side. Shock heating has also been suggested to be playing a role in this region (Ref. 15). Nevertheless we see the two transitions following each other quite well.

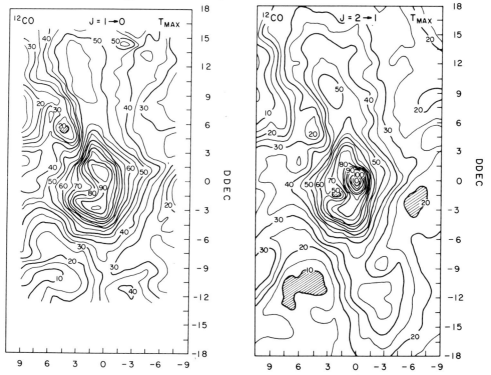

Fig. 2 - Contours of peak antenna temperature (corrected for atmospheric absorption and coupling efficiency) in two transitions of ^{12}CO in the Orion molecular cloud. The central position is R.A. (1950) = 5^h 32^m 47^s, Dec. (1950) = 5^0 24' 30"; the offsets are in minutes of arc.

It is remarkable how complex the structure of this cloud is in ^{12}CO - a far cry from a symmetric source heated by a single object. In particular, the extent of quite hot material (≥ 40K) over a region greater than 10' (1.5 pc) from the BN - KL nebula is difficult to understand at the present time.

A more extreme example of heating from a source that is hidden by a very large quantity of material is the molecular cloud/HII region complex

W33. A recent study of this source (Ref. 11) confirms the picture that the large HII region is located behind almost all of the molecular material, which has a peak hydrogen column density of 10^{23} cm^{-2}. In Figure 3 we show a map of the peak ^{12}CO J = 2 → 1 temperature obtained at the MWO. The broken curve is the extent of the 2K brightness temperature of the 8.6 GHz continuum emission (Ref. 16), while the cross marks the maximum of the continuum source G12°.8 - 0°.2. The correlation between the kinetic temperature, as traced out by the CO, and the HII region heat source is good but not perfect. This may be in part due to the complex density structure of this source (discussed in Ref. 11 and below.) Another interesting feature of this source is that neither of the OH sources W33A or W33B - the former associated with strong IR emission, but neither having significant radio continuum emission - is associated with enhanced temperature in the CO. Thus, despite the general agreement of different CO transitions which suggests that these optically thick lines do see into the molecular cloud and that the correct value of the kinetic temperature is being obtained, there clearly are subtleties in the temperature structure of giant molecular clouds that require further study.

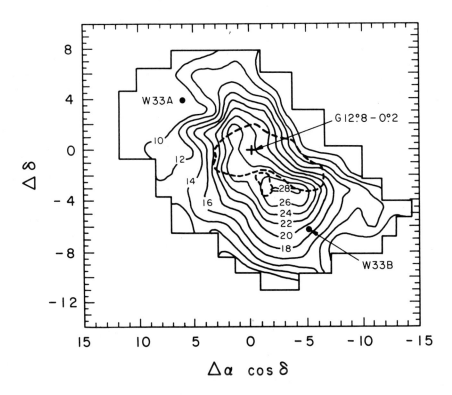

Fig. 3 - Contours of peak corrected antenna temperature for the J = 2 → 1 transition of ^{12}CO in W33. The cross indicates the peak of the radio continuum source G12°.8-0°.2 and the broken curve is the 2K brightness temperature contour of the 8.6 GHz continuum emission. The two OH sources W33A and W33B are also indicated; neither seems to be associated with any ^{12}CO temperature enhancement.

Observation of any ^{12}CO transition is potentially sensitive to the temperature distribution along the line of sight; thus it is relatively surprising that we do not more often see absorption of the relatively intense emission from distant hot sources (such as the gas immediately adjacent to the HII region in W33) by cooler gas lying along the line of sight. While this is surely giving us some information about the velocity field and radiative transfer within the cloud, the absence of "self reversals" or "absorption features" seems harder to understand than those cases where such effects are detected. Submillimeter observations have the ability to reveal cooler foreground gas that may escape detection in the lower transitions. This can be seen from eq. (2); if the foreground gas is cold, but not too cold, it can absorb more strongly in the higher transitions. For example, in thermal equilibrium at 10K, $\tau_{3\to2} \geqslant \tau_{1\to0}$. The behavior shown in Figure 1 for nonequilibrium situations is generally true for lower-kinetic temperatures as well, and thus moderate density cool gas can be seen much more easily in submillimeter than in longer wavelength transitions. This effect is in the spectra of a number of sources studied by Phillips et al. (Ref. 17). In Figure 4 we show profiles of 3 different ^{12}CO transitions from this study taken towards the source DR21. Obviously, there are significant differences in the line shapes, with the self reversal apparent only in the $J = 3 \to 2$ profile. Phillips et al. interpret their data in this and other sources to mean that the intensity of the $J = 1 \to 0$ transition, also, is being affected by the intervening cooler gas even though one does not see such a blatant effect as a clear dip in the center of the line profile. However, this issue is really a complex one depending on the radiative transfer and mass motions within the cloud. At present, the most promising solution to this problem would appear to be the existence of a velocity field with an intermediate correlation length - that is, longer than a photon mean free path but significantly smaller than the size of the cloud (Ref. 18). In this model (depending on the correlation length) the "self-reversals" can persist in the higher-J transitions but be absent in $J = 1 \to 0$. Velocity fluctuations between different portions of the cloud are a natural consequence of an intermediate correlation length and it seems plausible that density variations would be associated with this type of structure - which may be relevant to the question of cloud densities discussed in the following section.

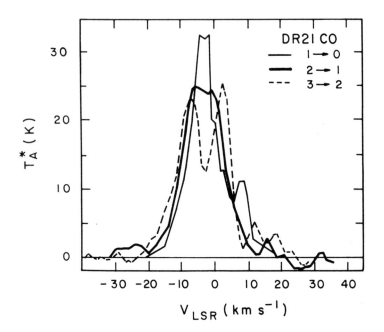

Fig. 4 - Profiles of the three lowest rotational transitions of ^{12}CO in the direction of DR21. From Ref. 17.

Thus while submillimeter observations have not yet definitively answered the question of molecular cloud temperatures, on the whole they have increased the confidence with which we can say that ^{12}CO can be used as a temperature probe of dense regions. Assuming that our ideas of temperature structure are correct leads to major difficulties in understanding the thermal balance of molecular clouds (cf. Ref. 11 and Ref. 19). While it is not possible to go into a discussion of thus issue here, it is possible to be quite optimistic that submillimeter data will help resolve this problem.

III. MOLECULAR CLOUD DENSITY

The density structure of molecular clouds contains critical information about the mass distribution, the dynamical state of these objects, and to a significant extent defines just what a molecular cloud is. A number of different techniques have been used to determine cloud densities, each having particular usefulness and pitfalls.

One technique is to choose a particular molecule or transition which is believed to map out only a certain type of region, e.g. a difficult-to-excite transition can be assumed to be a tracer of regions having a high hydrogen density. While useful, this method is difficult to put on a quantitative basis as densities and abundances are inextricably

interwoven, unless other assumptions are made. The present evidence is
that chemical abundance variations for most species are sufficiently
great in molecular clouds that taking a single molecular transition to
accurately trace the hydrogen density is not a satisfactory procedure.

Closely related to this approach is the idea that the column density
of a particular molecule can be relatively accurately measured; dividing
this by an assumed line of sight extent of the cloud gives a mean volume
density of this species. Then, the assumption of a certain fractional
abundance yields the mean hydrogen density. Since everything is averaged
in this technique certain types of variations will cancel out and the
mean results obtained should be fairly accurate. However, one has lost
sensitivity to variations along the line of sight and under certain conditions even the determination of total column density can have significant problems. An example of this is that for molecules other than CO,
the fractional population in any particular level, as well as the fractional abundance of the molecular species in question relative to
hydrogen, may depend quite sensitively on the total density. Thus, this
type of analysis should not be unquestioningly accepted.

A third technique, which I will discuss in more detail, is to use
multiple transitions of a given molecular species as a direct probe of
the excitation conditions and hence of the hydrogen density. Assuming
that the collisional cross sections are known (which now is the case for
a good number of astronomically interesting molecules, cf. Ref. 20), we
can solve the set of coupled rate equations to determine the level populations and thus the predicted intensity of each transition for a particular hydrogen density and kinetic temperature. It is obviously
advantageous to choose a molecular species for which the optical depth is
small, to minimize the complications of radiative transfer effects. In
Figure 5 we show the results of such a calculation for CO; the column
density has been chosen to keep all transitions optically thin, and might
be characteristic of $C^{18}O$ in a typical cloud. We see that the lower few
transitions are good densitometers for $10^2 \leqslant n(H_2) \leqslant 10^4$ cm^{-3} but lose
sensitivity to changes in density for densities higher than this. Using
submillimeter transitions [λ (J = 6 $\rightarrow$ 5) = 434 μm, for example] sensitivity to densities as high as 10^5 cm^{-3} can be obtained. These limits
come about as the populations of various levels with different A-
coefficients (eq. (1)) are thermalized; the critical density at which
this occurs can be estimated from a simple two-level model balancing
spontaneous decay and collisional excitation. Ignoring stimulated
emission we can solve for the hydrogen density required to obtain a given
excitation temperature

$$n(H_2) = \frac{A/C}{\exp[\frac{h\nu}{k}(\frac{1}{T_{EX}} - \frac{1}{T_{KIN}})] - 1} \qquad (3)$$

where A and C are the downwards radiative and collisional rates. A hydrogen density $n(H_2) \sim A/C$ produces $T_{EX} \sim h\nu/k$. The purpose of observing
pairs of transitions can be thought of in terms of determining the exci-

tation temperature of a pair of levels. This is most easily understood for two successive transitions in a linear molecule. In the optically thin limit, the line intensity of each transition is proportional to the population of its upper level, and thus the ratio of two intensities actually gives the ratio of the populations of the upper and lower levels of the higher frequency transition. In general, the ratio of the antenna temperatures of two optically thin rotational transitions A and B of a linear molecule is given approximately by

$$\frac{T_A(A)}{T_A(B)} = \frac{\nu_A^2}{\nu_B^2} \exp\left[\frac{E_u(A) - E_u(B)}{T_{EX}(A,B)}\right] \quad (4)$$

where $T_{EX}(A,B)$ is the excitation temperature describing the populations of the upper levels of A and B.

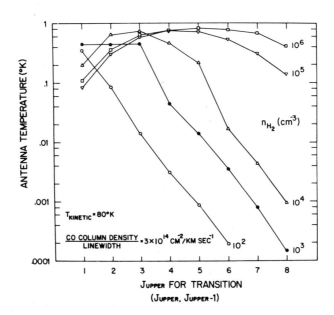

Fig. 5 - Antenna temperatures predicted from statistical equilibrium calculation of CO at 80K kinetic temperature. The low column density of CO ensures that all transitions are optically thin.

The $J = 2 \rightarrow 1$ and $J = 1 \rightarrow 0$ are by far the most accessible transitions in the CO molecule's rotational ladder, with observations of the higher transitions being increasingly hampered by atmospheric absorption and lower receiver sensitivity (although the latter situation is certain to be changing rapidly in the near future.) As can be seen in Figure 5, this pair of transitions can discriminate densities between ~ 10^2 cm^{-3} and 10^4 cm^{-3}. This is illustrated in Figure 6 where the ratio of the two transitions (for optically thin material) is plotted as a function of hydrogen density. The ratios, obtained from complete statistical equilibrium calculations, are in good agreement with the approximate formula from eq. (4):

$$\frac{T_A(J=2\to1)}{T_A(J=1\to0)} = 4 \exp[-11/T_{EX}(2,1)]. \tag{5}$$

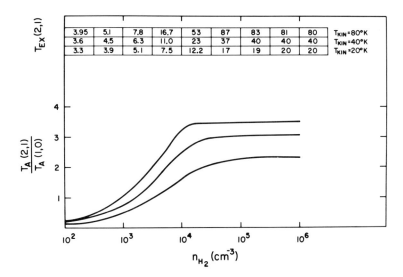

Fig. 6 - Ratio of the antenna temperatures of two lowest CO transitions for optically thin emission at three different kinetic temperatures.

Fully sampled maps of the Orion molecular cloud have been made in the $J = 2 \to 1$ and $J = 1 \to 0$ transitions of ^{13}CO to study the kinematics and density structure of this region (Ref. 21). A great wealth of detail is seen in this data. One view of this is shown in Figure 7, which is a contour map of ^{13}CO $J = 2 \to 1$ emission in the $10.5 - 11.5$ km s^{-1} velocity interval. The regions of strongest emission are in the northern part of the cloud, due to a large scale systematic shift from the central velocity of 9 km s^{-1}, and the enhanced emission lying just outside the optical bar, discussed previously. From either transition alone, it is not clear whether a density, temperature, or column density enhancement is responsible for the increased emission. In Figure 8 we plot the ratio of the $J = 2 \to 1$ and $J = 2 \to 1$ intensities, and we see that only in the region south and west of the BN/KL source at the (0,0) position is there an enhanced value of the intensity ratio.

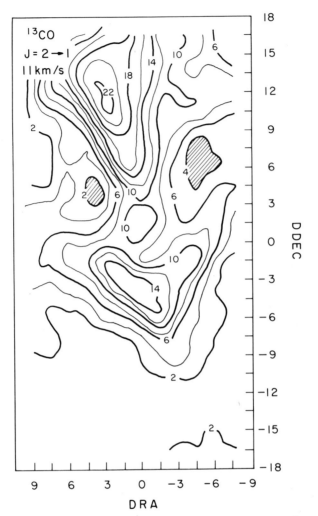

Fig. 7 - Contour map of the Orion molecular cloud in the average intensity of the J = 2 → 1 ^{13}CO transition between 10.5 and 11.5 km s^{-1}. The antenna temperatures have been corrected for atmospheric absorption and a coupling efficiency of 0.7.

We see from Figure 6 that a ratio of 3 implies $n(H_2) \geq 2 \times 10^4$ cm^{-3} and $T_{KIN} \geq 40K$ while a ratio of 3.5 indicates that $T_{KIN} \geq 80K$ with $n(H_2) \geq 10^4$ cm^{-3}. Thus we really cannot say whether the density might not be $\geq 10^5$ cm^{-3}; other density tracers must be used to answer this question. The region of highest temperature ratio actually lies entirely inside the area of strongest emission. This could be indicative of a real density gradient but could also be due to optical depth effects which are preventing us from seeing the high density material.

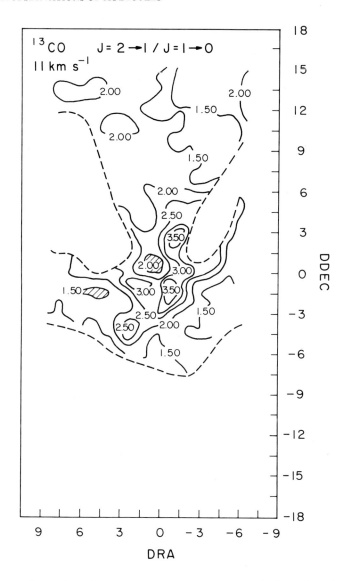

Fig. 8 - Map of ratio of antenna temperatures of $J = 2 \to 1$ and $1 \to 0$ ^{13}CO in Orion. For the area where the data is shown the rms of the fractional uncertainties of the two transitions is less than 0.2; outside the broken lines corresponding to this limit the data is not plotted.

Outside the region surrounding BN/KL the typical $J = 2 \to 1/J = 1 \to 0$ ratio is 2, indicative of $n(H_2) \simeq 6 \times 10^3$ cm^{-3} for $T \simeq 40K$. This is very close to the value 4×10^3 cm^{-3} obtained for the (0,0) position by Plambeck and Williams (Ref. 22). When compared to densities derived from other molecules (see below and Ref. 23) the CO result is quite low. It may be that it applies only to a large, low-density envelope and that we are <u>not</u> seeing the ratio characteristic of the denser core. Since there is general agreement that even ^{13}CO in molecular clouds is optically thick especially in its submillimeter transitions (Refs. 23, 24), we may indeed be missing something, but the temperature data discussed earlier suggests that in ^{12}CO this is not the case. An alternative is again that the density is uneven on a scale small compared to present beamwidths, and that the ^{13}CO emission is dominated by relatively low density material.

The greater spatial extent of molecular clouds in CO compared to that seen in other molecules, and the relative ease of excitation of this molecule reinforce the idea that this species does not trace out the dense portions of these objects. Thus, studies using other molecular species have been undertaken to determine cloud density and structure; some of the tracers that have been used are HCO$^+$, HCN and CS. All of these molecules have relatively large dipole moments and as the spontaneous emission rate scales as the square of the permanent dipole moment, these species are inherently sensitive to the presence of higher density material. Submillimeter observations play a particularly important role here as the higher-J transitions necessary for inter-line comparisons lie in this spectral range.

Several clouds have been studied in the $J = 3 \rightarrow 2$ ($\nu = 267.5$ GHz) and $J = 4 \rightarrow 3$ ($\nu = 356.7$ GHz) transitions of HCO$^+$ (Refs. 25 and 26). The absence of absorption features suggests that the high density regions are surrounded by regions insufficiently dense to excite HCO$^+$ levels above $J = 1$; HCO$^+$ is present in these "envelopes" as indicated by the self- (or foreground) absorption seen in the $J = 1 \rightarrow 0$ transition. The limited signal to noise ratio and differing beam sizes make direct intensity comparisons and density determinations difficult at the present time. HCO$^+$ also is conspicuously strong in the broad lines produced in high velocity flows. Density determinations are obviously important for understanding the dynamics of these events. The presence of HCO$^+$ $J = 3 \rightarrow 2$ emission in the line wings of the flow in NGC2071 has been used to set a minimum density of 10^5 cm^{-3} (Ref. 25). The higher angular resolution of submillimeter instruments presently under construction should allow determination of the density structure within these regions and hence a better understanding of their energy and momentum input to the surrounding molecular clouds.

Studies of the density in molecular cloud cores have increased in sophistication over the past few years. Three features of most work now in progress are first, a recognition of the importance of equal beam sizes for comparison of line intensities: this can be obtained by numerically convolving data to the lowest angular resolution instrument used, or by judicious selection of the instruments themselves. Second, radiative transfer codes, admittedly idealized and restricted in scope, are available for analysis of the data. Third, the general feeling is that many (e.g. more than two) transitions can be used to improve the overall accuracy of the derived densities.

In Figure 9 three maps of the Orion molecular cloud in CS are shown (Ref. 27). Although the spontaneous decay rate for $J = 5 \rightarrow 4$

(ν = 245 GHz) exceeds that for J = 1 → 0 by a factor of 170, the overall sizes of the emitting regions are quite similar - which is surprising in view of the large density, approximately 7×10^6 cm^{-3}, required to bring the excitation temperature of J = 5 → 4 reasonably close to the kinetic temperature (eq. 3). Limited observations of rare isotopic species suggest that optical depths are not large enough to make radiation trapping a major effect in the excitation of CS. The similarity of the line ratios at different positions argues against large scale, smooth density gradients. A best fit to the three lines at a typical position requires $n(H_2) \sim 0.5-5 \times 10^6$ cm^{-3}, where we have assumed that the ^{12}CO gives the kinetic temperature. The mass in the central 5' x 10' (.75 x 1.5 pc) region is then between 10^4 and 10^5 $M_\odot$.

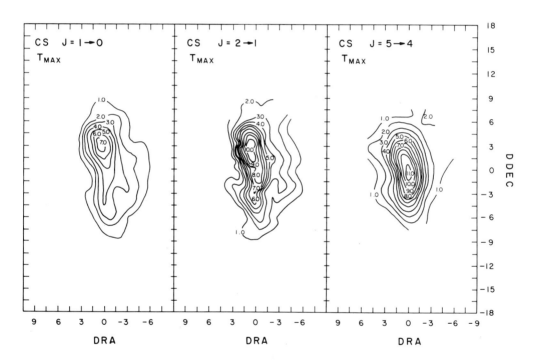

Fig. 9 - Maps of the peak corrected antenna temperatures of three rotational transitions of CS in the Orion molecular cloud. The central position is as in Figure 2, and all of the maps have been convolved to an angular resolution of 2'.1.

It does seem surprising that there is no stronger evidence for large scale density gradients - as indicated by modelling of dust continuum emission (Ref. 28) and correlation of the angular extent different species with different excitation requirements (Ref. 29). It is possible that the analysis of this data in terms of the LVG radiative transfer model is seriously misleading us and the isotopic data is part of the confusion as well.

The three transitions studied in Orion, NGC2264 and DR21 (Ref. 27) themselves suggest that a single density does not satisfactorily describe the emission at a given position and velocity, which is a fundamental aspect of the LVG model. The evidence is that different pairs of transitions give significantly different results - with the $J = 5 \rightarrow 4$ transition systematically indicating higher densities than indicated by the lower pair of transitions. This is reminiscent of the situation we are confronted with when comparing densities derived from different molecules (Ref. 30) but obviously in a more acute form since chemical differentiation cannot be used to escape from the difficulty. Again, significant density inhomogeneities may be present but detailed evidence or theoretical modelling to compare with observations is lacking.

Another study of CS has utilized the transitions shown in Table 1 (Ref. 31), and has resulted in maps of M17, NGC2024, and S140; some results have already been reported (Ref. 32). The spontaneous decay rates cover a range greater than a factor of 20. In fitting four transitions of the common isotopic species, it is clear that in some sources (particularly M17), a single density clearly does not result in a satisfactory fit, but in some sources (S140) it does. The densities from a least squares fit to the four transitions are between 1 and 10×10^5 cm^{-3}, quite similar to those obtained in the CS study discussed previously. It may be restating the obvious, but consistency with a single density model does not prove that this is the correct description of the cloud structure. These results will have to be compared with more detailed models to extract the most meaning from the data. Also, observation of higher - J transitions at shorter wavelengths in the submillimeter region, when receiver sensitivity and telescope technology permit, should improve our ability to determine to what extent density inhomogeneities are present.

Table 1 - Spectral Lines Observed

Species	Transition	Frequency (GHz)	n* (cm^{-3})	E_u (K)	Telescope	Beam Size (arcsec)
C^{34}S	2-1	96.41	5.4×10^5	7	FCRAO 14m	58
CS	2-1	97.98	5.7×10^5	7	FCRAO 14m	62
CS	3-2	147.96	1.5×10^6	14	NRAO 11m	55
C^{34}S	5-4	241.02	6.9×10^6	35	MWO 4.9m	72
CS	5-4	244.94	7.2×10^6	35	MWO 4.9m	76
CS	6-5	293.91	1.2×10^7	49	MWO 4.9m	72

IV. CONCLUSIONS

While the structure of molecular clouds is certainly imperfectly understood, submillimeter data is forcing us to deal with some of the crucial aspects which determine their dynamical state and likely evolution. It is dangerous to generalize but myopic not to attempt to do so. Thus some important results are

(1) submillimeter studies of optically thick lines support the idea that by some means we do "see into" molecular clouds - and detect the effects of highly obscured heating sources. The presence of cooler foreground material sometimes is detectable, but probably rarely significant.

(2) the picture of large-scale systematic motions being the agent that allows this to happen is seriously challenged by the inability to derive consistent densities for different molecules and even for multiple transitions of a single species at a given position within the cloud, and at a particular velocity.

(3) considerable information is available in the line profiles of submillimeter and millimeter lines. This has really not been interpreted adequately to date, due to the complexity and variety of cloud models and radiative transfer processes. It may be that points (1) and (2) can be satisfactorily reconciled in terms of partially turbulent (random) and partially systematic motions.

At present we can be alarmed at the complexity of this problem but I am excited at the prospect of pursuing it and about the potential contribution of submillimeter molecular spectroscopy to its resolution.

I would like to thank my collaborators at FCRAO - Richard Arquilla, Michael Brewer, Robert Dickman, Neal Erickson, and Ronald Snell - for their assistance and stimulation. This research has been supported by NSF grants AST81-21481 and AST82-12252. This is contribution #541 of the Five College Astronomy Department.

REFERENCES

(1) Blitz, L., and Shu, F.H. 1980, Ap.J., 238, 148.
(2) Solomon, P.M., and Sanders, D.B. 1980, in Giant Molecular Clouds in the Galaxy, P. Solomon and M. Edmunds (eds.), Oxford: Pergamon, 41.
(3) Goldsmith, P.F., and Langer W.D. 1978, Ap.J., 222, 881.
(4) Cheung, A.C., Rank, D.M., Townes, C.H., Thornton, D.D., and Welch, W.J. 1968, Phys. Rev. Lett., 21, 1701.
(5) Solomon, P.M., Jefferts, K.B., Penzias, A.A., and Wilson, R.W. 1971, Ap. J., 168, L107.
(6) Linke, R.A., Cummins, S.E., Green, S., and Thaddeus, P. 1982, in Regions of Recent Star Formation, R.S. Roger and P.E. Dewdney (eds.), Dordrecht: Reidel, 391.
(7) Goldsmith, P.F., Krotkov, R., Snell, R.L., Brown, R.D., and Godfrey, P. 1982, submitted to Ap.J.

(8) Walmsley, C.M., and Ungerechts, H. 1982, preprint.
(9) Penzias, A.A. 1975, in Atomic and Molecular Physics and the Interstellar Matter (Les Houches Session XXVI), Amsterdam: North Holland, 374.
(10) Goldsmith, P.F. 1972, Ap.J., 176, 597.
(11) Goldsmith, P.F., and Mao, X., 1983, Ap.J., in press.
(12) White, G.J., Phillips, J.P., and Watt, G.D. 1982, in Regions of Recent Star Formation, R.S. Roger and P.E. Dewdney (eds.), Dordrecht: Reidel, 237.
(13) Phillips, T.G., Huggins, P.J., Wannier, P.G., and Scoville, N.Z. 1979, Ap.J., 231, 720.
(14) Koepf, G.A., Buhl, D., Chin, G., Peck, D.D., Fetterman, H.R., Clifton, B.J., and Tannenwald, P.E. 1982, Ap.J., 260, 584.
(15) Schloerb, F.P., and Loren, R.B. 1982, in Symposium on the Orion Nebula to Honor Henry Draper, A.E. Glassgold, P.J. Huggins, and E.L. Schucking (eds.), New York: New York Acad. Sci., 32.
(16) Bieging, J.H., Pankonin, V., and Smith, F.H. 1978, Astron. Astrophys., 64, 341.
(17) Phillips, T.G., Knapp, G.R., Huggins, P.J., Werner, M.W., Wannier, P.G., Neugebauer, G., and Ennis, D. 1981, Ap.J., 245, 512.
(18) Dickman, R.L., and Kleiner, S.C. 1982, B.A.A.S., 14, 931, and private communication.
(19) Evans, N.J. II, Blair, G., Harvey, P., Israel, F., Peters, W.L. III, Scholtes, M., deGraauw, T., and Vanden Bout, P., 1981, Ap.J., 250, 200.
(20) Green, S., and Chapman, S. 1978, Ap.J.Suppl., 37, 169.
(21) Goldsmith, P.F., Arquilla, R., Schloerb, F.P., and Scoville, N.Z. 1982, in Regions of Recent Star Formation, R.S. Roger and P.E. Dewdney (eds.), Dordrecht: Reidel, 295.
(22) Plambeck, R.L. and Williams, D.R.W. 1979, Ap.J., 227, L43.
(23) Goldsmith, P.F., Plambeck, R.L., and Chiao, R.Y. 1975, Ap.J., 196, L39.
(24) White, G.J., Watt, G.D., Cronin, N.J., and vanVliet, A.H.F., 1979, M.N.R.A.S. 186, 107.
(25) Sandqvist, A., Loren, R.B., Wootten, A., Friberg, P., and Hjalmarson, A. 1982, in Proc. ESA Workshop on The Scientific Importance of Submillimeter Observations, Noordwijkerhout, Netherlands, 131.
(26) Padman, R., Scott, P.F., and Webster, A.S. 1982, M.N.R.A.S. 200, 183.
(27) Brewer, M., Goldsmith, P.F., and Linke, R.A. 1983, in preparation.
(28) Westbrook, W.E., Werner, M.W., Elias, J.H., Gezari, D.Y., Hauser, M.G., Lo, K.Y., and Neugebauer, G. 1976, Ap.J. 209, 94.
(29) Goldsmith, P.F., Langer, W.D., Schloerb, F.P., and Scoville, N.Z., 1980, Ap.J., 240, 524.
(30) Evans, N.J. II 1980, in Proc. IAU Symposium 87, Interstellar Molecules, Dordrecht: Reidel, 1.
(31) Snell, R.L., Mundy, L.G., Goldsmith, P.F., Evans, N.J. II, and Erickson, N.R. 1983, in preparation.
(32) Vanden Bout, P.A., 1982, in Proc. ESA Workshop on The Scientific Importance of Submillimeter Observations, Noordwijkerhout, Netherlands, 109.

ENERGETIC OUTFLOWS, WINDS AND JETS AROUND YOUNG STARS

Charles J. Lada

Steward Observatory, University of Arizona

1. INTRODUCTION

During the last decade the development of millimeter-wave and infrared astronomical spectroscopy has enabled astronomers to explore and directly probe the cold component of the galaxy, a component which contains 50-90% of the total mass of the interstellar medium and is rich in astrophysical complexity, activity and mystery. The result of this exploration has been a revolution in our understanding of star formation and the interstellar medium. Millimeter-wave CO observations, for example, led to the recognition of the giant molecular clouds (GMCs). With sizes often in excess of 100 parsecs and masses between 10^5 and 10^6 $M_\odot$, the GMCs are the largest and most massive structures in the Milky Way. Studies of these clouds have helped elucidate spiral structure in our galaxy and refined, through improvements in the rotation curve, the total mass of the Milky Way system (Blitz 1979, Bok 1983). Moreover, infrared observations have shown GMCs to be the primary sites of star birth in our galaxy. Indeed, molecular-line observations of these clouds have given us the first detailed glimpses into the physical processes associated with the earliest stages of stellar evolution and perhaps star formation itself.

In this paper I am going to discuss a recent and exciting development of such molecular cloud research: the recognition and study of an unexpected, and important stage of early stellar evolution. Millimeter-wave CO observations have revealed the existence of massive and energetic outflows of cold molecular gas emanating from very young stars. The discovery of such energetic and massive outflows in star formation regions appears to unite several seemingly unrelated astrophysical phenomena such as water masers, Herbig-Haro Objects, and molecular hydrogen emission regions as manifestations of the same underlying physical process: a new stage of early stellar evolution characterized by energetic mass loss. Detailed observations of these energetic outflows are not only contributing important information to our understanding of early stellar evolution and star formation but they are also providing us with unique opportunities to investigate the physics of supersonic gas flows in space as well as the formation of interstellar jets. In the remainder of this paper I will review the important observational characteristics of this phenomena and discuss the astrophysical implications as presently perceived.

2. THE HIGH VELOCITY MOLECULAR FLOWS (HVMFs)

2.1 Hypersonic Outflow

The dynamical state of the gas in molecular clouds can be investigated by study of molecular line profiles observed toward these regions. CO and occasionally NH_3 observations indicate kinetic temperatures of about 10 K for most material in a GMC. In regions of active star formation, usually a small fraction of area of a GMC, temperatures can be considerably higher, 20-65 K. The sound speed in such molecular gas is only a few tenths of a kilometer per second. Yet, observed linewidths from such regions are typically between 1-3 km s^{-1} (full width at half-power). In regions of very active star formation such linewidths can reach values as high as 6-8 km s^{-1}. Consequently the dynamical state of molecular gas in GMCs is one characterized by supersonic bulk motion. Since the lifetimes of such clouds (i.e. > 3×10^7 years, Blitz and Shu 1980) are considerably greater than the corresponding free fall time scales (~ 3×10^6 yrs, for n_{H_2} = 200 cm^{-3}) the supersonic velocity fields of GMCs are probably not due to global collapse of the clouds. The large linewidths are most likely the result of turbulence. Maintaining such a supersonic turbulent state for a GMC would be a difficult proposition if the turbulence had to be generated in a free-fall time. The total kinetic energy of a 10^5 M$_\odot$ GMC is about 3×10^{48} ergs and would have to be dissipated and regenerated in a few million years. This has been a long-standing problem for molecular cloud studies. However, since the Alfven speed in GMCs could be comparable to the observed linewidths, turbulent dissipation times may be considerably longer than a free-fall time (Arons and Max 1975) and perhaps even approach the cloud lifetime. However in the absence of any information on the nature of magnetic fields in GMCs, the generation of supersonic turbulence must still be considered a major problem.

In addition to the supersonic velocity fields which characterize an entire GMC, bulk gas motion greatly in excess of the general supersonic turbulence has been observed toward many localized regions of star formation (e.g. Bally and Lada 1983, Rodriguez et al. 1982). Invariably centered on embedded young stellar objects these hypersonic velocity fields are confined to regions between 0.1 and 3 parsecs in size. These high velocity molecular flows (HVMFs) are identified by intense emission wings on molecular line profiles. The full velocity extent of such high velocity emission is typically between 20 km s^{-1} and 100 km s^{-1}! Figure 1 shows the ^{12}CO J = 1→0 spectrum observed toward the Orion A molecular cloud. Clearly, ^{12}CO emission is observed over 100 km s^{-1} in velocity extent. In a recent survey of young stellar objects Bally and Lada (1983) found 27 sources with localized hypersonic vel-

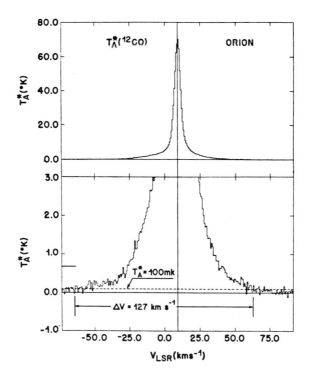

Figure 1. ^{12}CO (1-0) spectrum of Orion showing broad emission wings (from Bally and Lada 1983).

ocity fields characterized by linewidths of 20 km s^{-1} or greater. A summary of these results is shown in Figure 2.

Maps of more than a dozen of these HVMFs have been made and all but one flow (V645 Cyg) has been resolved, enabling a determination of flow extents (i.e., 0.1-3.0 pc). These relatively large extents coupled with the high velocities of the hypersonic flows suggests strongly that they are not gravitationally confined (Lada and Harvey 1981). For a typical source a central mass of order $10^4 M_\odot$ would be required to gravitationally confine the flow, yet the total mass of these central regions is observed to be only about 100 $M_\odot$. Therefore, the HVMFs represent high velocity <u>outflows</u> of cold molecular material from around young stellar objects. The most convincing and direct evidence for the outflow interpretation is found for the Orion and L1551 flows where proper motions measured for H$_2$O masers (Orion) and Herbig-Haro objects (L1551) associated with the flows unambigously indicate outflow motions (Genzel <u>et al.</u> 1981, Cudworth and Herbig 1980). Asymmetries in ^{12}CO emission profiles and association of many HVMFs with hot molecular hydrogen also support an outflow interpretation as discussed in detail by Bally and Lada (1983).

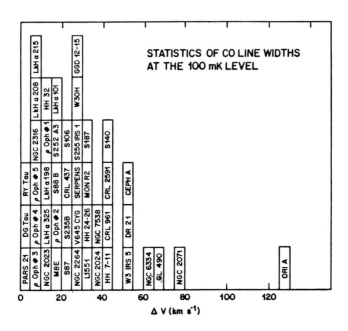

Figure 2. Histogram of sources studied in the Bally and Lada (1983) survey of young stellar objects. The number of sources is plotted against the width of ^{12}CO emission, (at the 100 mK level) observed toward each source with the 7m Bell Labs antenna. Sources with $\Delta V > 20$ km s^{-1} are almost certainly outflow sources.

2.2 A New and Energetic Stage of Stellar Evolution

Determination of the size and velocity extent of the HVMFs can be used to determine the dynamical time scales (i.e. $\tau_d = R/V$) of the flows. These range between 10^3 and 10^5 years (Bally and Lada 1983), but for most sources, $\tau_d \approx 10^4$ years. As first discussed by Lada and Harvey (1981) these short dynamical time scales lead to a surprisingly high local formation rate of HVMFs (i.e. 3×10^{-4} yr^{-1} Kpc^{-2}). The total galactic rate of formation is about 0.1 yr^{-1} (Bally and Lada 1983). The birth rate of such flows is comparable to the formation rate of massive stars, and may be much higher than estimated since an exhaustive search for such objects has not been made. Many more certainly remain to be found. The association of many HVMFs with low mass stars (e.g. Snell and Edwards 1981, Edwards and Snell 1982, Frerking and Langer 1982) also suggests a higher birthrate, a birthrate comparable to the total star formation rate. Their high formation rate and association with newly formed stars implies that HVMFs are a consequence of star formation and represent a hitherto unrecognized stage of early stellar evolution.

The astrophysical importance of this phase of early stellar evolution is only fully appreciated when one considers the masses and energetics of the phenomenon. Molecular spectroscopy is a powerful tool for not only determining the velocity structure of these flows but also for determining abundances and masses of the swiftly moving material. Observation of the J = 1-0 ^{13}CO transition and the J = 2-1 ^{12}CO transition toward many HVMFs indicate that the ^{12}CO emission is optically thick (Lada and Harvey 1981; Bally 1982; Ho, Rodriguez and Moran 1983; Lada 1983; Plambeck, Snell and Loren 1983). ^{13}CO emission is, however, optically thin and can be used to estimate fairly reliable gas column densities and masses (Lada 1983). To determine a reliable flow mass the spatial extent of the flow as well as the gas column density must be well determined. Therefore, masses are only determined for the dozen or so HVMFs which have been well mapped. Self-consistent mass determinations have been made for all the well studied objects (Lada 1983), and masses in the range between 2 and 70 $M_\odot$ have been determined for the various sources. These large masses indicate that the HVMFs represent very energetic outflows of cold gas. The momenta and kinetic energies of these flows are enormous. For example, the kinetic energies of the three most energetic flows (Mon R2, S140 and NGC 2071) exceed 2×10^{47} ergs. This is nearly 0.1% of the energy of a supernova. However, the range in energetics of the flows studied to date is large. For example, the flow associated with T Tauri has a kinetic energy estimated to be approximately 4×10^{43} ergs (Edwards and Snell 1982), a thousand times less than the NGC 2071 flow.

Understanding the energetics of the HVMFs is a critical step toward understanding their origin and nature and their significance for both the evolution of molecular clouds and young stars. The observed quantities of mass (M), velocity (V) and size (R) can be used to estimate the force required to drive a flow, $P = \dot{M}V = MV/\tau_d = MV^2/R$ (momentum supply rate), and the mechanical luminosity of the flow, $L_{HVMF} = E_K = 1/2 \, MV^2/\tau_d = MV^3/2R$ (energy supply rate), if we assume that the flows have been steady over their dynamical lifetime. The range in $\dot{M}V$ and L_{HVMF} for the observed flows is a few orders of magnitude and is much greater than can be accounted for by any uncertainties in determinations of M, V or R. The differences in the energetics therefore represent differences in the energetics of the driving engines

In Figure 3 the mechanical energies of the flows are plotted against the radiant luminosities (L_*) observed for the young stellar object or objects at the centers of the outflows. A correlation appears to exist between the two quantities (L_{HVMF} and L_*) when all the flows are considered (Lada 1983, Rodriguez et al. 1982, Bally and Lada 1983). The energetics of a molecular outflow therefore is related to either the mass or luminosity of the central star, as might be expected if the central stars are the source of energy for driving the flows. The relation $L_{HVMF} = L_*$ is also plotted in Figure 3. That all flows fall below this line suggests that over their dynamical lifetimes, the central sources have radiated considerably more luminous energy than is now present as kinetic energy in the outflow. In other words, the

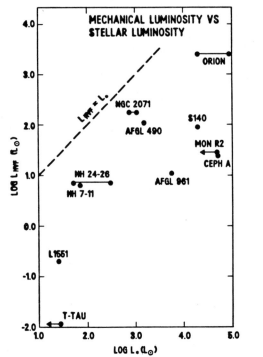
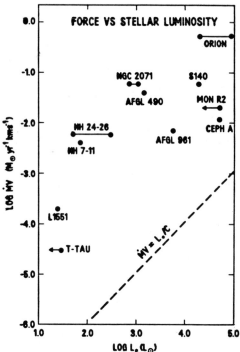

Figure 3. Mechanical luminosity of molecular outflows plotted against the total radiant luminosity of embedded central objects. All sources fall below the relation $L_{HMVF} = L_*$, although a few sources are very close to that line (Bally and Lada 1983).

Figure 4. Force needed to accelerate molecular outflows plotted against the bolometric luminosity of the central object (Bally and Lada 1983).

energy in the radiation field is sufficient to drive the flows if they have been steady over their dynamic lifetimes. However, the relatively high values of the ratio of mechanical to radiant luminosity (i.e. 0.2-0.002) indicates an extremely efficient conversion of radiant to mechanical energy if the energy source for the molecular outflows is the radiation fields of the central objects.

To investigate further the possibility of a radiation driven engine, let us consider the relation between $\dot{M}V$ and L_* for the flows. This is plotted in Figure 4. Again, there is a correlation between the force needed to drive the flows and the central source luminosity. However, all the sources lie well above the $\dot{M}V = L_*/C$ relation. The quantity L_*/C is the force that radiation pressure could supply to the flow if every photon from the central source were completely absorbed or scattered at least once before escaping. Since for all the flows $\dot{M}V/L_*/C \sim 100-1000$, radiation pressure would not be effective unless each photon was scattered a hundred times or more. This might be possible in some cases if the material experienced all its acceleration very close to an embedded central star where dust opacities, even in the near-infrared could be high (Phillips and Beckman 1980). Whether such conditions actually exist in the circumstellar regions around the central objects remains to be determined. For T-Tauri, a visible star, such a situation is unlikely.

In summary, molecular-line observations suggests that the HVMFs are very massive outflows of cold material from young stellar objects. The energies of these flows are substantial. The correlation of mechanical luminosity and force required to drive the flows with stellar radiant luminosity suggests that energetics of the outflow are determined in some way by the mass or luminosity of the central objects. That the correlations exist over many decades in source luminosity perhaps suggest that the physics underlying the phenomenon is similar for all sources. Apparently radiation pressure is not responsible for driving the flows. If the origin of these massive outflows is an energetic wind from near the surface of the embedded objects, then infrared spectroscopy with high spatial and velocity resolution may provide a most unique and powerful tool to directly observe the physical conditions at the heart of driving engine.

3. OTHER MANIFESTATIONS OF MASS OUTFLOW FROM YOUNG STARS

3.1 T-Tauri Stars and Steady Mass Loss

The idea that at least some stars go through a phase of mass-loss during their formative years is not new. Optical observations of T-Tauri stars have suggested that these young solar-mass stars may experience episodes of both steady mass-loss (e.g. Herbig 1962, Kuhi 1964) and eruptive mass-loss (i.e. Herbig 1977). T-Tauri stars are emission-line, variable stars which appear to be contracting toward the main sequence. Analysis of high resolution $H\alpha$ emission line profiles from T-Tauri stars often reveals broad, asymmetric, double-peaked lines. In about 60% of all T-Tauri stars, there is double-peaked structure which is due to an absorption feature on the blue side of the $H\alpha$ emission line profile (Kuhi 1978). Figure 5 shows such a typical spectrum for the T-Tauri star AS 205. Both the large blue displacement of the absorption feature (-162 km s^{-1}) and the large velocity width of the resolved profile are suggestive of mass outflow from this particular star. Similar structures in line profiles in other T-Tauri stars has been interpreted as evidence of early episodes of steady mass loss from these objects (e.g. Herbig 1962, Kuhi 1964, DeCampli 1981). Indeed, 5% of all T-Tauri stars exhibit classical P-Cygni profiles (where the blueward absorption feature dips below the continuum) a circumstance which is often regarded as unambiguous evidence for mass outflow (Kuhi 1978).

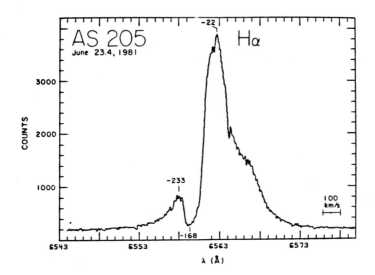

Figure 5. High resolution Hα spectrum of AS 205 obtained by R. Mundt (1983) with the MMT.

Estimating the magnitude of the mass loss and the energetics of the inferred T-Tauri winds, however, is exceedingly difficult. In many stars the half-widths of the Hα profiles and the magnitude of the blueward displacement of the absorption dips are less than the surface escape speed of the underlying star. Indeed, about 5% of all T-Tauri stars exhibit inverse P-Cygni profiles suggesting mass infall, with some stars showing evidence for both infall and outflow at different times (Kuhi 1978, Ulrich 1978). Consequently, determining the terminal velocities of T-Tauri winds is a difficult task. In addition, using emission line luminosities to estimate masses of such winds can result in uncertainties of three orders of magnitude in mass-loss rates from typical T-Tauri stars (DeCampli 1981). Although emission-line studies of T-Tauri stars have long suggested the existence of stellar winds from young stars, determining the energetics and therefore significance of such winds for early stellar evolution from these observations has proved difficult.

Additional evidence of winds from T-Tauri stars has recently been found from observation of the centimeter-wave radio continuum spectra of ultracompact H II regions around a number of T-Tauri stars (Cohen, Bieging and Schwartz 1982). The radio emission spectra are consistent with electron densities decreasing away from the star with the inverse square of the distance. This density law is what one would expect from a uniform spherical outflow of an ionized wind (Wright and Barlow 1975). Furthermore, the available ultraviolet radiation from T-Tauri stars is not believed to be sufficient to produce such compact H II regions via photo-ionization. Consequently the presence of such ionized gas is thought to be the result of collisional ionization which could occur in a stellar wind (e.g. Krolik and Smith 1981). Limits to the mass-loss rates can be derived from radio observations. These limits are generally consistent with upper limits to such rates derived from optical data (i.e. ~ 10^{-7} $M_\odot$ yr^{-1}). Observations of molecular flows around young stars probably give the most accurate estimates of wind energetics obtainable. The total kinetic energies and momenta derived from molecular-line data place strong constraints on the total energy and momentum input into the surrounding gas during these episodes of early mass loss. In many sources, particularly for more massive stars, mass loss rates inferred from CO observations greatly exceed that derived from radio data. This suggests that mass loss from T-Tauri stars has not been steady and most sources are presently relatively quiet (Lada and Gautier 1982, Bally and Lada 1983, Bally and Predmore 1983).

3.2 FU Ori Stars and Eruptive Mass-Loss

That young stars experience episodes of non-steady, eruptive outbursts is strongly suggested by the FU Ori phenonemon. In a six-month period between 1936 and 1937 the star FU Ori flared, increasing its brightness by more than 5 visual magnitudes. It has remained near its elevated brightness through the present epoch. Since FU Ori flared, two other stars, V1057 Cyg and V1515 Cyg, have also been observed to experience similar eruptive outbursts. Figure 6 shows the light curve for V1057 Cyg. Figure 7 shows a high resolution spectrum of V1057 Cyg recently obtained by R. Mundt with the MMT. Multiple, blue-shifted

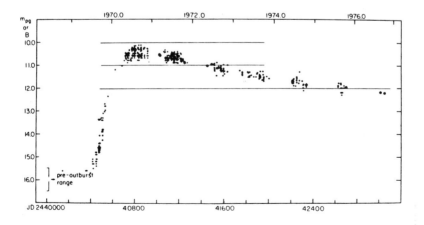

Figure 6. An example of FU Ori outburst: the light curve for V1057 Cyg (from Herbig 1977).

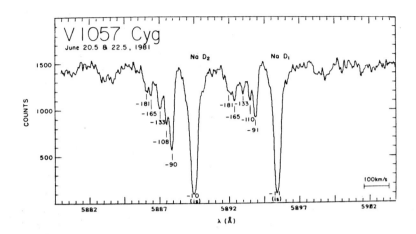

Figure 7. High resolution spectrum of V1057 Cyg obtained by R. Mundt (1983) with the MMT.

components of the sodium lines are present, suggesting that mass ejection has accompanied the outburst of light. Spectra obtained of V1057 Cyg before its outburst indicated that it was a T-Tauri type star, providing the first strong evidence that FU Ori type outbursts were associated with early stellar evolution. The observation of three such outbursts in 80 years suggests that T-Tauri stars experience multiple FU Ori type outbursts during their early evolution (Herbig 1977). The mechanism responsible for these eruptions is unknown and the energetics of the ejected gas are not well determined, consequently the relationship of this phenonemon to T-Tauri winds and massive molecular outflows is not certain.

3.3 Herbig-Haro Objects

The recent recognition of Herbig-Haro objects as manifestations of strong stellar winds has provided important additional evidence for energetic mass loss from young stars (e.g. Schwartz 1975). Herbig-Haro objects are semistellar emission nebulae with spectra containing strong low excitation emission lines such as [OI], [SII], [NI], and [FeII]. Originally thought to be protostellar condensations, it is now obvious that their spectra are best explained by shock excitation resulting from the interaction of an energetic stellar wind with ambient molecular gas (Schwartz 1975, Raymond 1979, Dopita 1978). Observations of Herbig-Haro radial velocities also suggest an association with mass outflow. Probably the most convincing evidence that Herbig-Haro objects are associated with mass outflow derives from their proper-motion measurements. Large proper motions have been detected for H-H objects in L1551 and HH 1-2 in Orion (Cudworth and Herbig 1979, Herbig and Jones 1981). Large tangential velocities have been detected (200 -350 km s^{-1}) suggesting a very energetic ejection of material from nearby young stellar objects. Most Herbig-Haro objects have radial velocities that are significantly blue-shifted with respect to ambient molecular gas in which they are embedded (Stom, Grasdalen and Strom 1974; Canto 1981). Since Herbig-Haro objects are optical emission knots in molecular clouds, extinction probably prevents us from readily observing Herbig-Haro objects which embedded stars ejected away from us (i.e. deeper into the cloud). We tend to see those objects near the cloud's surface and most of those are likely to be directed toward us.

Clearly the velocities of the winds needed to accelerate H-H objects must be high, 100-400 km s^{-1}! The mass-loss rates needed to provide sufficient excitation of the emission lines are between 10^{-5} and 10^{-6} M$_\odot$ yr^{-1} (Schwartz 1975) if the winds are isotropic. Although comparable to the rates inferred for molecular flows, these wind parameters for Herbig-Haro objects may be overestimates since Herbig-Haro objects may be accelerated by anisotropic winds (see Section 4.).

3.4 H_2O Masers

Evidence of mass loss from higher mass stars is suggested by observations of H_2O maser sources at centimeter wavelengths. Approximately 70% of all luminous H_2O maser sources have high velocity (i.e. V > 10 km s^{-1}) components (Jaffe, Güsten and Downes 1981). These features can be explained as an interaction of ambient gas with strong stellar winds (Strelnitsky and Sunyaev 1972, Norman and Silk 1979, Rodriguez et al. 1980). Indeed, H_2O maser sources and Herbig-Haro objects may be very similar objects observed at different stages of evolution in an intense stellar wind (Norman and Silk 1979). As with the Herbig-Haro objects, the most conclusive demonstration of the relationship of H_2O masers to outflows from young stars derives from proper motion measurements. VLBI proper motion measurements of the maser sources associated with the energetic molecular outflow in Orion, have detected a directed expansion of maser sources away from the infrared source IRC2 (Genzel et al. 1981). Accurate determinations of the physical parameters of maser knots cannot be easily made and deriving wind parameters from maser source observations alone is very uncertain. The coincidence of the maser outflow and the molecular outflow in Orion, however, indicates that the two outflow phenomena are related and probably driven by the same underlying mechanism.

4. BIPOLAR FLOWS AND INTERSTELLAR JETS

Perhaps the most remarkable and intriguing property of the high velocity molecular flows is their tendency to appear bipolar (e.g. Lada and Harvey 1981, Bally and Lada 1983). More than a dozen molecular flows have been resolved and mapped with existing millimeter-wave telescopes. With only a few exceptions such data show molecular outflow sources to be bipolar in nature. That is, they often consist of two spatially separate lobes of emission, with one lobe containing predominantly blue-shifted gas and the other predominantly red-shifted gas. Furthermore, the two separating lobes are more or less symmetrically situated about an embedded infrared source or young star.

The prototypical bipolar molecular flow is in L1551 (Snell, Plambeck and Loren 1980), and is shown in Figure 8. This flow contains

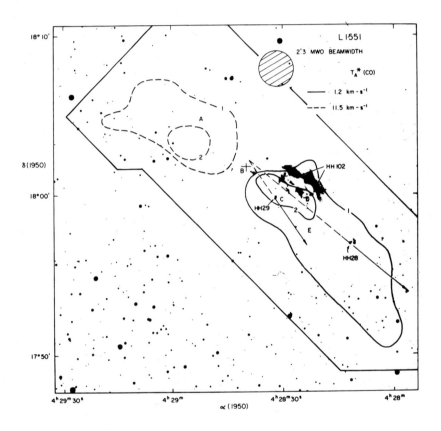

Figure 8. Map of high velocity CO emission from L1551 obtained by Snell, Plambeck and Loren (1980).

conspicuous Herbig-Haro objects associated with the blue-shifted lobe of CO emission. Not surprisingly, the radial velocities of the H-H objects are large and blue-shifted with respect to the molecular gas (Strom, Grasdalen and Strom 1974). In addition two of the Herbig-Haro objects have large proper motions (Cudworth and Herbig 1979). When these motions are projected back in time, their vectors intersect very near an embedded infrared source at the center of the bipolar CO flow.

Clearly gas is being ejected from the central star in two well-collimated, oppositely directed streams or jets. In fact, close inspection of deep photographs of the L1551 area reveal a very collimated optical jet extending from the vicinity of the infrared source to the blue-shifted CO lobe (e.g. Konigl 1982, Mundt 1983). In addition, high resolution VLA maps of the ionized gas show that radio continuum emission within a few arcseconds of the infrared source is elongated along the same jet axis (Cohen, Bieging and Schwartz 1982). This suggests that the collimation of the outflow is produced in the vicinity of the central object driving the flow. It also reinforces the idea that the physical process which produces the Herbig-Haro objects also produces the cold molecular outflows.

Collimated, bipolar gas ejection is not only an apparent characteristic of molecular flows, but is also observed in numerous Herbig-Haro objects. For example, proper motion measurements of H-H 1 and 2 show them to be rapidly moving away from each other along a well defined axis which contains a young T-Tauri star, suspected of driving the flow (Herbig and Jones 1981). Recent observations of the T-Tauri star AS 353 has revealed two Herbig-Haro objects on opposite sides of the star moving in opposite directions along the line of sight at speeds of about 400 km s^{-1} relative to the star (Mundt, Stocke and Stockman 1983). A narrow bridge of emission line radiation appears to connect the blue-shifted H-H object with the stellar image of AS 353, suggesting a very highly collimated jet-like flow emanating from very near the surface of the star.

Probably the most striking and extremely collimated jet-like flow known is associated with the Herbig-Haro objects HH 46 and 47 in the Gum Nebula (Dopita, Schwartz and Evans 1982). Figure 9 shows a photograph of this interesting object taken by Bart Bok in 1978. The H-H objects are knots in the narrow jet of material which appears to have been ejected from the dark globule. A fainter H-H knot (HH 47C) has been discovered about two arcminutes southwest of HH 46-47. The new H-H object falls almost exactly on the axis of the HH 46-47 jet but on the opposite side of the globule. Measured radial velocities of these objects obtained by Graham and Elias (1983) are shown in Figure 10. The HH 46-47 stream is blueshifted with velocities around -150 km s^{-1} while HH 47C on the opposite end of the globule is redshifted with a velocity of around 100 km s^{-1}! Apparently an object embedded in the globule has produced a highly collimated bipolar jet of gas which has pierced through the cloud. The blueshifted jet of gas is clearly observed at the near edge of the cloud while the receding jet of gas is almost entirely obscured by the globule. A possible candidate for the central source driving the jets has been found by Graham and Elias (1983) near HH 46 and is probably a heavily obscured star of low luminosity. Other known Herbig-Haro objects (e.g. HH 7-11 and HH 12) also appear to be aligned in linear strings and may be similarly highly collimated jets observed at the surfaces of other molecular clouds. The porponderance of blueshifted radial velocities of H-H objects is consistent with such an idea. These other objects may also be bipolar

Figure 9. Deep photograph of a globule with jet-like Herbig-Haro objects taken by Bart Bok on Valentine's night in 1978.

in nature with the receeding material buried in molecular gas and dust and obscured from view. Certainly the association of HH 7-11 with the blueshifted lobe of a bipolar <u>molecular flow</u> would support that point

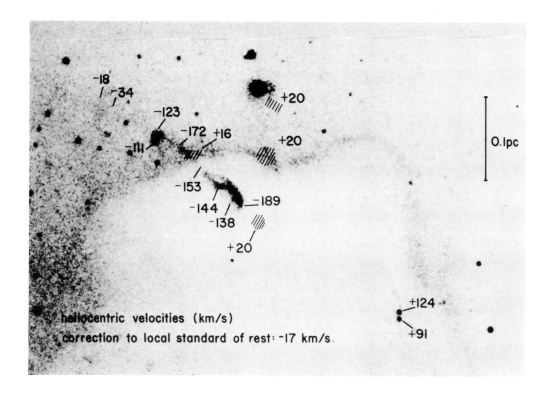

Figure 10. Velocities obtained by J. Graham and J. Elias (1983) along the Herbig-Haro jet shown in Figure 9.

of view (Snell and Edwards 1981). Deep optical imaging might reveal the redshifted counterparts if a bipolar ejection characterizes these and other Herbig-Haro objects.

Unlike the Herbig-Haro flows, most cold molecular flows, though bipolar, are not very well collimated (Bally and Lada 1983). Indeed, at least one source, S140, shows no evidence of bipolar spatial structure (Lada and Wolf 1983). This is perhaps a very fortunate circumstance since it indicates that the mechanism which produces bipolar streams and collimated jets does not always operate for molecular flows. Consequently, comparing the physical conditions in and around highly collimated flows with those of poorly collimated flows may lead to an understanding of the mechanism which gives rise to jet formation.

The poor collimation of many molecular flows has led to suggestions that the collimation and bipolarity are produced at large distances from the central object due to large-scale density inhomogeneities in the ambient molecular clouds (Ho, Rodriguez and Moran 1983). Torrelles et al. (1983) have suggested that bipolar outflows are focused by large interstellar toroids (tenths of parsecs in size) and recent interferometer observations of molecular SO emission toward Orion support this idea (Plambeck et al. 1982). Since the large scale density structure is expected to vary from cloud to cloud, a large variation in collimation may be expected for molecular flows in this model.

However, observations of the Herbig-Haro objects discussed earlier suggest that highly collimated flows and jets are formed very close to the surface of the stars driving the flows. The lack of many well collimated molecular flows could be the result of one of two possibilities. First, the degree of collimation is related to the mass of the underlying central object. Low-mass objects have much more collimated flows than high mass sources. (Most known molecular flows occur around higher mass stars, while H-H objects are a phenomenon associated with low mass stars.) Perhaps the large energies in winds from luminous stars (Bally and Lada 1983) tend to quench the jet formation mechanism. Yet there are some luminous stars with very collimated molecular flows (e.g. NGC 2071). A second possible explanation for poorly collimated flows proceeds as follows: near the surface of the driving stars the flows are focused into supersonic jets which appear as Herbig-Haro objects, but when the jets interact with the surrounding inhomogeneous ambient molecular cloud they become defocused. The more inhomogeneous the cloud, the less collimated the molecular flows appear. I think the observations of the H-H 7-11 outflow support this kind of picture. In that flow the H-H objects appear highly collimated while the large-scale molecular flow does not. Moreover, the morphology of the molecular lobes is correlated with the distribution of dense ambient gas in the region (Schwartz, Waak and Smith 1983). The real solution to this problem may contain elements of both possibilities however, and more detailed analyses of outflow-cloud interaction will be required to critically test these ideas. Studies of the interaction between flows and ambient molecular gas could also shed some light on the long standing problem of giant molecular cloud longevity. Because of their frequency and energetics, molecular outflows may represent a significant source of internal energy for molecular clouds. Whether or not such flows could support molecular clouds is at this time very uncertain (e.g. Bally and Lada 1983).

The unexpected discovery of bipolar outflows and, in particular, collimated jets around young stellar objects is an exciting and important development for astrophysics. It suggests that supersonic jet formation is a widespread phenomenon in the universe. Could the underlying physical mechanism which produces pre-main-sequence jets be similar to that which produces jets in radio galaxies? Konigl (1981) has proposed a unified interpretation of molecular flows, Herbig-Haro objects and other observed manifestations of outflow in the context of a jet formation model which is very similar to models originally developed for radio galaxy jet formation. What is truly exciting is the prospect for detailed observational tests of such models. The proximity of young stellar jets enables us to study them with a wide range of observational techniques in detail which is not possible for extragalactic objects. Future study of young stellar outflows will undoubtably be a virgorous and productive astronomical endeavor, and optical and infrared spectroscopy will play a critical role in deciphering the nature of this most fascinating phenomenon.

ACKNOWLEDGMENTS

I thank Reinhardt Mundt and John Graham for generously providing data in advance of publication and Chris McKee for enlightening discussions. This work is supported, in part, by the Alfred P. Sloan Foundation.

REFERENCES

Arons, J. and Max, C. E.: 1975, Astrophys. J., 196, L77.
Bally, J.: 1982, Astrophys. J., 261, pp. 558.
Bally J. and Predmore, R.: 1983, Astrophys. J., 265, pp. 778.
Bally, J. and Lada, C. J. 1983, Astrophys. J., 265, pp. 824.
Blitz, L.: 1979, Astrophys. J., 231, L115.
Blitz, L. and Shu, F.: 1980, Astrophys. J., 238, pp. 148.
Bok, B. J.: 1983, Astrophys. J., in press.
Canto J.: 1981, in "Investigating the Universe", ed. F. D. Kahn, pp. 95, Dordrecht: D. Reidel Publ. Co.
Cohen, M., Bieging, J. H. and Schwartz, P. R.: 1982, Astrophys. J., 253, pp. 707.
Cudworth, K. M. and Herbig, G. H.: 1979, Astron. J., 84, pp. 548.
DeCampli, W.: 1981, Astrophys. J., 244, pp. 124.
Dopita, M. A.: 1978, Astrophys. J. Suppl., 37, pp. 117.
Dopita, M. A., Schwartz, R. P. and Evans, I.: 1982, Astrophys. J., 263, L73.
Edwards, S. and Snell, R. L.: 1982, Astrophys. J., 261, pp. 151.
Frerking, M. A. and Langer, W. D.: 1982, Astrophys. J., 256, pp. 523.
Genzel, R., Reid, M. J., Moran, J. M. and Downes, D.: 1981, Astrophys. J., 244, pp. 884.
Graham, J. A. and Elias, J. H.: 1983, preprint.
Herbig, G. H.: 1962, Adv. Astr. Ap., 1, pp. 47.
Herbig, G. H.: 1977, Astrophys. J., 217, pp. 693.
Herbig, G. H. and Jones, B. F.: 1981, Astron. J., 86, pp. 1232.
Ho, P. T. P., Rodriguez, L. F. and Moran, J. M.: 1983, Astrophys. J., in press.
Jaffe, D. T., Gusten, R. and Downes, D.: 1981, Astrophys. J., 250, pp. 621.
Konigl, A. 1982, Astrophys. J., 261, pp. 115.
Krolik, J. H. and Smith, H. A.: 1981, Astrophys. J., 249, pp. 628.
Kuhi, L. V.: 1964, Astrophys. J., 140, pp. 1409.
Kuhi, L. V.: 1978, in "Protostars and Planets", ed. T. Gehrels, pp. 708. Tucson: The University of Arizona Press.
Lada, C. J.: 1983, in preparation.
Lada, C. J. and Gautier, N.: 1982, Astrophys. J., 261, pp. 161.

Lada, C. J. and Harvey, P. M.: 1981, Astrophys. J., 245, pp. 58.
Lada, C. J. and Wolf, G.: 1983, in preparation.
Mundt, R.: 1983, unpublished data.
Mundt, R., Stocke, J. and Stockman, H. S.: 1983, Astrophys. J., 263, L73.
Norman, C. A. and Silk, J.: 1979, Astrophys. J., 228, pp. 197.
Phillips, T. G. and Beckman, J.: 1980, Mon. Not. R. Astr. Soc., 193, pp. 245.
Plambeck, R. L., Snell, R. L. and Loren, R. B.: 1983, Astrophys. J., in press.
Raymond, J.: 1979, Astrophys. J. Suppl., 44, pp. 379.
Rodriguez, L. F., Moran, J. M., Ho, P. T. P. and Gottlieb, E. W.: 1980, Astrophys. J., 235, pp. 845.
Rodriguez, L. F., Carral, P., Ho, P. T. P. and Moran, J. M.: 1982, Astrophys. J., 260, pp. 635.
Schwartz, P. R., Waak, J. A. and Smith, H. A.,: 1983, preprint.
Schwartz, R. D.: 1975, Astrophys. J., 195, pp. 631.
Snell, R. L., Loren, R. and Plambeck, R.: 1980, Astrophys.J., 239, L17.
Snell, R. L. and Edwards, S.: 1981, Astrophys. J., 251, pp. 103.
Strelnitsky, V. S. and Sunyaev, R. A.: 1972, Astron. Zh., 49, pp. 704.
Strom, S. E., Grasdalen, G. L. and Strom, K. M.: 1974, Astrophys. J., 191, pp. 111.
Torrelles, J. M., Rodriguez, L. F., Canto, J., Carral, P., Marcade, J., Moran, J. M. and Ho, P. T. P.: 1983, preprint.
Ulrich, R. K.: 1978, in "Protostars and Planets", ed. T. Gehrels, pp. 718. Tucson: University of Arizona Press.
Wright, A. E. and Barlow, M. J.: 1975, Mon. Not. R. Astr. Soc., 170, pp. 41.

NEAR-IR SPECTROSCOPY OF H_2 AND CO IN MOLECULAR CLOUDS

Donald N B Hall

Space Telescope Science Institute
Homewood Campus
Baltimore, Maryland 21218 USA

1. Introduction

   The near infrared vibration rotation (v-r) bands of CO and H_2 provide a powerful tool for the study of physical conditions in molecular clouds. They cluster individual v-r transitions, encompassing a wide range of excitation energy and line strength, into a relatively narrow frequency interval which can be observed with uniform instrumental techniques, often with a single instrument. Precise line frequencies and transition probabilities are available for both CO and H_2. The v-r lines are not seen in emission from cool (T $\lesssim$ 200K) cloud material because their upper states (E/k $\gtrsim$ 3000K) are not collisionally populated at such temperatures. However, the H_2 transitions are emitted by shocked regions where T is $\gtrsim$ 2000K and the CO transitions are seen in absorption against embedded continuum sources.

   In Section 2 of the present paper, the properties of the CO and H_2 v-r transitions are reviewed with special emphasis on their application to determination of physical conditions in molecular clouds using observations with current instruments. In sections 3 and 4, the application of these to a particularly well studied case, OMC-1 in Orion, is used to illustrate the information which can be obtained from v-r band observations. In section 5, the future prospects for these are briefly discussed.

2. Properties of CO and H_2 Vibration Rotation Transitions

   The properties of v-r transitions of diatomic molecules are treated in detail by Herzberg (1950). They involve photon emission or absorption through a transition in which the molecule remains in the same electronic state but <u>both</u> the vibrational quantum number, v, and the rotational quantum number, j, vary.

In both CO and H_2 the vibrational level energies are much larger than the rotational (a factor $\sim$70 for H_2 and $\sim$2,000 for CO). The vibrational level structure, determined by the radial separation potential which is approximately a simple harmonic oscillator for low lying levels, consists of nearly equally spaced levels of energy E_v:

$$E_v = \omega_e (v + 1/2) \qquad (1)$$

Each of these breaks up into a rotational fine structure, approximated by that of a rigid rotator with rotational energy E_j:

$$E_j = B_e\, j\, (j + 1) \qquad (2)$$

The effects of vibration-rotation interaction and modest departures from a simple harmonic oscillator require a power series expansion

$$E_{v,j} = \sum_{l,m} Y_{lm} (v + 1/2)^l (j(j+1))^m \qquad (3)$$

where Y_{lm} are the Dunham coefficients (Herzberg, 1950).

Under molecular cloud conditions, only v-r transitions within the ground electronic states of CO and H_2 are anticipated. The Dunham expansion is valid for CO under these conditions but there may be substantial departures for high rotational levels ($j > 7$) of H_2. Both CO and H_2 have $^1\Sigma$ electronic ground states and so have no fine structure within rotational levels. Molecular constants for CO are given by Mantz et al (1975) and for H_2 by Fink, Wiggins and Rank (1965) and Bragg, Brault & Smith (1982).

The selection rule for a harmonic oscillator is $\Delta v = \pm 1$ and the $\Delta v = 1$ transitions are the strongest v-r bands. The 1-0 band origins occur near 2140 cm^{-1} for CO and 4161.2 cm^{-1} for H_2. Departures from a pure harmonic vibrational potential allow overtone v-r bands but only the 2-0 transition of CO, which is centered on 4260 cm^{-1} and has transitions nearly two orders of magnitude weaker than the 1-0, has to date been observed in molecular cloud sources (the H_2 2-0 and CO 3-0 transitions are at higher frequencies where markedly increased obscuration hinders their detection).

For a heteronuclear molecule such as CO, the rotational dipole transition selection rule is $\Delta j = \pm 1$ for all vibrational transitions. With the small CO rotational constant ($Y_{01} \sim 1.9$ cm^{-1}), the CO 1-0 and 2-0 bands at cloud temperatures consist of well defined P and R branches of nearly equally spaced lines (Fig. 1). Line strengths for all of these transitions have been measured directly (Young & Eachus, 1966) and are also well fitted by both parametrizations of larger sets of experimental data (Tipping, 1976) and ab-initio calculations (Kirby-Docken & Liu, 1978). All of the parameters necessary to reduce observed line profiles within a band to rotational termperature, CO column density and radial velocity are thus available.

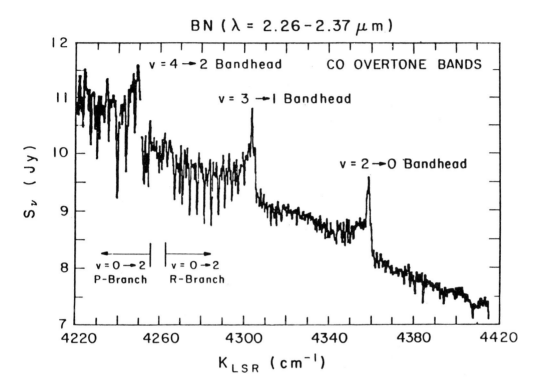

Figure 1. The spectrum of the Becklin-Neugebauer (BN) source in the vicinity of the CO first overtone bands. Terrestrial absorption features have been removed by ratioing to an AOV star. Low j 2-0 cloud lines are seen in absorption; the resolution is inadequate to resolve the two velocity components. The overtone bandheads prominent in emission are localized on BN.

Vibration rotation transitions of CO arising in cloud material have, to date, only been observed in absorption against embedded continuum sources. In this case, the physical conditions are probed in a column of intervening cloud material whose size is determined by the dimensions of the embedded source, i.e. typically on scales $\lesssim$ a few AU. Several dozen sources embedded in molecular clouds are accessible to observations at spectral resolutions from 5 to 20 km s^{-1} with current Fourier Transform Spectrometers (FTS) which have limiting magnitudes $\sim$ +7 at 2.3 µm and $\sim$ +1 at 4.6 µm for such observations. The high resolution and stable instrumental profile of the FTS are well suited to these observations because the CO absorption lines are closely spaced (~4 cm^{-1}), often narrow ($\lesssim$ 5 km s^{-1}), confined to a limited frequency interval ~50-100 cm^{-1} and often blended with terrestrial absorption.

For a homonuclear molecule such as H_2, dipole transitions are not allowed and the much weaker quadrupole transitions must be observed. The selection rule for these is $\Delta j = \pm 2$, 0 (0-0 not allowed). With the much larger rotational constant of H_2 (~60 cm^{-1}), the O and S transitions cover much wider frequency intervals than the CO P and R branches, with only the Q branch lines clustered near the band origin. The transition probabilities for the H_2 v-r transitions have been calculated by Turner, Kirby-Docken & Dalgarno (1977) and have been measured directly for some low energy transitions by Bragg, Brault and Smith (1982).

Despite the high abundance of H_2 in cloud material, the weakness of H_2 quadrupole absorption features places them beyond present detection capability. However, H_2 v-r transitions are seen in emission in many clouds and are now believed to originate in regions of shocked gas with T ~2000K. In most such regions the emission is optically thin and vibration/rotation temperatures, column of hot emitting gas and radial velocity information can be derived. Because the observed H_2 $\Delta v = 1$ cloud transitions cover an octave of frequency (from 2500 - 5000 cm^{-1}) and arise in regions which are heavily obscured by intervening cloud material, the scaling and frequency dependence of the extinction law affect the interpretation of the H_2 emission features. The latter can, in turn, be used to investigate the extinction law; sets of O, Q and S transitions which arise from the same upper state and encompass large frequency ranges are particularly well suited for this.

The H_2 emission lines are observed at modest resolution ($\lambda/\Delta\lambda$ ~ 100 - 1000) with filter wheel and grating spectrometers and at high resolution ($\lambda/\Delta\lambda$ ~ 10^4) with Fabry Perot and Fourier Transform spectrometers. The lower resolution instruments are well suited to spatial mapping of integrated line fluxes and the higher to obtaining resolved line profile information at selected locations. In the K band, the FTS achieves a multiplex advantage and can provide observations of all lines between 2.0 and 2.4 μm in time comparable to that required to observe a single profile with a dispersive spectrometer. At longer wavelengths, however, cooled dispersive spectrometers have substantial advantages becauses of their reduced thermal background noise.

3. The CO 1-0 and 2-0 Bands in OMC-1

Vibration-rotation CO features were first detected in OMC-1 material seen in absorption against the continuum of the Becklin-Neugebauer (BN) source by Hall et al (1978). The 1-0 transitions were resolved into doublets, one at $+8.9 \pm 0.5$ km s^{-1} LSR which was associated with undisturbed OMC-1 material and a second at -17.7 ± 0.3 km s^{-1}; the derived CO column densities of 6 ± 2 and 4 ± 1.5 $\times 10^{18}$ cm^{-2} respectively. The authors interpreted the blue shifted component as expanding material in the radio 'plateau' source (Zuckerman & Palmer, 1975) and, noting the 27 km s^{-1} velocity difference was supersonic, suggested the shock front between the two CO components as the location of the H$_2$ emission.

Extensive spectra of much higher quality which have since been obtained are presented by Scoville et al (1983). These more recent spectra allow resolution of five separate CO velocity components (Fig. 2), the two OMC-1 absorption systems detected earlier, an emission system at 20.2 km s^{-1} LSR, and absorption system at 30.2 km s^{-1} LSR (both of the latter are associated with circumstellar rather than cloud material on the basis of their 600K rotational temperatures) and a weak system at -3.0 km s^{-1} LSR.

The 2-0 line profiles constrain the average rotational temperature in the undisturbed OMC-1 material to 150 ± 10K (Fig. 3) and in the expanding material to 150 ± 30K; the measured radial velocities for the two systems are $+8.5 \pm 0.1$ km s^{-1} LSR and -17.8 ± 0.1 km s^{-1} LSR and the derived CO column densities are $7.5 \pm 0.3 \times 10^{18}$ cm^{-2} and $5.2 \pm 0.3 \times 10^{18}$ cm^{-2}.

The 1-0 CO line profiles observed in OMC-1 in absorption against the BN object are heavily saturated. However, $^{13}C^{16}O$ 1-0 transitions are observed with strengths comparable to the $^{12}C^{16}O$ 2-0 features; the decreased ^{13}C isotopic abundance approximately compensates the increased transition probabilities in the fundamental. So long as the same column of CO is seen at 2100 cm^{-1} and 4260 cm^{-1}, $^{12}C/^{13}C$ isotope abundance ratios of 96 ± 6 and 95 ± 9 are derived for the OMC-1 and expanding gas. Weak $^{12}C^{18}O$ 1-0 features are also detectable at a strength which implies a $^{16}O/^{18}O$ isotope abundance ratio of 420 ± 150. All of these values are consistent with the hypothesis that the $^{12}C/^{13}C$ and $^{16}O/^{18}O$ isotope ratios are the same in OMC-1 and the expanding material as in primitive solar system material.

The BN spectra also exhibit emission lines of atomic hydrogen corresponding to the $5 \to 4$ (Br α), $7 \to 4$ (Br γ), $7 \to 5$ (Pf β) and $8 \to 5$ (Pf γ) transitions. If these are produced within a compact HII region associated with BN then the line flux ratios indicate a visual reddening to BN of $A_V = 25$ magnitudes (assuming the van der Hulst #15 reddening law - Johnson, 1968). The total CO column density of $12.6 \pm 0.5 \times 10^{18}$ cm^{-2} then corresponds to a ratio of CO column density to visual extinction of $N_{CO}/A_V = 5 \times 10^{17}$ cm^{-2} mag^{-1}.

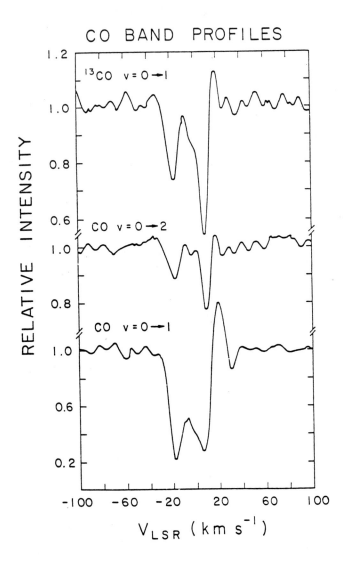

Figure 2. Composite line profiles for the $^{13}C^{16}O$ 1-0 band and the $^{12}C^{16}O$ 1-0 and 2-0 bands. Each was obtained by averaging all rotational transitions observed in each band. The velocity resolution is 7 km s^{-1} in all cases.

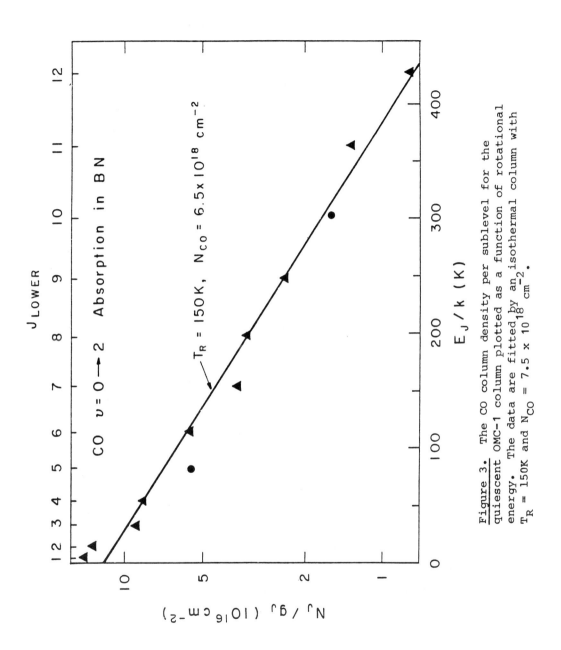

Figure 3. The CO column density per sublevel for the quiescent OMC-1 column plotted as a function of rotational energy. The data are fitted by an isothermal column with $T_R = 150K$ and $N_{CO} = 7.5 \times 10^{18}$ cm^{-2}.

Similar spectra of a number of other sources embedded in molecular clouds have been observed. The CO rotational temperatures of cloud material seen against UOA 27 is comparable to that in OMC-1, as is the N_{CO}/A_V ratio. However, the material seen against sources such as Lick Hα 101 and MWC 297 is characterized by rotational temperatures ~10K and by a markedly reduced ratio of N_{CO}/A_V. A plausible mechanism for the change in N_{CO}/A_V is the 'freezing out' of CO on grains at low enough cloud temperatures.

4. Near Infrared H_2 Emission in OMC-1

The detection by Gautier et al (1976) of seven lines of the vibration-rotation quadrupole bands of H_2 in emission from OMC-1 stimulated numerous further observations; these have resulted in broad observational constraints on the emission mechanism. Although H_2 quadrupole emission has since been detected from numerous other sources (Shull and Beckwith, 1982), the observations of the OMC-1 region are by far the most complete.

Many properties of the H_2 emission now rest on a firm observational basis. All subsequent observations have confirmed the original conclusion of Gautier et al (1976) that the H_2 quadrupole emission lines in OMC-1 are optically thin. All recent high resolution spectroscopic observations have shown broad line profiles with emission typically extending over more than 100 km s^{-1} (Nadeau and Geballe, 1979; Scoville et al, 1982; Nadeau, Geballe & Neugebauer, 1982). There is also ample evidence of marked spatial structure on the 5" scale of the highest resolution mapping to date (Beckwith et al, 1978).

However, estimates of the extinction to the H_2 emitting region vary from $A_{2.1\mu}$ ~ 1.2 to 4 magnitudes, implying total H_2 luminosities and masses which differ by more than an order of magnitude. The values $A_{2.1\mu}$ ~ 4 magnitudes obtained by Beckwith, Persson & Neugebauer (1979) and Simon et al (1979) were based on observed 1-0 Q(3)/S(1) line ratios. Because of the modest frequency separation of these lines (4123 cm^{-1} - 4713 cm^{-1}), accurate determination of $A_{2.1\mu}$ requires accurate observed and theoretical values for the line intensity ratio and is sensitive to the reddening law adopted. Beckwith, Persson & Neugebauer (1979) obtained 1-0 Q(3)/S(1) line flux ratios which agree with later observations but used an incorrect theoretical line ratio. This ratio was fortuitously supported by observations of the modestly reddened H_2 emission in NGC 7027 because of high Pfund series H line contamination in the spectral reference channels. The Q(3)/S(1) line ratio observed by Simon et al (1979) has not been reproduced in later observations.

Much lower estimates of extinction have since been obtained by Scoville et al (1982). Observations of the 1-0 Q(3)/S(1) line ratio at high angular ($3.8''$) and spectral (20 km s^{-1}) resolution at 12 locations in the Orion H$_2$ emission yield A$_{2.1\mu}$ extinctions between 1.2 and 2.7 magnitudes. A key result is that the extinction varies markedly over a scale $\sim 5''$; such variations may be responsible for much of the observed spatial structure in the H$_2$ emission. The value of 1.25 magnitudes at Peak 1 agrees well with the measurements of Beckwith et al (1983). The value of 2.9 ± 0.2 magnitudes obtained with a $17''$ beam by Davis, Larson & Smith (1982) is also consistent with the Scoville et al (1982) values averaged over the larger beam.

The derivation of total extinction from line ratio excesses requires specification of the reddening law and, in the case of the 1-0 Q(3)/S(1) ratio, the derived extinction is sensitive to the form of the law. Use of the longer wavelength O Branch and pure rotation lines (Knacke and Young, 1980; 1981) reduces this sensitivity and allows a test of the form of the reddening law. Beckwith et al (1983) find the observed line ratios are consistent with either the van der Hulst #15 (Johnson, 1968) law or a λ^{-1} law. Interpretation of these results is complicated by probable internal extinction within the H$_2$ emitting region. The relative geometry of the source and obscuring material also calls into question the assumption that photons are simply scattered out of the line of sight beam. Despite these uncertainties, the total H$_2$ luminosity is clearly $\sim 200 L_\odot$ rather than the earlier estimates $\sim 10^3 L_\odot$.

The temperature structure of the H$_2$ emitting region is more complex than indicated by early 1-0 observations which yielded a uniform temperatures ~ 2000K. Observations of the 0-0 S(2) line by Beck, Lacy & Geballe (1979) demonstrated the existence of a substantial column of H$_2$ at T $\ll$ 2000K whilst Beckwith et al (1983) derive a v-r rotational temperature $T_R \sim 3000$K for the v=2 level. The H$_2$ observations are clearly probing gas with an appreciable thermal distribution.

The evidence in favor of shock excitation of the H$_2$ is extremely compelling. The breadth of the observed line profiles and the range of observed H$_2$ temperatures are consistent with the shock mechanism and the H$_2$ emission occurs in a region where both millimeter and infrared CO observations show evidence for outflow centered within the infrared cluster. Scoville et al (1982) have shown the kinematics implied by their observed H$_2$ line profiles are in good agreement with 'plateau' source kinematics deduced from millimeter and infrared CO observations.

The eventual goal in observations of the OMC-1 H$_2$ emission is undoubtedly the mapping of line profiles at high spatial and spectral resolution over the entire emitting complex. The capability to do this in pairs of lines selected to optimize information on parameters such as reddening or temperature would permit observational testing of detailed models of processes within the H_2 emitting region.

5. Summary

The near infrared bands of CO and H_2 allow us to probe the physical conditions in both quiescent and energetic regions of molecular clouds. Observations of these transitions have provided a wealth of information about these regions in OMC-1 and are now being applied to a number of other molecular clouds. Forseeable improvements in observational techniques will substantially increase both the number of sources accessible to such observations and also the level of sophistication of observation of the brigher sources.

REFERENCES

Beck, S.C., Lacy, J.H., and Geballe, T.R. 1979, Ap. J. (Letters), 234, L213.
Beckwith, S., Persson, S.E., and Neugebauer, G. 1979, Ap. J., 227, 436.
Beckwith, S., Persson, S.E., Neugebauer, G., and Becklin, E.E. 1978, Ap. J., 223, 464.
Beckwith, S., Evans, N.J., III, Gatley, I., Gull, G., and Russell, R.W. 1983, Ap. J., 264, 152.
Bragg, S.L., Brault, J.W., and Smith, W.H. 1982, Ap. J., 263, 999.
Davis, D.S., Larson, H.P., and Smith, H.A. 1982, Ap. J. 259, 85.
Fink, U., Wiggins, T.A., and Rank, D.H. 1965, J. Molec. Spectrosc., 18, 384.
Gautier, T.N., III, Fink, U., Treffers, R.R., and Larson, H.P. 1976, Ap. J. (Letters), 207, L129.
Hall, D.N.B., Kleinmann, S.G., Ridgway, S.T., and Gillett, F. 1978, Ap. J. (Letters), 223, L47.
Herzberg, G. 1950, "Spectra of Diatomic Molecules" (Princeton: Van Nostrand).
Johnson, H.L. 1968, in "Stars and Stellar Systems, Vol 7, Nebulae and Interstellar Matter" ed B.M. Middlehurst and L.H. Allen (Chicago, University of Chicago Press), p 167.
Kirby-Docken, K. and Liu, B. 1978, Ap. J. Suppl., 36, 359.
Knacke, R.F. and Young, E.T. 1981, Ap. J. (Letters), 249, L65.
1980, Ap. J. (Letters), L183, 131.
Mantz, A.W., Maillard, J.P., Roh, W.B., and Rao, K.N. 1975, J. Molec. Spectrosc., 57, 155.
Nadeau, D., and Geballe, T.R. 1979, Ap. J. (Letters), 230, L169.
Nadeau, D., Geballe, T.R., and Neugebauer, G. 1982, Ap. J., 253, 154.
Scoville, N.Z., Hall, D.N.B., Kleinmann, S.G., and Ridgway, S.T. 1982, Ap. J., 253, 136.
Scoville, N.Z., Kleinmann, S.G., Hall, D.N.B., and Ridgway, S.T. "The Circumstellar and Nebular Environment of the Becklin-Neugebauer Object: $\lambda = 2-5$ Micron Spectroscopy" Ap. J., Submitted.

Shull, J.M., and Beckwith, S. 1982, Ann. Rev. Astron. Astrophys., 20, 163.
Simon, M., Righini-Cohen, G., Joyce, R.R., and Simon, T. 1979, Ap. J. (Letters), 230, L175.
Turner, J., Kirby-Docken, K., and Dalgarno, A. 1977, Ap. J. Suppl., 35, 381.
Tipping, R.H. 1976, J. Molec. Spectrosc., 61, 272.
Young, L.A., and Eachus, W.J. 1966, J. Chem. Phys., 44, 4195.
Zuckerman, B., and Palmer, P. 1975, Ap. J. (Letters), 199, L35.

SPECTROSCOPY OF HII REGIONS IN THE 1–15 μm REGION

J.H. Lacy

University of California, Space Sciences Laboratory, Berkeley

Infrared spectroscopy of HII regions can provide valuable information about heavy ions in regions too heavily obscured to be studied optically. Three such areas of study are discussed: elemental abundances in the Galaxy, excitation of compact, young, HII regions, and the structure and dynamics of compact HII regions.

1. Introduction

Why do astronomers observe ionized gas in the infrared? As is well known, the visible portion of the spectrum contains a much denser collection of ionic lines and is much more easily observed. Of course, interstellar extinction prevents optical observations of many sources, but extinction is not negligible at the shorter infrared wavelengths, and the VLA now allows high sensitivity and high resolution maps of free-free continuum and recombination line radiation at wavelengths where extinction is virtually absent. The better sensitivities of optical and radio instruments have forced infrared observers to concentrate on the questions which only they can study. These are primarily studies of ionic species which cannot be observed in the radio, from nebulae which are too heavily obscured to allow optical measurements.

This review concentrates on astronomical problems which require the special capabilities of the infrared. In Section II, I discuss observations of elemental abundances in the Galaxy. This study requires observations of several ionic lines in objects which are often obscured due to their large distances in the galactic plane. I then discuss the excitation of young HII regions in Section III. Studies of nebulae which are still obscured by their parent molecular clouds have presented some problems regarding their ionization structure, and promise to add to our understanding of the spectra of early-type stars. The third topic of this review (Section IV) is the structure and dynamics of compact HII regions. In this case, radio observations provide very similar information to that from infrared measurements. Nevertheless, infrared spectroscopy has been important here, especially when studies of structure, excitation, and abundances are combined.

No attempt is made in this review to mention every near infrared observation of an HII region. Rather, I discuss what I feel are the most interesting and promising types of observations, and a few representative papers.

II. Elemental Abundances

To measure the abundance of an element, one would like to be able to measure emission lines from all of the relevant ionic species of that element. This is not possible with only infrared observations, for any element other than hydrogen, but the dominant ionization states are observable for several elements. Only one ionic state of neon, Ne', is observable, but it is the dominant ion for $25,000 \text{ K} \lesssim T_{eff} \lesssim 35,000 \text{ K}$, where T_{eff} is the effective temperature of the exciting stars of the nebula. The two observable sulfur ions, S^{+2} and S^{+3} dominate for $25,000 \text{ K} \lesssim T_{eff} \lesssim 50,000 \text{ K}$. Similarly argon is well represented by Ar^+ and Ar^{+2} for $20,000 \text{ K} \lesssim T_{eff} \lesssim 40,000 \text{ K}$. Helium has not been extensively studied in the infrared, but HeI and HeII recombination lines are observable, although weak.

Two groups have attempted extensive infrared surveys of the abundances of neon, sulfur and argon in the HII regions and two groups have studied abundances in planetary nebulae. The studies of planetary nebulae were made by Dinerstein (1980) and Beck et al. (1981). These and other observations of planetary nebulae are reviewed by Dinerstein (1982) and Aitken (this volume) and so will not be discussed further here. Lester (1979), Herter et al. (1981; 1982) and Pipher et al. (1983) studied abundances in HII regions.

Studies of abundances in HII regions are of particular interest for the information they can provide about nucleosynthesis in the Galaxy and the dependence of metal abundances on galactocentric distance. Most of the work in this field has been done at radio and optical wavelengths. Radio observations have demonstrated the presence of a gradient in the electron temperatures of HII regions, with T_e decreasing by $\sim 400 \text{ K kpc}^{-1}$ toward the Galactic Center (e.g. Churchwell et al. 1978). The electron temperature in an HII region is determined primarily by the abundances of the cooling elements, especially oxygen and nitrogen. Consequently the observed electron temperature gradient has been interpreted to imply an oxygen abundance gradient of $d(\log O/H)/dr \sim -0.03 \text{ kpc}^{-1}$ (Panagia 1979).

Optical studies have also shown the presence of gradients in the abundances of heavy elements. Optically observed gradients are well established in several external galaxies and until recently over a rather small range in galactocentric distance in our Galaxy; see the review by Pagel and Edmunds (1981). A recent study by Shaver et al. (1983) uses radio observations of T_e to interpret optical line observations, and includes a large number of HII regions with galactocentric distances as small as 3.5 kpc. They find abundance gradients $\sim -0.07 \text{ dex kpc}^{-1}$ for

O, N, and Ar, but significantly less for S and Ne.

Abundances of neon, sulfur, and argon in several nebulae measured in the infrared by Herter et al. (1981; 1982), Lester (1979), and Lester, Dinerstein, and Rank (1979) are shown in Table I. The agreement between the different measurements is not impressive and indicates that the uncertainties are substantial. Nevertheless, there appear to be real differences between the different sources. Most notable is G29.9 at R_G = 5 kpc which appears overabundant in all three elements, and Sgr A in which argon appears overabundant (Lester et al. 1981). Surprisingly, however, W33 at R_G = 6 kpc may show low abundances, and W3 at R_G = 12 kpc may have high abundances. Certainly no abundance gradient has been convincingly observed; rather it appears that several interesting objects have been found.

Table I
IONIC ABUNDANCES RELATIVE TO "COSMIC" ATOMIC ABUNDANCES

	R_G	He$^+$	Ne$^+$	S^{+2}	S^{+3}	Ar$^+$	Ar^{+2}	ref
Sgr A	0		1.0		<0.02	1.8	0.2	
G29.9	5	0.7	2.7	1.8	<0.2	2.4	1.5	H81
			1.6		0.16		1.4	L79
W33	6		0.45	0.34	0.12	0.3	0.4	H81
G333.6	7	1.05	1.8	0.9	0.1		1.4	R78
G45.1	7.5		0.41	0.53	0.22	0.3	0.6	H81
			0.28	0.81	0.04		0.16	L79
S106	10		0.62	1.0	<0.1	1.2	<0.16	H82
			0.64	3.9	0.06		0.37	L79
M42	10		0.41*	1.0*	0.27	<0.6	0.5	H82
			0.26	0.75	0.60			LDR
G298.2	11	0.8	0.09		0.5		0.35	R78
S156	12		0.48	1.7		<2.3	0.20	H82
			1.6	0.6	0.05		2.5	L79
W3A	12		0.48	1.1	0.44	0.1	1.3	H81
			0.36	0.8	1.06		5.5	L79

*Ne^{+2}: 0.55, S$^+$: 0.02

Assumed Cosmic Abundances:
Ne/H = 1.0×10^{-4}
S/H = 1.6×10^{-5}
Ar/H = 6×10^{-6}
He/H = 0.1

References: H81: Herter et al. 1981
H82: Herter et al. 1982
L79: Lester 1979
LDR: Lester, Dinerstein and Rank 1979
R78: Rank et al. 1978

The abundance of helium can also be measured from its infrared emission lines. The strongest helium lines, at 2.06μm and 1.08μm are severely affected by radiation trapping, but more easily interpreted recombination lines exist. The $(n,1) = (5,4) \to (4,3)$ transition of the HeI has been observed by Lacy et al. (unpublished). Some of these results are shown in Table I. The apparent low abundance of helium in G29.9 is probably due to the presence of unobserved neutral helium. The low abundance in G298.2 is more remarkable, however, as its high excitation rules out He°. G298.2 may also be underabundant in sulfur, neon, and argon, although unobserved higher ionization states of these elements may be important.

Infrared emission line studies, like those described above, should be particularly well suited for the measurement of elemental abundances in the Galaxy. So far they have only partly fulfilled this promise, largely because of the difficulty of the observations and the uncertainties in their interpretation. The primary uncertainty in measuring abundances from infrared emission lines is in the extinction correction. The ionic lines discussed here have extinctions of ~0.03 - 0.10 A_V and are most often compared to HI Brα with A ~0.03 A_V or the unobscured radio free-free continuum. As the extinctions to the sources of interest range up to A_V ~50 and are uncertain by factors ~2, the derived abundances can be very uncertain. Several methods are used to determine the extinction. Most frequently the shape of the dust continuum near 10μm is used with the Gillett et al. (1975) case I or II fitting procedure to determine the 9.7μm extinction. A standard extinction curve (e.g. Becklin et al. 1978) is then used to determine the extinction at the observed lines. Unfortunately the case I and II results differ by $A_{9.7}$ ~1 mag and neither method is sensitive to a gray component of the extinction. The ratio of $A_{9.7}/A_V$ is uncertain by up to a factor of 2. An alternative approach is to observe Brγ (2.17μm), Brα (4.05μm), and/or the radio continuum to determine $A_{2.17}$, $A_{4.05}$, or $E_{2.17-4.05}$. A standard extinction curve is then used to extrapolate to longer wavelengths. Lester (1979) compared $A_{9.7}$ from case II continuum fits with $A_{4.05}$, and found a reasonably good correlation. A systematic difference between the case I $A_{9.7}$ and the true $A_{9.7}$ is not ruled out, however.

The best solution for the extinction uncertainty is to compare ionic lines with nearby hydrogen lines. This has been done in a few cases with ArII (7.0μm) and Pfα (7.45μm), (Lester et al. 1981), NeII (12.8μm) and Huα (12.4μm), and HeI and HI Brα (4.05μm) (Lacy et al. unpublished). For these lines relatively accurate abundances should be obtainable, but there are no strong hydrogen lines near ArIII (9.0μm) or SIV (10.5μm) for which the extinction is most uncertain.

III. Excitation

The excitation, or level of ionization, of a nebula is most directly measured by comparison of lines of different ionization states of an element. Sulfur and argon each have two ionization states which are observable in the infrared, and Herter, Helfer, and Pipher (1983) used the results of Herter et al. (1981, 1982) and Pipher et al. (1983) to study the excitation of HII regions in this way. The difficulty with this method is that [SIII] and [ArII] are measured from the KAO with focal plane apertures ~20 - 30", whereas [SIV] and [ArIII] are measured from ground based telescopes with much smaller beams. This problem can be overcome by comparing [NeII], [ArIII], and [SIV], but then the relative elemental abundances must be assumed. Lester (1979) and Lacy, Beck and Geballe (1982) used this latter approach. With either method, the extinction correction introduces a substantial uncertainty. In any case, the results differ so markedly with expectations that the uncertainties are relatively unimportant.

Each of these studies of excitation of HII regions has found systematically lower excitation than is predicted for nebulae ionized by single main-sequence stars with luminosities sufficient to ionize the nebulae. The most extreme case is G333.6 (Rank et al. 1978) for which the relative ionic abundances require a spectral class of O9-B0 while the luminosity corresponds to that of an O4V star. A total of 80 B0V stars or 4 O9I stars would be required to ionize the nebula. In HII regions where the exciting stars have been observed directly (M42, W3A) one star produces most of the ionizing radiation, and it seems improbable that the more obscured nebulae could be ionized by sufficiently populous clusters with only relatively late-type O stars, as implied by the observations. Several explanations have been suggested to explain this discrepancy. Dust mixed with the ionized gas could soften the radiation field in an HII region if it were to absorb sufficiently strongly shortward of the HeI ionization edge. Herter, Helfer, and Pipher (1983), have successfully reproduced integrated ionic emission spectra of many HII regions in this way. There is no other evidence that interstellar grain material has the required absorption spectrum, however, and selective absorption of short wavelength ultraviolet radiation by grains should produce a central high excitation region which is not seen in G333.6, for example. Alternatively, the ionizing stars may lie above the main sequence. Such a situation could occur early in the life of an O star, but it then seems unlikely that it would be so commonly observed. It is perhaps more likely that the assumed stellar spectra are at fault. Kurucz (1979) or similar LTE line-blanketed model atmospheres were used in the above studies. Mihalas (1972) non-LTE, non-line-blanketed atmospheres are much brighter in the ionizing ultraviolet, and so produce even larger discrepancies with the observations. Models including non-LTE, line blanketing, and perhaps non-planar geometries might be required to produce better fits to the data.

Rubin, Hollenbach, and Erickson (1983) have made detailed models of the ionic distributions in G333.6 which also indicates that modifications in the model stellar atmospheres are required. They were able to fit most of the observations with a power-law density distribution and an ionizing star with T_{eff} = 34,000 K and R = 60 $R_\odot$. The very large radius of this star was required to supply the required ionizing luminosity while leaving about half of the helium neutral. Even this model predicts much more S^{+3} and O^{+2} than is observed, however. To fit these ions, they found it necessary to suppress the stellar flux beyond 35 eV by a factor of ~20.

IV. Structure and Dynamics

Radio interferometers, notably Westerbork, Cambridge, and now the VLA, have dominated the study of the spatial distribution of ionized gas in HII regions. In addition, aperture synthesis of radio recombination lines now allows studies of the motions of the ionized gas as well. Nevertheless infrared observations have contributed significantly to our knowledge of the structure and dynamics of HII regions, and promise to continue to do so.

Infrared observation have several advantages in this field. First, several different ions can be observed so that information about the excitation and density can be obtained at each position in a map. Second, thermal broadening does not obscure systematic motions in heavy ions as it does with hydrogen. And third, near-infrared recombination lines can be used to study much denser, and so more optically thick, regions than can radio lines (see Wynn-Williams, this volume).

The most detailed infrared study of the structure and dynamics of an HII region is that of the Galactic Center (Lacy et al. 1980). In this case, spectral resolution of different velocity components allowed separation of spatially overlapping clouds of gas. The cloud velocities were then used to infer the stellar mass distribution (see Gatley, this volume).

Maps of HII regions in several ionic lines, but without velocity resolution, were made by Geballe et al. (1981) and Lacy, Beck, and Geballe (1982). Of those nebulae, G333.6 and G29.9 appear to be rather smooth, centrally peaked distributions of ionized gas. Both have been explained as face-on blister sources (see Hyland et al. 1980). Curiously, these two nebulae also have very similar ionization, being very luminous, but of low excitation. The other observed sources all have rather irregular, clumpy, distributions of ionized gas. It seems likely that this results from clumping present in the parent molecular clouds. The existence of clumping is probably indicative of the youth of the nebulae, but even so, clumps would not be expected to be observed if they were to dissipate in the sound-crossing time for ionized gas. Neutral condensations within the HII region are probably required to replenish the clumps. An important conclusion from the combined excitation and structure study is that the clumps do not each surround an ionizing star. The excitation was found to be constant over regions containing many clumps, indicating that the clumps are all bathed in a uniform radiation field. Radio observations of these nebulae could easily have reached the erroneous conclusion that the luminosity of each clump reflects that of one star.

References

Beck, S.C., Lacy, J.H., Townes, C.H., Aller, L.H., Geballe, T.R., and Baas, F. 1981, Ap. J., 249, p.592.
Becklin, E.E., Matthews, K., Neugebauer, G., and Willner, S.P. 1978, Ap. J., 220, p.831.
Churchwell, E., Smith, L.F., Mathis, J., Mezger, P.G., and Huchtmeier, W. 1978, Astr. Ap., 20, p.719.
Dinerstein, H.L. 1980, Ap. J., 237, p.486.
Dinerstein, H.L. 1983, IAU Symposium No. 103, "Planetary Nebulae".
Geballe, T.R., Wamsteker, W., Danks, A.C., Lacy, J.H., and Beck, S.C. 1981, Ap. J., 247, p.130.
Gillett, F.C., Forrest, W.J., Merrill, K.M., Capps, R.W., and Soifer, B.T. 1975, Ap. J., 200, p.609.
Herter, T. et al. 1981, Ap. J., 250, p.186.
Herter, T., Helfer, H.L., Pipher, J.L., Briotta, D.A., Forrest, W.J., Houck, J.R., Rudy, R.J., and Willner, S.P. 1982, Ap. J., 262, p.153.
Herter, T., Helfer, H.L., and Pipher, J.L. 1983, preprint.
Hyland, A.R., McGregor, P.G., Robinson, G., Thomas, J.A., Beckler, E.E., Gatley, I. and Werner, M.W. 1980, Ap. J., 241, 709.
Kurucz, R.L. 1979, Ap. J. Suppl., 40, p1.
Lacy, J.H., Beck, S.C., and Geballe, T.R. 1982, Ap. J., 255, p.510.
Lacy, J.H., Townes, C.H., Geballe, T.R., and Hollenbach, D.J. 1980, Ap. J., 241, p.132.
Lester, D.F. 1979, Ph.D. Thesis, University of California, Santa Cruz.
Lester, D.F., Bregman, J.D., Witteborn, F.C., Rank, D.M., and Dinerstein, H.L. 1981, Ap. J., 248, p.524.
Lester, D.F., Dinerstein, H.L., and Rank, D.M. 1979, Ap. J., 232, p.139.
Mihalas, D. 1972, Non-LTE Model Atmospheres for B and O Stars, NCAR-TN/STR-76.
Pagel, B.E. and Edmunds, M.G. 1981, Ann. Rev. Astron. Astrophys., 19, 77.
Panagia, N. 1979, Mem. S.A. It.
Pipher, J.L., Helfer, H.F., Herter, T., Briotta, D., Houck, J.R., Willner, S.P., and Jones, B. 1983, preprint.
Rank, D.M., Dinerstein, H.L., Lester, D.F., Bregman, J.D., Aitken, D.K., and Jones, B. 1978, M.N.R.A.S., 185, p.179.
Rubin, R.H., Hollenbach, D.J., and Erickson, E.F. 1983, Ap. J., 265, p.239.
Shaver, P.A., McGee, R.X., Newton, L.M., Parks, A.C., and Pottasch, S.R. 1983, preprint.

FAR-INFRARED SPECTROSCOPY OF HII REGIONS

R J Emery and M F Kessler

Astronomy Division, Space Science Department,
European Space Agency,
Noordwijk, The Netherlands

I. INTRODUCTION

Interest has developed rapidly in the astrophysics associated with far-infrared line emission from ionised regions, following the development of spectroscopic instruments and observing facilities appropriate to those wavelengths. The first detection of a far-infrared fine structure line was by Ward et al. (1975) with the 88 μm line of O^{++}. Subsequent instrumental development has improved the limiting sensitivity by a factor of about 1000, principally through the use of cooled Fabry-Pérot etalons. These observations have sufficient spectral resolution to match the width of lines from exceptional regions, such as the galactic centre, equivalent to a velocity range of 100 km/sec. A resolution about 10 times better than this is required to match the lines from normal galactic HII regions and this performance was approached by Baluteau et al. (1976) using a Michelson interferometer.

The fact that this is a relatively new topic stems from the observational difficulties caused by the absorption and emission of the atmosphere in this wavelength range. Figure 1 shows the zenith transmission spectrum of the atmosphere from a 4 km altitude site, and the emission spectrum of a 300 K temperature body to represent the background contribution from a warm telescope. Observations are therefore only generally possible using an airborne facility, such as the NASA Kuiper Airborne Observatory or the University College London balloon-borne telescope. Improvements in sensitivity have been achieved by narrowing the instrumental spectral bandwidth to reduce the background emission from the atmosphere and telescope.

Foundations for interpreting far-infrared fine structure observations were set out by Petrosian (1970) and Simpson (1975). This and more general work covering HII regions are given in Osterbrock (1974), Spitzer (1978) and more recently by Herter et al. (1983a). Far-infrared observations and their interpretation are now at the stage where the need for specific developments in theoretical and laboratory work have been identified. A summary is given in this volume by M. Harwit. The need is also apparent for the development of models dealing with more realistic astrophysical situations.

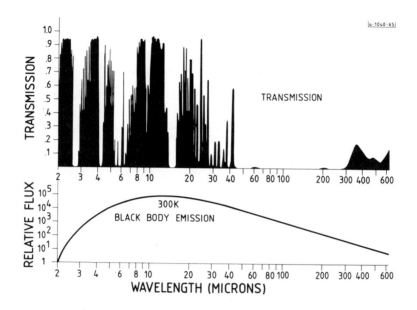

Figure 1: Calculated atmospheric absorption spectrum at the zenith angle from a 4 km altitude site and the emission from a 300 K body.

Figure 2 shows graphically the scheme which was followed to define the subject area for this paper and adjoining topics, particularly in relation to the papers by M. Harwit and D. Watson. In the overlap region, D. Watson treats observations of C^+ which has an ionisation energy of 11.26 eV compared with the value of 13.6 eV for hydrogen. M. Harwit also discusses the general significance of these observations. In defining the wavelength range as far-infrared, wavelengths longer than 15 μm are adopted to match up with the paper by J. Lacy.

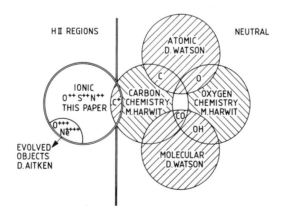

Figure 2: Scheme for subdividing topics on the spectroscopy of interstellar matter.

II. DEVELOPMENT OF FAR-INFRARED IONIC SPECTROSCOPY

Far-infrared spectroscopy of HII regions was pioneered by Ward et al. (1975), who measured the [OIII] 88 μm line in M17. The detection of the first pair of infrared lines was achieved when Melnick et al. (1978) found the companion [OIII] 52 μm line in the Orion nebula. Over the next few years other lines were observed: [SIII] 18.7 μm by Baluteau et al. (1976); [NIII] 57 μm by Moorwood et al. (1980b); and [SIII] 33 μm tentatively by Moorwood et al. (1980a) in M17 and definitely by Herter et al. (1982a) in M42.

Long wavelength fine structure line studies may be considered to have come of age in 1979 with the mapping of various sources with 1 to 3 arc minute resolution. McCarthy et al. (1979) detected [SIII] 18.7 μm emission from many positions in M17, while Storey et al. (1979) mapped three sources both in the [OIII] 88 μm line and also in the adjacent continuum emission. A complete map of the Orion nebula in three far-infrared lines was made by Furniss et al. (1983).

At the beginning of the 1980's this work developed with individual sources being studied in detail and also with the application of infrared results to other astrophysical problems, e.g. abundance gradients. Moorwood et al. (1980a) observed M17 in five far-infrared lines in up to 14 positions with high spectral resolution. New [SIII] 18.7 μm observations were presented by Herter et al. (1981) as part of a compilation of spectroscopic data in the 2-30 μm wavelength range on six HII regions at different galactocentric distances. The application of infrared lines to abundances has been continued by Lester et al. (1983) who used the [NIII] and [OIII] lines to investigate the variation of the N/O ratio as a function of distance from the galactic centre.

Table II.1 indexes all published detections by line and by object. The vast majority of these results have been obtained using the NASA Lear Jet or the Kuiper Airborne Observatory. The exceptions are Greenberg et al. (1977), who used a ground-based telescope on Mauna Kea and Moorwood et al. (1980b), Furniss et al. (1983) and Emery et al. (1983) who made their observations with the University College London balloon-borne system.

TABLE II.1: FAR IR FORBIDDEN LINES OBSERVED IN HII REGIONS

ASTRONOMICAL OBJECT	[SIII] 18.7 μm	[SIII] 33.5 μm	[OIII] 52 μm	[OIII] 88 μm	[NIII] 57 μm
M17	8, 11	11	7,11,12,14,17	1,4,9,11,12,14,17	11,12,14,17
M42	2,6,8,16,19	16	5,7,15,23	2,4,6,9,15,23	15,23
G333.6 - 0.2	3		12	12	
G29.9 - 0.0	20				
W51	8		7,12,17	4,9,12,17	12,17
W49	13		13,17	9,13,17	
M8			13	4	
W43			17,23	23	23
W3	6,18,20	18		4, 9	
NGC7538/S158 A&G	13,18,20	18	13,17	9,13,17	
NGC6357			7,12	4, 12	
NGC2170 Mon R2	19				
Sgr A	21		10	4,22	
G45.1 + 0.1	20				
S 106	19				
NGC2024	8				
G75.84 + 0.4	18, 20	18		23	23
G12.8 - 0.2(W33)	20				
NGC6334			12		
S 156	19				
S 88B	19				

1. Ward et al.   (1975)
2. Baluteau et al.   (1976)
3. Greenberg et al.   (1977)
4. Dain et al.   (1978)
5. Melnick et al.   (1978)
6. Moorwood et al.   (1978)
7. Melnick et al.   (1979)
8. McCarthy et al.   (1979)
9. Storey et al.   (1979)
10. Watson et al.   (1980)
11. Moorwood et al.   (1980a)
12. Moorwood et al.   (1980b)
13. Baluteau et al.   (1981)
14. Emery et al.   (1983)
15. Furniss et al.   (1983)
16. Herter et al.   (1982a)
17. Watson et al.   (1981)
18. Herter et al.   (1982b)
19. Herter et al.   (1982c)
20. Herter et al.   (1981)
21. Herter et al.   (1983b)
22. Genzel et al.   (1983)
23. Lester et al.   (1983a)

Upper Limits have also been set to the [FeII] (87 μm and 35 μm); [ClII] (37 μm); [NeIII] (36 μm); [SiII] (35 μm) and [PII] (33 μm) lines in Orion A and W3 IRS 1 by Moorwood et al. (1978).

III. ANALYSIS OF FAR-INFRARED LINE OBSERVATIONS OF HII REGIONS

Within the HII region, a balance is established between the energy input to the region from the ionising star or stars, and the various emission mechanisms involving continuum emission and lines. The region shows differentiation into zones for the various ionisation states of the elements, the zones reflecting the change in the ionising radiation as it travels out from the exciting star. The far-infrared lines involve transitions within the ionic ground states since these states are easily excited by collisional interaction with the magnetic fields of the ambient electrons, which have a kinetic temperature typically around 8000 K. Spontaneous line radiation from the excited state is competitive with collisional de-excitation. This introduces a sensitivity to the electron density, c.f. M. Harwit, this volume. The far-infrared line emission is relatively insensitive to the electron kinetic temperature, however, since the excited ionic ground state energies are much lower than the kT of the electrons. In this respect, the information derived from the far-infrared lines is complementary to optical wavelength measurements which are temperature dependent.

Galactic HII regions are generally optically thin to far-infrared line emission, so that the observations relate to the total column density of emitting material. Analysis is frustrated by the lack of density information along the line of sight, and so assumptions about the 3-dimensional structure of the region must be made. Substantial improvement to this situation must await observations with high spectral resolution (say equivalent to better than 15 km/sec velocity resolution) to support modelling of the structure of the emitting region.

Far-infrared fine structure lines result from magnetic dipole transitions between states in the ground configuration of the ion. Those species with a p^2 or a p^4 electron configuration (e.g. S^{++} and Ne^{++}) have a triplet ground state with two detectable infrared lines. Those with p^1 or p^5 configurations (e.g. N^{++} and Ne^+) have a doublet state and therefore only one infrared line. The details of various lines are shown in Table III.1.

Starting from the early measurements involving one line, a significant step forward in interpretation was made with the ability to measure both lines arising from ions with triplet ground states. This allows the density of the electrons which are coexisting with the ion to be determined (as discussed by M. Harwit in this volume or Watson and Storey, 1980). It is these electrons which are responsible for the collisional excitation and de-excitation of the ions. When the density of the emitting region varies along the line of sight, it must be noted that this derived electron density represents an average value weighted in a rather complicated way.

Problems with astrophysical interpretation are also eased by the identification of quantities derived from two or more lines, such as the N/O ratio, which modelling has shown to be fairly insensitive to the structure of the region.

TABLE III.1: SOME SELECTED FAR-INFRARED LINES ARISING IN HII REGIONS
($\lambda > 15\mu m$)   (after Watson and Storey, 1980)

Species	$\lambda(\mu m)$	Transition	$A(s^{-1})$	Excitation Potential (eV)	Ionisation Potential (eV)	Critical Density (cm^{-3})	Detected
N^{++}	57.30	$2P^1: {}^2P_{3/2} \to {}^2P_{1/2}$	4.77×10^{-5}	29.601	47.448	2.0×10^3	YES
O^{+++}	25.91		5.18×10^{-4}	54.934	77.413	–	(YES)*
N^+	203.9	$2P^2: {}^3P_1 \to {}^3P_0$	2.13×10^{-6}	14.534	29.601	4.1×10^0	NO
	121.7	${}^3P_2 \to {}^3P_1$	7.48×10^{-6}			2.6×10^2	NO
O^{++}	88.356	${}^3P_1 \to {}^3P_0$	2.62×10^{-5}	35.117	54.934	6.7×10^2	YES
	51.815	${}^3P_2 \to {}^3P_1$	9.75×10^{-5}			4.9×10^3	YES
Ne^{++++}	24.28	${}^3P_1 \to {}^3P_0$	1.3×10^{-3}	97.11	126.21	1.0×10^5	(YES)*
	14.33	${}^3P_2 \to {}^3P_1$	4.6×10^{-3}			4.2×10^5	(NO)
S^{++}	33.47	$3P^2: {}^3P_1 \to {}^3P_0$	4.72×10^{-4}	23.33	34.83	4.1×10^3	YES
	18.713	${}^3P_2 \to {}^3P_1$	2.07×10^{-3}			4.3×10^4	YES
Ne^{++}	15.55	$2P^4: {}^3P_1 \to {}^3P_2$	5.99×10^{-3}	40.962	63.45	2.8×10^5	(YES)*+
	36.04	${}^3P_0 \to {}^3P_1$	1.15×10^{-3}			4.2×10^4	NO
Ar^{++}	21.83	$3P^4: {}^3P_0 \to {}^3P_1$	5.19×10^{-3}	27.629	40.74	–	NO

* Detected in Planetary Nebulae.   + Detected by IRAS.

Another factor to be considered is the dust which is often associated with HII regions, sometimes partly situated within the HII region but mainly outside it. The dust can have condensed from material ejected by the exciting star, as in the example of planetary nebulae, or reside in the surrounding neutral material such as in star formation regions. The dust often comes within the line of sight causing reddening or even total optical extinction. Measurements at infrared wavelengths can probe these regions, often allowing interpretation to be made without major corrections for extinction, although the uncertainty in the value of the extinction coefficient is likely to be larger than at visible wavelengths.

Following the notation given by D. Watson in this volume, the line intensity I (power per unit area and solid angle) is given by

$$I = \frac{hc}{4\pi\lambda} A_{kj} \int f_k n f d l \tag{1}$$

where A_{kj} is the Einstein A-coefficient for the radiative transition from state k to state j, n is the density of the emitting species, f is the ionisation fraction, f_k is the fraction of these ions occupying the state k, and the integration is along the line of sight l. The fractional populations f_k are given under steady state conditions by:

$$\sum_{j \neq k} n_e f_j \gamma_{jk} + \sum_{j > k} f_j A_{jk} = \sum_{j \neq k} f_k n_e \gamma_{kj} + \sum_{j < k} f_k A_{kj} \tag{2}$$

where γ_{jk} is the electron collisional rate coefficient and n_e the electron density. The left hand side of equation 2 deals with transitions populating level k and the right hand side with transitions depopulating level k. For the far-infrared emission lines, only the ground state levels need be considered. The fractional populations f_k are therefore given by solving the two or three relevant equations of the form of equation 2, together with the condition $\sum_k f_k = 1$.
Examples of solutions have been given by Storey et al. (1979) for O^{++} and Watson et al. (1981) for O^{++} and N^{++}.

At low densities when collisional de-excitation of the upper levels is significant, f_k is proportional to n_e, making the line intensity proportional to $n_e \cdot n$. Assuming a constant ionic abundance within the emitting region, this makes the intensity proportional to n^2. On the other hand, when the density has increased to the point where the line is thermalised, f_k is proportional to the statistical weights of the levels, making the line intensity proportional to n. This shows that the measured emission will increasingly emphasize the high density regions as lines with higher critical density are selected.

Figures III.1, 2 and 3 show, in the form of flow diagrams, various schemes which have been used to derive astrophysical information from fine structure lines. With analysis of a single line, an important aspect is the determination of ionic abundances, as shown in Figure III.1. The advantages of measuring ionic abundances via infrared lines are that the uncertainty due to dust extinction is reduced; that the lines have little intrinsic temperature dependence; and that the lines become thermalised at lower densities than their optical counterparts. The last point is particularly important because it enables an accurate ionic column density to be derived independent of electron density and temperature.

The general method to convert a measured line intensity, I, to an average ionic abundance $<n_i/n_e>$ starts by re-writing equation (1) as

$$I = \int j \cdot dl \tag{3}$$

where j, the line emissivity, is given by

$$j = \frac{hc}{4\pi\lambda} A_{kj} f_k nf \tag{4}$$

Thus without any assumptions an accurate column density for the ions in the upper state may be found. If the line is thermalised then f_K is not sensitive to n_e and an accurate ionic column density is derived. To go from this to a number density requires some modelling of the source to derive or assume an emitting length. This assumption can be made explicitly as in the left-hand branch of Figure III.1 or implicitly by using the radio emission measure. In this case, equation (3) may be expanded to

$$I = \int (j/n_i n_e) \cdot \left(\frac{n_i}{n_e}\right) \cdot n_e^2 \, dl \tag{5}$$

Plots of $(j/n_i n_e)$ as a function of n_e are given by Simpson (1975). Average ionic abundances are derived from,

$$<\frac{n_i}{n_H}> = \left(\frac{n_e}{n_H}\right) \frac{\int \frac{n_i}{n_e} dl}{\int dl} \tag{6}$$

which, as discussed by Baluteau et al. (1981), may be approximated to

$$<\frac{n_i}{n_H}> = \left(\frac{n_e}{n_H}\right) \frac{I}{\int (j/n_i n_e) n_e^2 \, dl} \tag{7}$$

Figure III.1. Emission line analysis with one observation

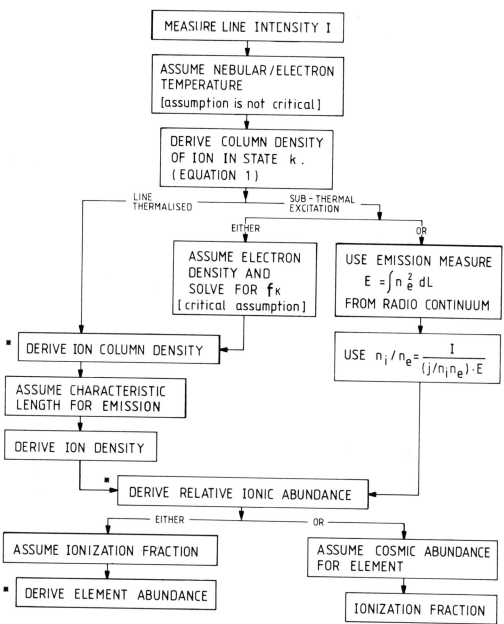

f_k is the fraction of ions in state k.

n_i and n_e are the ionic and electron densities respectively and j is the emissivity of the line as used by Simpson (1975).

* shows commonly required astrophysical quantities.

If an average value for $(j/n_i n_e)$ is used, as shown by the right-hand branch in Figure III.1, equation (7) further reduces to

$$<\frac{n_i}{n_H}> \cong \frac{I}{<\frac{j}{n_i n_e}> E} \cdot \left(\frac{n_e}{n_H}\right) \qquad (8)$$

where E is the emission measure as found from the radio continuum and the equation may be further simplified by using $n_e \approx n_H$. In equation (8), the assumptions about the length of the emitting region are hidden in I and E.

Figure III.2 shows the analysis of observations of two infrared lines arising within a triplet ground state, leading to a measure of the electron density. The analysis for this is discussed in detail by Watson and Storey (1980). Comparison between this and the emission measure allows an estimate of the clumping of the region to be made, as discussed by Herter et al. (1982b). This electron density value is given directly from the line measurements whereas values derived from radio continuum data require the source region to be modelled for a characteristic emission length. In comparison with optical determination, the infrared values are less affected by dust extinction and uncertainty in the electron temperature, but the observations have a much lower spatial resolution. The low critical density values for most infrared lines also makes them less biased towards the very small-scale dense emission structures seen in the visible.

By measuring the spatial distribution of emission from one or more ions, information is obtained about the ionisation structure of the region, and hence, about the ionising radiation field. Certain ions e.g. S^{++} and O^{++} (Herter et al., 1982b) and N^{++} and O^{++} (Genzel et al., 1983) have been shown by models to be spatially co-extensive and thus measurements of lines from these species permit determination of very accurate relative ionic abundances. This is shown in Figure III.3, together with the development by Watson et al. (1981) which aims at deriving the effective temperature of the exciting star.

IV. RESULTS

In section III and summarised in Figures III.1, 2 and 3, the astrophysical quantities derivable from fine structure lines are outlined and in this section, some results will be summarised.

From Table II.1, it is clear that sufficient observations of the bright HII regions M42 and M17 exist to make an assessment of the data repeatability. After correcting for beam sizes, the measurements on similar positions on M42 in the [SIII] 18.7 μm, [OIII] 52 μm and 88 μm lines agree to within a factor of 2. The agreement on M17 in these three lines and the [NIII] 57 μm line is within a factor 4. However the

Figure III.2. *Emission line analysis when observations of both lines arising from a triplet ground state can be obtained.*

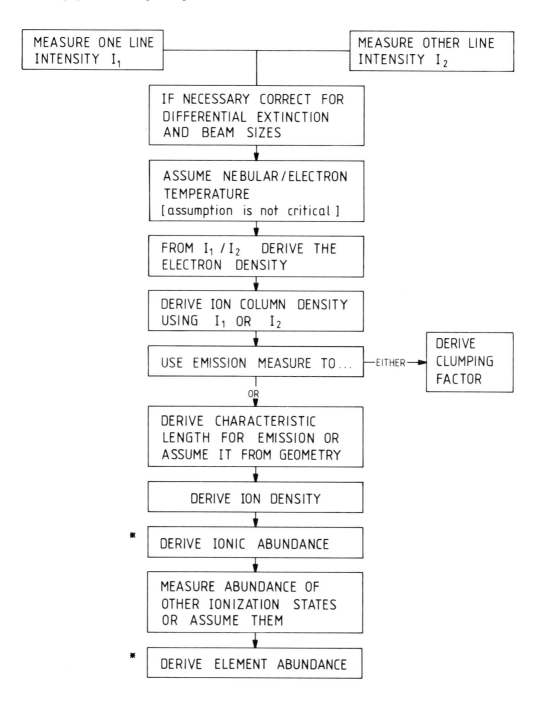

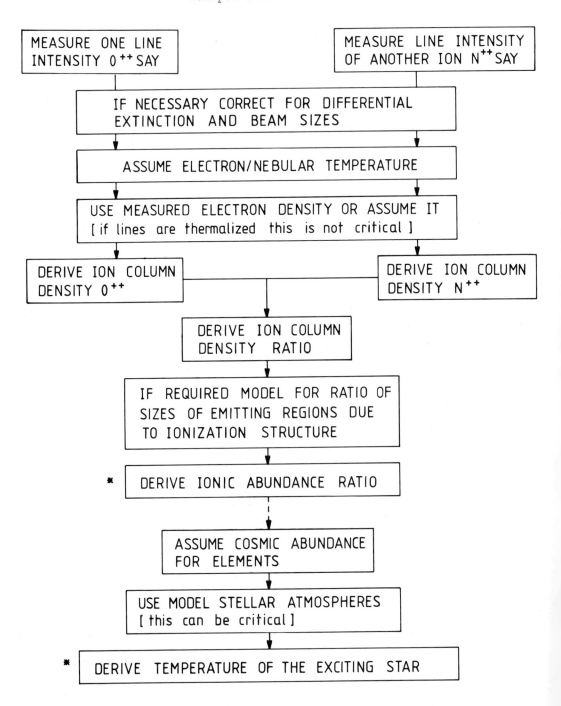

Figure III.3. *Emission line analysis with observations of different ions. Example with O^{++} and N^{++}*

process of correcting for beam sizes by simple scaling is sensitive to the density and ionisation structure of the source and this may account for most of the variation. When the large beam measurements on M17 are excluded, the various measurements agree more closely, the remaining discrepancy being x2 for the [OIII] and [SIII] lines and x3 for the [NIII] line. Thus, reported fluxes ought, in general, to be reliable to within a factor of about two.

Measured line strengths have been used to date in the following areas:

(i) Comparison with model predictions

(ii) Determination of ionic abundances, leading to information on the ionisation structure of the nebula and on the effective temperature of the ionising radiation

(iii) Determination of electron number densities, leading to density modelling of the source

(iv) Positional information from maps

(v) Extraction of velocity information.

IV. (i) Comparison with Predictions

Measured line intensities were initially compared with predictions based upon simple models (Petrosian, 1970; Simpson, 1975) or with the results of calculations based upon radio measurements (e.g. Dain et al., 1978). Two general facts emerged. Firstly, the effect of differential dust absorption must be considered, especially for the sulphur lines (Greenberg et al., 1977) and even for the oxygen lines in the direction of very obscured objects (Melnick et al., 1978). Secondly, the predicted intensities depend critically on assumed elemental abundances, assumed ionisation fractions and assumed density structure. Thus after the initial few papers, the arguments were inverted and the data used to derive or comment upon these quantities.

IV. (ii) Ionic Abundances

Abundances have been calculated from measurements of the [SIII], [OIII] and [NIII] lines via the methods summarised in section III. Although the calculation of the S^{++}/H ratio from the infrared lines is complicated in some cases by dust extinction (e.g. NGC 2024, SgrA, McCarthy et al., 1979), it has been measured in a dozen or so HII regions (e.g. Herter et al., 1981, 1982c). Values for the S^{++}/H ratio have been found ranging from 0.4 to 2.2 times the cosmic elemental sulphur abundance ratio of 1.6×10^{-5}. To reduce the problem of assuming an average emissivity (i.e. approximation from equation (7) to (8)), Moorwood et al. (1980a) used their three-component density model of M17 to improve their estimate of S^{++}/H. This resulted in an increase of around 50% compared to the average-emissivity value. For NGC 7538 Baluteau et al.

(1981), used a two-component model, which decreased their initial S^{++}/H value by a small amount. Herter et al. (1982b) took density structure into account and increased their previously published (Herter et al., 1981) abundances in G75.84 + 0.4 and W3 IRS1 by a factor of 2.

There is, at present, insufficient data to make any definite comparison between S^{++}/H ratios derived in the optical and in the infrared. For example, McCarthy et al. (1979) find that their measurements of S^{++}/H in M42 are consistent with the optical values of Peimbert and Torres-Peimbert (1977) while Moorwood et al. (1980a) find a peak value in M17 about 50% higher than the cosmic S/H ratio, but still less than Peimbert and Costero's (1969) S/H determination. In NGC 7538, Baluteau et al. (1981) find S^{++}/H about 3 times higher than the optical value of Talent and Dufour (1979) but Herter et al. (1981) find an infrared value more consistent with the optical.

Column densities or ionic abundances have been calculated for most of the nebulae listed in Table II.1. Storey et al. (1979) and Watson et al. (1981) report column densities with the former authors emphasising the difficulty of proceeding to ionic abundances. This can be quantified by looking at the measured O^{++}/H ratios for M17. Storey et al. (1979) find an ionic abundance ratio of 1×10^{-4} and conclude that it is not necessary to increase the commonly accepted oxygen abundance. This is in contrast to Melnick et al. (1979) who find $O^{++}/H = 3.8 \times 10^{-4}$ and thus need to increase Simpson's (1975) O abundance by a factor of 2. They support this argument by quoting the optical work of Peimbert and Torres-Peimbert (1977). After correcting for density effects, Moorwood et al. (1980a) find a value of $\sim 4.6 \times 10^{-4}$ in agreement with the optical line work of Peimbert and Costero (1969). Moorwood et al. (1980b) give a result of 2.4×10^{-4} agreeing with the uncorrected value of Moorwood et al. (1980a). Finally, Lester et al. (1983a) measure 4×10^{-4} and summarise the situation by saying that the error in this value is probably a factor of two or three.

Results for the N^{++}/H ratio have only been explicitly given for M42 (Furniss et al., 1983) and M17 (Moorwood et al., 1980a). The latter find, after modelling, an ionic abundance much greater than the optically determined elemental abundance (Peimbert and Costero, 1969) but still consistent with the cosmic nitrogen abundance.

Lester et al. (1983a) argue strongly the case for using measurements of the [NIII] 57 μm and [OIII] 52 μm and 88 μm lines to determine N^{++}/O^{++}. These lines are intrinsically insensitive to temperature, and due to their similar critical densities, their ratio is also insensitive to density. Thus, even in the absence of any information about the electron temperature and density, N^{++}/O^{++} can be determined to within a factor of 2. A further refinement is to use the [OIII] lines to determine n_e, a value which will also be appropriate for [NIII] since the two ions are expected to be co-extensive. On the basis of models, Lester et al. (1983a) further argue that N^{++}/O^{++} will be indicative of N/O, a ratio important for nucleosynthesis theory. Infrared measurements will enable the galactic abundance gradient of N/O to be investigated over a much larger distance than from work in the optical. Lester et al. (1983a) present new results for N^{++}/O^{++} for W43, Orion A and G75.84+0.4 together with data for M17 and W51 from Watson et al. (1981). Their W51 data agree with measurements of Moorwood et al. (1980b) but their M17 results are higher by about a factor of 1.5. The N^{++}/O^{++} ratios of the five HII regions of Lester et al. (1983a) are substantially higher than the currently accepted solar neighbourhood values for N/O. The value for W43, which is closest to the galactic centre, has the highest value of all. However, the infrared ratios are about x2 higher than optical ones and until this discrepancy is resolved, any conclusions about a galactic N/O gradient must be considered tentative.

Information on the ionisation structure of a nebula may be derived from the spatial variation of the ionic abundance if it is assumed that this reflects changes in the ionisation excitation rather than in the elemental abundances. Various authors have used the presence or absence of different ionisation states, or the strength or weakness of certain lines to make general statements about the ionising radiation field. The weakness of [OIII] emission from the galactic centre was interpreted by Dain et al. (1978) to be due to a lack of hard UV photons. This source was further investigated by Watson et al. (1980), who used [OIII] measurements to conclude that the effective temperature of the ionising source was between 32000 K and 40000 K. They concluded by suggesting that the effective temperature will be better determined from the [NIII] (57 μm)/[OIII] (52 μm) line ratio.

Moorwood et al. (1980a) conclude that the ionisation structure of M17 is complex. The ratio O^{++}/O decreases by a factor of 5 from a peak on M17N towards the ionisation front in the SW. The rough symmetry of the [SIII]/[OIII] and [NIII]/[OIII] line ratios around M17S are interpreted as consistent with the heavily reddened star cluster of Beetz et al. (1976) being the dominant ionising source. They find a minimum of N^{++}/O^{++} on M17S and thus imply the presence of a small N^{+++} zone and hence an effective temperature of > 40000 K for the radiation field. However, the latter conclusion is not supported by the results of Watson et al. (1981), who present extensive measurements of the oxygen and nitrogen lines across M17 and find no minimum on M17S. They compare their results with the model of Icke et al. (1980) and find good agreement.

IV.(iii) Electron Densities

Following the analysis shown in Figure III.2, differences between electron number densities derived from radio and optical measurements and those from either [OIII] lines (e.g. Melnick et al., 1979; Moorwood et al. 1980b; Watson et al. 1981) or [SIII] lines (e.g. Herter et al. 1982a; Herter et al. 1982b) have been interpreted as indicative of the amount of clumping in the source. This may be expressed as a filling factor, ϕ, which is defined as the ratio of the volume occupied by fully ionised high density clumps to the total volume of the HII region. Moorwood et al., (1980a), using the above concept, proposed an idealised three-component density model for M17 to account for the radio and infrared results. The components have densities of 300 cm^{-3} with ϕ = 0.9, 3000 cm^{-3} with ϕ = 0.1 and 3×10^4 cm^{-3} with ϕ = 0.001. Similar modelling, but with two components, was done by Baluteau et al. (1981) for NGC7538 and W49. Both oxygen lines have been measured from the galactic centre (52 µm by Watson et al., 1980; 88 µm by Genzel et al., 1983). These results lead to an electron density for the O^{++}-containing plasma of > 6000 cm^{-3}, consistent with densities deduced for the [NeII]-emitting clouds. No evidence is seen for a medium density halo as deduced from radio observations.

IV.(iv) Positional Information from Far-Infrared Line Maps

The variation of far-infrared line strengths across various sources has been measured and compared with that of the radio or far-infrared continuum. McCarthy et al. (1979) found that their [SIII] 18.7 µm fluxes from M17 followed the radio (Schraml and Mezger, 1969) fairly closely. The distribution of the [OIII] 88 µm line in M17, NGC 7538 and W51 was measured by Storey et al. (1979) and, in each case, the fine structure peak was co-incident with the radio continuum peak but displaced from the far-infrared continuum peak. Models of star formation regions (Icke et al., 1980) predict the displacement between the radio and far-infrared continuum peaks. A general correspondence between fine structure and radio peaks is expected because the relevant ionisation zones will be largely co-centered. However, ionisation structure, with the lower ionisation zones extending further from the exciting star(s) will affect this coincidence. Higher spatial resolution mapping than the current $\sim$ 1 arc minute is likely to be required to investigate such effects. This is demonstrated by the shorter wavelength measurements of [NeII] 12.8 µm and [SIV] 10.5 µm by Lester et al. (1983b) where a displacement of 2 arc seconds was observed between the peak positions of these lines at the ionisation front in M17.

Emery et al. (1983) presented a map of electron density across M17 derived from the [OIII] lines. When compared to a radio map (Wilson et al., 1979) substantial differences were found which could derive from the ionisation structure of the source. The Orion nebula, which can be modelled as a blister viewed face-on, was mapped by Furniss et al. (1983) who, within positional errors, found the expected co-incidence between the [OIII] 52 µm and 88 µm peaks, the radio continuum peak and the [OIII] - derived electron density peak.

IV.(v) Velocity Information

In the absence of laboratory determinations, the most accurate values of the wavelength of some infrared fine structure lines have been made through astrophysical measurements. Moorwood et al. (1980b) report rest wavenumbers of $192.996 + 0.002$ cm^{-1} and $113.178 + 0.003$ cm^{-1} for the [OIII] lines and 174.43 ± 0.01 cm^{-1} for the [NIII] line. Best values for the [SIII] lines are 534.39 ± 0.01 cm^{-1} (Moorwood et al. 1980a) and 298.78 ± 0.18 cm^{-1} (Herter et al. 1982a).

Limited velocity information has been gained to date. Moorwood et al. (1978) using a Michelson Interferometer observed that the most blue-shifted [OIII] 88 μm emission from M42 came from a region centred over the bright bar and including the trough between the bar and the Trapezium, agreeing with visible [OIII] 5007 Å measurements. In NGC 7538, Baluteau et al. (1981) observed a systematic variation of the [OIII] line centres across the nebula with the NE redshifted with respect to the SW. This agrees with optical Hα measurements.

The [SIII] 18.7 μm line has been resolved in M42 (Baluteau et al., 1976; Moorwood et al., 1978), in M17 (Moorwood et al., 1980a, in W49A (Baluteau et al., 1981) and marginally in the galactic centre (Herter et al., 1983b). The results are generally consistent with results obtained at other wavelengths. For example, the [SIII] line-width over the Trapezium is comparable to that obtained from radio recombination lines in a slightly larger beam, while further north it is a factor of two narrower, consistent with optical line widths. In short, present instrumentation yields only marginal velocity information.

V. CONCLUSION

The papers reviewed here show the development of far-infrared spectroscopy to the point where it is now making important contributions to general astrophysics. One example of this is the work on galactic abundance gradients. The present pace of observation is restricted by the very limited availability of airborne telescope facilities to get above the worst of the atmospheric absorption. The sensitivity of present day instrumentation, as described in some detail by D. Watson in this volume, is appropriate for a range of observations throughout most of the Galaxy. It is also reasonable to expect continued significant development in sensitivity of these systems over the next few years and the availability of heterodyne systems at these wavelengths. In the longer term, however, it is only with the development of space facilities that infrared observations will achieve sensitivities in keeping with those currently available at most other wavelengths. For this, the German Infrared Laboratory (GIRL) is expected to operate from the Shuttle in about 1987, the European Space Agency now has an approved Infrared Space Observatory (ISO) mission for operation in the 1990's and NASA is continuing with plans for a Space Infrared Telescope Facility (SIRTF).

ACKNOWLEDGEMENTS

The authors are grateful to their many colleagues who made results available in advance of publication

REFERENCES

Baluteau, J.-P., Bussoletti, E., Anderegg, M., Moorwood, A.F.M., and Coron, N., 1976, Ap. J. 210, L45-L48.
Baluteau, J.-P., Moorwood, A.F.M., Biraud, Y., Coron, N., Anderegg, M., and Fitton, B., 1981, Ap. J. 244, 66-75.
Beetz, M., Elsässer, H., Poulakos, C. and Weinberger, R., 1976, Astron. Astrophys. 50, 41-46.
Dain, F.W., Gull, G.E., Melnick, G., Harwit, M., and Ward, D.B., 1978, Ap. J. 222, L17-L21.
Emery, R.J., Naylor, D.A., Fitton, B., Furniss, I., Jennings, R.E., and King, K.J., 1983, Ap. J. 268, 721-726.
Furniss, I., Jennings, R.E., King, K.J., Lightfoot, J.F., Emery, R.J., Naylor, D.A., and Fitton, B., 1983, MNRAS 202, 859-865.
Genzel, R., Watson, D.M., Townes, C.H., Dinerstein, H.L., Hollenbach, D., Lester, D.F., Werner, M., and Storey, J.W.V., 1983, Ap. J. (submitted).
Greenberg, L.T., Dyal, P., and Geballe, T.R., 1977, Ap. J. 213, L71-L74.
Herter, T., Briotta, D.A. Jr., Gull, G.E., Shure, M.A., and Houck, J.R., 1982a, Ap. J. 259, L109-L112.
Herter, T., Briotta, D.A. Jr., Gull, G.E., Shure, M.A., and Houck, J.R., 1982b, Ap. J. 262, 164-170.
Herter, T., Briotta, D.A. Jr., Gull, G.E., Shure, M.A., and Houck, J.R., 1983b, Ap. J. 267, L37-L40.
Herter, T., Helfer, H.L., and Pipher, J.L., 1983a, Astron. Astrophys. Suppl. Ser. 51, 195-212.
Herter, T., Helfer, H.L., Pipher, J.L., Briotta, D.A. Jr., Forrest, W.J., Houck, J.R., Rudy, R.J., and Willner, S.P., 1982c, Ap J. 262, 153-163.
Herter, T., Helfer, H.L., Pipher, J.L., Forrest, W.J., McCarthy, J., Houck, J.R., Willner, S.P., Peutter, R.C., Rudy, R.J., and Soifer, B.T., 1981, Ap. J. 250, 186-199.
Icke, V., Gatley, I, and Israel, F.P., 1980, Ap. J. 236, 808-822.
Lester, D.F., Dinerstein, H.L., Rank, D.M., and Wooden, D.H., 1983b, Ap. J. (submitted).
Lester, D.F., Dinerstein, H.L., Werner, M.W., Watson, D.M., and Genzel, R.L., 1983a, Ap. J. (accepted).
McCarthy, J.F., Forrest, W.J., and Houck, J.R., 1979, Ap. J. 231, 711-719.
Melnick, G., Gull, G.E., and Harwit, M., 1979, Ap. J. 227, L35-L38.
Melnick, G., Gull, G.E., Harwit, M., and Ward, D.B., 1978, Ap. J. 222, L137-L140.
Moorwood, A.F.M., Baluteau, J.-P., Anderegg, M., Coron, N., and Biraud, Y., 1978, Ap. J. 224, 101-108.

Moorwood, A.F.M., Baluteau, J.-P., Anderegg, M., Coron, N., Biraud, Y., and Fitton, B., 1980a, Ap. J. 238, 565-576.
Moorwood, A.F.M., Salinari, P., Furniss, I., Jennings, R.E., and King, K.J., 1980b, Astron. Astrophys. 90, 304-310.
Osterbrock, D.E., 1974, Astrophysics of Gaseous Nebulae, pub. W.H. Freeman and Co.
Peimbert, M., and Costero, R., 1969, Bol. Obs. Tonantzintla y Tacubaya, 5, 3-22.
Peimbert, M., and Torres-Peimbert, S., 1977, MNRAS 179, 217-234.
Petrosian, V., 1970, Ap. J. 159, 833-846.
Schraml, J., and Mezger, P.G., 1969, Ap. J. 156, 269-301.
Simpson, J., 1975, Astron. Astrophys. 39, 43-60.
Spitzer, L., Jr., 1978, Physical Processes in the Interstellar Medium, pub. John Wiley and Sons.
Storey, J.W.V., Watson, D.M., and Townes, C.H., 1979, Ap. J. 233, 109-118.
Talent, D.L., and Dufour, R.J., 1979, Ap. J. 233, 888-905.
Ward, D.B., Dennison, B., Gull, G., and Harwit, M., 1975, Ap. J. 202, L31-L32.
Watson, D.M., and Storey, J.W.V., 1980, Int. J. IR and MM Waves 1, 609-629.
Watson, D.M., Storey, J.W.V., Townes, C.H., and Haller, E.E., 1980, Ap. J. 241, L43-L46.
Watson, D.M., Storey, J.W.V., Townes, C.H., and Haller, E.E., 1981, Ap. J. 250, 605-614.
Wilson, T.L., Fazio, G.G., Jaffe, D., Kleinman, D., Wright, E.L., and Low, F.J., 1979, Astron. Astrophys. 76, 86-91.

INFRARED SPECTROSCOPY OF LATE TYPE STARS

Stephen T. Ridgway

Kitt Peak National Observatory

The uses of infrared spectroscopy in the study of late-type stars are introduced. Topics include instrumentation, observing procedures, characteristics of the infrared stellar spectrum, and applications of low and high resolution spectroscopy.

Introduction

This review serves as an introduction to the general area of infrared stellar spectroscopy, and a review of recent results. The topics selected represent a compromise between those in which the most work has been done, and those which seem most important (currently or potentially). I have concentrated on results published since the last review of this subject (Merrill and Ridgway 1979). There are numerous exceptions, however, as some older results are cited for illustrative purposes, and some recent work falls outside the topics chosen for emphasis.

1. Spectrometers for Infrared Spectroscopy

The area of low resolution infrared spectrophotometry is currently dominated by techniques based on the circular variable filter (CVF). With this technique, an instrument with a single detector can cover several octaves of spectrum, providing quite a lot of information about the thermal characteristics of a source. However, the limitation to resolutions ≈ 100, and to signal-to-noise ≈ 100 means that the technique is limited to study of continua and to relatively strong emission or absorption features.

Grating spectroscopy is of course possible in the infrared. It is not widely used, in part because of the complications associated with maintaining the grating and other optics at cryogenic temperatures. However, the grating technique is more readily adapted to operation at higher resolutions, and detector arrays now under development will yield direct gains in observing speed when used in the dispersed spectrum of a grating device. The cryogenic grating spectrometer has a promising future in infrared astronomy.

At intermediate and high resolutions, typically 200 to 10^6, the Fourier transform spectrometer (FTS) has many applications. Most high resolution work in the 1 to 5 μm range has been with this instrument. For spectral measurements covering a substantial (> 1%) region of spectrum with a single detector, the FTS method is as fast or faster than other instrumental techniques. Numerous other factors, such as throughput, wavelength precision, achievable noise levels and available resolution generally favor the FTS method.

For moderately high resolutions ≈20000 the Fabry-Perot (FP) has been used effectively in applications where a small region of spectrum is required. Particularly in the vicinity of 10 μm where the thermal background levels are so high, the FP can provide good resolution to fainter limits than available with a contemporary FTS.

1.1 Observing Techniques in Infrared Spectroscopy

Since instrumentation for infrared spectroscopy is becoming generally available to the community, the potential user should be aware of some of the observing techniques used in the acquisition of infrared spectra. At wavelengths beyond about 2.5 μm the telescope optics and sky are always bright. Thus some type of differential measurement is required to extract the spectrum of the astronomical source from the foreground emission. This is generally built into the instrument in a fundamental way. Techniques similar to photometry often use rapid beam switching. FTS methods usually have a separate aperture off the source, and compensate the sky emission internally in the interferometric mixing.

Another area, where the responsibility falls squarely on the observer, is in the acquisition of appropriate standards. Throughout the infrared, molecular absorption from the terrestrial atmosphere is a fact of life. In practice it is possible to obtain data through spectral regions heavily obscured by terrestrial absorption, provided some care is exercised.

In the choice of standard, it is preferred that the source be free of intrinsic spectral features in the region of interest. The choice of standard will then depend on wavelength. A number of suggestions are collected in Table 1.

TABLE 1

λ	Comparison	Comments
1.0-4.0 μm	O or A star	Except near atomic H, He lines
	Sun	Lines generally sparse
Brackett gamma	IRC+10 216	
All Brackett lines	GIII star	Weak H, some weak atomic lines
⩾ 3.0 μm	Moon, mercury	
3.4-4.0 μm	IRC+10 216	A few weak features

A few generalities to remember are that F and G type stars tend to have few strong spectral features in the infrared. A type stars are very line free except for the atomic hydrogen lines, which are broad and smooth except at the core.

When working at high spectral resolution, $\lambda/\delta\lambda \geqslant 100000$, so that terrestrial line wings are resolved, it is possible to measure the atmospheric spectrum at one zenith distance and scale it to another zenith distance with confidence, as Lambert's law of absorption is approximately valid monochromatically. Since the atmospheric absorption can be scaled for other zenith distances, it is sometimes desirable, when working at high resolution, to record the standard for the terrestrial spectrum at large zenith distance and scale it to small zenith distance. In this way, the noise level is reduced (yes, it really works). When observing bright sources, this may be valuable, as the comparision is not always brighter.

When working at low resolution, the situation is quite different. It is not possible to scale the terrestrial absorption accurately - instead it is important to record the comparison source at the same zenith distance as the object, and if the object was observed over a range of zenith distances, proper allowance must be made in observing the comparison. In a source with weak features near terrestrial absorption, careful comparison acquisition is of utmost importance. With care, however, excellent results can be obtained. An example, with valuable accompanying discussion, is in Hall et al (1981).

2. An Overview of the Infrared Spectrum with Low Resolution and Wide-Band Spectrophotometry

By convention the term infrared spectroscopy is taken to include low resolution spectrophotometry. Low resolution spectrophotometric measurements in the infrared are of immense value in determining the radiation temperatures of cool sources (T < 5000K) since the peak of the Planck function is included in the infrared region. Furthermore, astrophysical sources typically exhibit a range of temperatures, generally decreasing in the outward radial direction, with the result that lower temperature regions of greater spatial extent can leave a distinguishing signature in the longer wavelength regions of a spectrum without obscuring the shorter wavelength emissions dominated by hotter parts of the source.

A convenient introduction to the infrared region of the stellar spectrum is available through consideration of the uses of wide-band spectral coverage observations of a survey nature. Such work has been done from the ground, and more recently from aircraft, often employing CVF techniques. The spectral regions covered have been throughout the infrared. The region from 1 to 5 μm is particularly useful for determining the total flux from a stellar photosphere, or the photospheric spectral distribution, both of which are useful in inferring the kinetic temperature conditions in the photosphere. The spectral region from 2 to 40 μm and beyond is useful for detection and diagnosis of circumstellar emission or absorption from dust.

An important contribution in this area came from Merrill and Stein (1976a,b,c). The systematic implementation of an established observational technique, insistence on high signal-to-noise, attention to calibration, and survey of a range of objects, provided an excellent set of observational material. Representative spectra from a more recent publication (Merrill 1977) are reproduced here to provide an overview of the 2 to 12 μm spectral region. Figure 1 shows the development of the 2 to 4 μm spectral region for a sequence of spectral types, from early KIII at the top to cool M, S and C at the bottom. The most prominent spectral features have been identified from high resolution studies. In the K2III and M3III spectra the depression longward from 2.3 μm is due to the CO molecule. In fact this molecule is abundant in all cool photospheres, and the CO depression may be discerned in all of the spectra in Figure 1. This is the first overtone of a vibration-rotation sequence, and the fundamental near 5 μm should be much stronger, though for a variety of reasons it is generally less obvious in low resolution spectra. The late M spectral types, M6 and cooler, show strong depression due to H_2O both at 2 μm, and at 2.2 to 3.8 μm. The S type star shows CO only at this resolution. The C type stars show a more complex spectrum in the 2 to 2.5 μm region where numerous electronic CN bands, not individually distinct, depress the

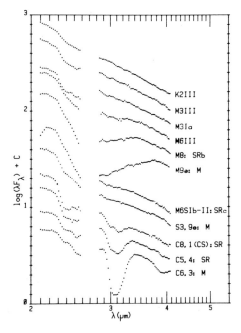

Figure 1. 2-4 μm spectra of late-type stars, $\lambda/\delta\lambda \approx 100$, from Merrill (1977).

spectrum. A prominent band at 3 to 3.3 μm is due primarily to the polyatomic molecules HCN and C_2H_2, which are seen only in spectra of cool C stars.

These qualitative characteristics of the infrared spectra are understood in terms of the photospheric abundances of carbon and oxygen. The importance of the C/O abundance ratio lies in the tendency, in giant stars cooler than $T \approx 4500K$, for CO to form until one of the elemental constituents is strongly depleted. Stars of the types K and M are understood to have an oxygen abundance exceeding the carbon abundance. Thus after formation of CO, substantial oxygen remains to form oxygen bearing molecules such as OH, SiO, and H_2O. In C type stars, with a carbon abundance exceeding the oxygen abundance, formation of CO leaves a significant carbon abundance for the formation of CN, C_2, and the polyatomic carbon molecules. The S type stars, finally, are thought to have carbon and oxygen abundances roughly equal, so that little carbon or oxygen remains, and except for CO, none of the molecular species mentioned is present in abundance.

An extended spectral range is shown in Figure 2. The 8 to 13 μm coverage picks up the most prominent emission bands of the circumstellar dust associated with many very cool or very high luminosity stars.

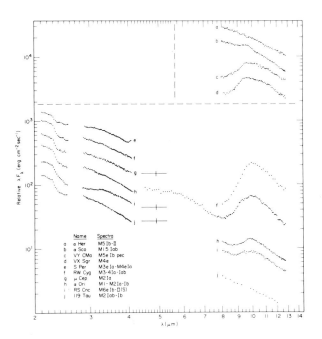

Figure 2. 2-12 μm spectra of cool M type stars, $\lambda/\delta\lambda \approx 100$, from Merrill and Stein (1976).

Based on plausibility arguments, and the approximate match of the infrared emission to laboratory or computed spectra, the dust material has been tentatively identified. The plausibility arguments rest on the expected availability of elemental species after formation of CO. The dust material in the shell of oxygen rich stars is though to be in the form of silicates. The carbon rich stars are thought to form dust including silicon carbide grains and some other carbon condensate.

From Figure 3 it is possible to see at a glance which regions of the infrared spectrum are available from the ground. The correspondence between regions of low average atmospheric transmission and gaps in the stellar spectrum is approximately indicative of the available regions for ground based spectroscopic work. Remember, however, that at high spectral resolution there are both windows in holes and holes in windows.

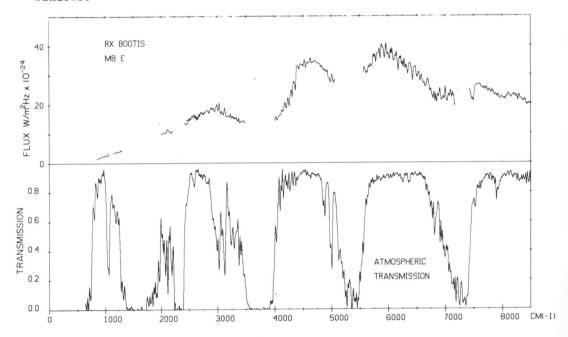

Figure 3. FTS spectra covering the range 700 - 2600 cm^1 and 2400 - 8500 cm^1 have been combined to give an overview of the entire region. Resolution ≈ 30 cm^1, or $\lambda/\delta\lambda \approx 150$. The stellar spectrum has been reduced photometrically to give absolute fluxes. The noise level in the region $\leqslant 3500$ cm^1 is high.

By working from a high altitude observatory the hindrance of the terrestrial atmosphere is reduced. An airborne telescope can cover almost all of the infrared. Aircraft spectra covering part of the infrared range, and at least some of the spectral region unobservable from the ground, have been reported in the last few years. Strecker et al (1979) observed a group of K and early M giants. Strecker et al (1978) obtained spectra of several Mira type variables. Goebel et al (1980, 1981) observed a number of C type stars. An example is the spectrum of the C star YCVn, from 1.2 to 30 μm. This is reproduced as Figure 4. The importance of molecular bands in determining the flux distribution of such a star is evident from the appearance of the spectrum, and the authors note the association between many of the strong absorption features and the bands expected for several molecules known to play a role in C star atmospheres. Data such as this serves two purposes: it provides the most complete overview of the spectrum of a star of this type; and it provides quantitative information for anyone bold enough to attempt an analysis.

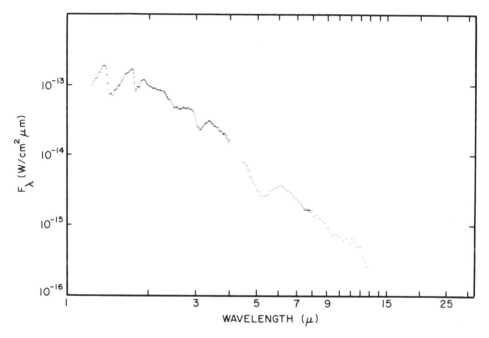

Figure 4. Spectra of the C star Y CVn from aircraft altitude, $\lambda/\delta\lambda \approx 100$, from Goebel et al (1980).

An important use of broadband infrared spectrophotometry such as this is in the selection and evaluation of model atmospheres. For the K and M type stars, a comparison of observations to theoretical models is possible now or in the near future. Initial efforts in this direction have been described by Tsuji (1978a, 1981), Scargle and Strecker (1979), Augason et al (1980), and Manducca et al (1981). The case of carbon stars is more complex, as the stars are cooler and the opacities more difficult.

2.1 Higher Resolution

The appearance of late-type stellar spectra changes rapidly with resolution, especially for low resolutions. The next three figures show spectra with resolution only very slightly higher than available from the CVF technique, but it will be seen that the information content is very much greater as a result of partially resolving some of the dominant spectral structure.

Figure 5 presents a series of FTS spectra for a range of cool giants, K0 to M7. The resolution, $\lambda/\delta\lambda \approx 150$, might be suitable for spectral classification. In these spectra, the CO overtone bands around 4000 - 4360 cm^1 and 5800 - 6400 cm^1 are partially resolved, so that bandheads of the rotational sequences are prominent. In addition, numerous atomic lines are evident, often as blends of several especially strong features.

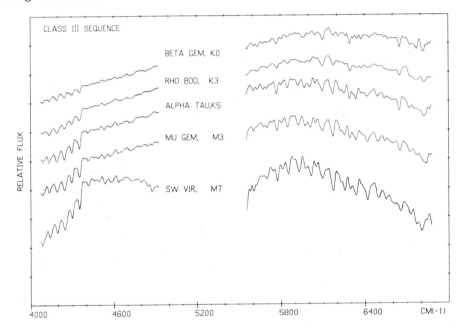

Figure 5. A sequence of K and M giant spectra with resolution ≈ 30 cm^1, or $\lambda/\delta\lambda \approx 150$. The data has been reduced photometrically to give correct relative fluxes. The author's unpublished data.

Examples of less common cool spectral types appear in Figure 6. Dramatic differences between the M type and C type stars are attributed to the strong line blanketing by CN and C_2. The Ballick-Ramsey bands of C_2 provide the depression near 5600 cm^1. The peculiar shape of the C star continuum is due primarily to CN. Note that near 6400 cm^1 the flux is depressed to less than 50% of the "continuum" level near 5800 cm^1. But in fact, the true continuum is substantially higher than would be guessed from the low resolution spectrum.

A series of spectra of the cooler M spectral types is shown in Figure 7, at a slightly higher resolution $\lambda/\delta\lambda \approx 250$. Even the hottest of these has strong depression of its spectrum due to H_2O. The spectra of such stars as these and those of Figure 6 represent part of the terra

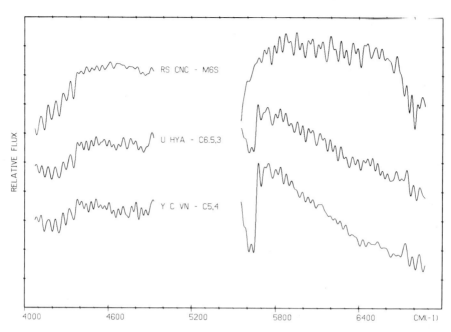

Figure 6. 4000-6800 cm^{-1} spectra of MS and C stars, $\lambda/\delta\lambda \approx 250$. The spectra have been reduced photometrically. The author's unpublished data.

incognita of infrared spectroscopy. The spectra contain thousands of spectral lines, and only a small fraction have been identified. (So far as I know, no one has even made a serious attempt).

In the study of stars, low resolution spectra such as those described above are used primarily for determination of approximate spectral types. Several recent topics are reviewed below.

2.2 Recent Results from Infrared Spectrophotometry of Late-Type Stars

Certain classes of stellar objects will be studied in the infrared because they are only detectable at those wavelengths. These include heavily obscured stars and very cool stars.

Obscuration may arise in conventional interstellar extinction, as in observing distant supergiants through the galactic disk. It may also arise more locally to the source of interest, as in the dense molecular clouds which cloak the youngest stellar objects. In both cases, the typical wavelength dependence of extinction, proportional to $1/\lambda$, insures that heavily obscured stars are more readily detected and studied in the infrared than in the visible. As an example, an M0 supergiant at the distance of the galactic center with no extinction would have stellar magnitudes of $M_V \approx 8$, and $M_K \approx 5$, whereas the

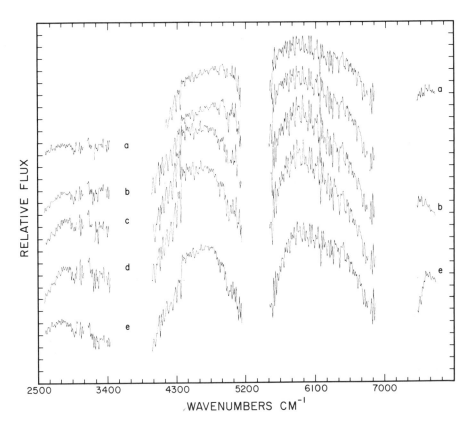

Figure 7. A selection of very cool late-type stars. The resolution is ≈20 cm^{-1}. The data has been reduced photometrically to give correct relative fluxes. a) RLyr; (b) SwVir, M7; (c) RXBoo, M7-8; (d) and (e) RLeo, M7-9.

presence of 30 visual magnitudes of extinction (estimated) would result in stellar magnitudes of $M_V \approx 38$, and $M_K \approx 8$. This is an example of substantial current interest. Low resolution spectrophotometry has been used to identify cool stars in the galactic center (Neugebauer et al 1976). Sources with CO absorption are presumed to be high luminosity cool stars or clusters of lower luminosity cool stars. More recently, followup studies with moderate resolution FTS spectroscopy (Wollman et al 1982, Hall et al 1982, Lebofsky et al 1982) have provided more detailed observational results which allow improved determination of spectral types. Measurement of radial velocities may yield estimates of the mass distribution from virial theorem arguments.

Cool stars in close association with hotter stars may be difficult to detect, and more difficult to identify. A now well established example is the case of the symbiotic stars. A discussion of these stars is given by Aitken (1983).

The study of possible precursors to planetary nebulae remains a very current topic, in which the outcome is still unclear. Infrared CVF and FTS spectroscopy of the source HM Sagittae (Puetter et al 1978, Thronson and Harvey 1981) show evidence of CO and H_2O absorption characteristic of a cool star, suggesting that in this case the transition from cool star with shell to planetary nebula (Kwok 1981) is still in the future. Analogous arguments apply to IRC + 10 216.

Some of the most interesting applications of infrared spectroscopy for diagnostic purposes may be in the inspection of very cool stellar objects. Two examples are the detection of cool companions to T Tauri (Dyck et al 1982) and to ζ Aquarii B (McCarthy et al 1982). These were detected only because of their binary effects or proximity to other interesting objects. Both are among the early results of infrared speckle interferometry, and may be the forerunners of whole new classes of objects. A determination of the nature of these sources, when it is done, will be based on infrared spectroscopy.

Stars may be obscured in yet another manner, when they eject such great quantities of gas and dust that they are effectively hidden by large extinction. In this case, however, it is not true that the infrared regime may be used to penetrate the extinction. If the dust shell has a large optical depth in absorption, it will also have a large effective emissivity, and since the dust is heated by proximity to the star, thermal emission at infrared wavelengths increases more strongly than the opacity decreases. Thus the infrared reveals the veiling dust rather than the underlying star. In fact the best wavelength for directly detecting the stellar radiation is in the very near infrared, around 1 μm, where the scattering is moderate, but the dust emission is still small. Spectroscopy in this region is rarely used, although new detectors may change that situation.

3. A Closer Look with High Resolution Spectroscopy

Spectra discussed in the previous sections were of low resolution, generally sufficient to distinguish some molecular bands and groups of atomic features. We will now move directly to resolutions sufficient to resolve individual molecular and atomic lines. This will mean, for typical giant type stars, a resolution $\lambda/\delta\lambda \approx 50000$ or greater. It should not be supposed that such resolution is limited to a small number of stars. With a 4 meter telescope, and current FTS techniques, such resolution is available on stars of magnitude $M_K \approx +4$. (This is a magnitude fainter than the deepest infrared sky survey). Thus bright representatives of many important stellar types are accessible to study. Further, it may be expected that developments in IR detector and spectrometer technology will extend the magnitude limit by 2 or 3 within the decade.

The infrared spectrum will normally be required for one of the following reasons: the molecular bands of CNO; the CNO isotope ratios; some difficulty in working in the shorter wavelength regions, such as faintness of the source or excessive line blanketing. Be warned, however, that little work has been done with the infrared spectrum. Atomic lines have not been selected for abundance studies, many atomic and molecular line parameters are not known, and model atmosphere techniques are not obviously optimized for study of the infrared spectrum.

Few examples of high resolution infrared spectra have been published. I have selected several examples to give an idea of what can be done. Figure 8 shows, from the top, spectra of α Orionis, α Tauri, and the terrestrial atmospheric transmission. The spectra of the stars have been corrected for the terrestrial absorption and Doppler corrections applied. These spectra are in the atmospheric window near 4 μm. The first thing to note is the sparseness of the spectra. Line blending is not a serious problem, nor is location of the continuum. Lines of the fundamental bands of OH and NH are noted below the top spectrum. Below the middle spectrum are noted the strongest unidentified spectral lines. These are mostly atomic lines, and we are investing some effort in identifications.

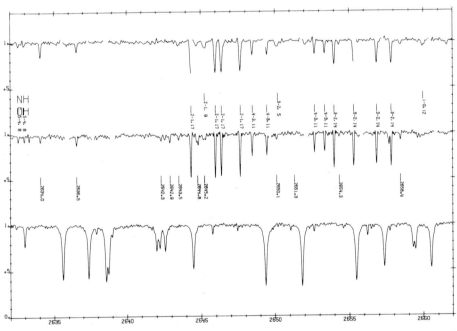

Figure 8. From the top, α Orionis, α Tauri, and the transmission of the terrestrial atmosphere. Resolution is $\approx .05$ cm^1, $\lambda/\delta\lambda \approx 50000$. The stellar spectra have been corrected for terrestrial absorption, with gaps in the spectrum to indicate where the correction was in doubt. The Doppler shifts have been removed. The author's unpublished data.

Figure 9 shows a nearby spectral region for three very different stars. From the top, these are o Ceti (Mira), R And (an S type Mira star), and TX Psc (a C type star). The first thing we notice is the higher density of strong lines. This is attributed primarily to the much lower atmospheric temperatures. The lower temperature results in lower continuum opacity and increased molecular association. From the discussion earlier of the M, C, and S star differences, it may be expected that these spectra will show substantial differences, and this is indeed the case. Here the simplest spectrum is that of TX Piscium. Lines of the CH fundamental band are noted scattered through the spectrum. Also the overtone band of CS is prominent near 4 µm. Most of the remaining spectral lines, and most of the lines in the figure, are probably due to CN. The density and strength of lines are about as predicted for CN. No accurate list of line positions is available, however, so a line by line study would require some work. In R And, a number of HCl fundamental lines are noted, with P Cygni type profiles. In addition, SiO appears near 4 µm. Otherwise, the lines remain unidentified. The spectrum of o Ceti is heavily blanketed with lines, presumably H_2O. Most likely surprises lurk in each of these spectra.

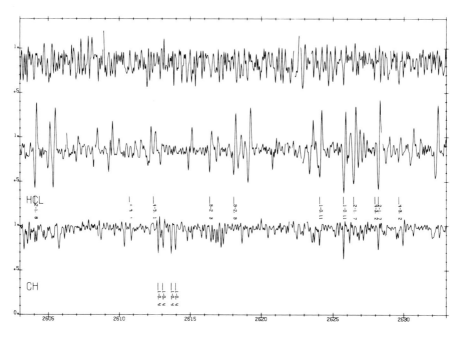

Figure 9. From the top, o Ceti, R Andromeda, and TX Piscium.

Similar spectra could be shown of any part of the infrared, and the discussion would be very similar - a few prominent features identified, much of the spectrum still unscrutinized. Rather than show further such examples, Figure 10 represents a slightly different type of information. Here we plot the "local apparent continuum", determined from the high points in high resolution spectra, but plotted at low resolution. Thus it is as close as one can come to a true observational continuum for these stars. The obvious disturbance near 4000 cm^1 is the

CO bands, and the gap from 4800 - 6600 cm^1, is due to terrestrial absorption. In these spectral distributions, the shift of the Planck peak into the region can be noted, and also the development of the H$^-$ opacity minimum near 6000 cm^1. Such information may be used as a diagnostic for model studies, or to assist in interpolation of a continuum over large ranges of spectrum.

In the following sections, we will consider the high resolution infrared spectroscopic studies recently reported. The topics covered will be temperature structure, mass motions, magnetic fields, pulsation, mass loss, elemental abundances, and the role of the sun in infrared astronomy.

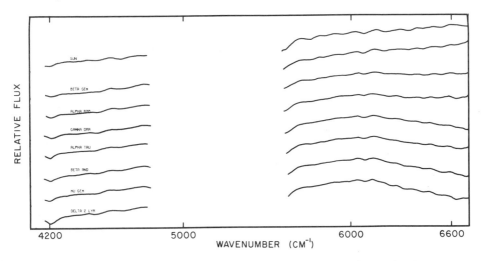

Figure 10. Local apparent continuum determined from high resolution spectra (resolution ≈.03 cm^1). Result smoothed slightly and plotted at low resolution. Author's unpublished data.

3.1 The Role of the Sun

Before continuing, it is perhaps appropriate to insert a short aside regarding the importance of the sun in the development of infrared stellar spectroscopy. As will be noted below in specific cases, study of the sun is critical in most applications of high resolution spectroscopic techniques. Many atomic and molecular transitions do not have strengths or other parameters well determined from laboratory studies. The sun often serves as a calibrator, so that analysis may be done differentially to the sun, which is often the preferred result in any case. The interpretation of line formation phenomena, originating in convection, chromospheric effects, turbulence and inhomogeneities of any kind must begin with the example of the sun, where these effects are present in abundance, though different degree and proportion. It is quite clear in surveying the literature for infrared results that without reference to the sun for baseline, zeropoint or concept, results for other stars would enjoy a low level of confidence.

3.2 Temperature and Temperature Structure of the Stellar Atmosphere

Theoretical models of stellar atmospheres are generally indexed by a few standard parameters, typically (T, log g) and, perhaps, an abundance or turbulence parameter. For the infrared spectroscopist, the suitability of a particular model must be considered in doubt until considerable attention has been given to its verification. This is in part because of the developmental status of the theory and practice of cool star models. The selection of a model temperature T would normally proceed with an attempt to match the broadband spectral distribution, especially in the infrared. Detailed discussion of this problem is presented by Manducca et al (1981). It is then wise to consider whether the temperature structure of the model represents that of the star.

Beginning with the easier cases first, consider the use of the infrared CO bands as an indicator of the temperature structure of the solar photosphere. Tsuji (1977) and Ayres and Testerman (1981) found that the CO fundamental band was unexpectedly shallow, suggesting higher temperatures than expected in the outer photosphere. They concluded that a multicomponent model is required, as also suggested by other studies. In the case of the sun, this may be a small correction, but in the case of much cooler, much higher luminosity stars it may be otherwise. Use of the CO fundamental as a diagnostic for a temperature rise in the upper atmosphere of α Bootis (K2III) by Heasley et al (1977) failed to show the temperature rise predicted on the basis of other observations, suggesting inhomogeneity with substantially different temperature structures in different spatial regions. A dramatic temperature rise relatively deep in the photosphere has been proposed to account for the flux distribution and emission line spectrum of T Tauri stars (Calvet-Cuni 1981). So with cool stars generally, the rule should be to proceed with caution.

The further specification of a model atmosphere for a star requires an evaluation of the pressure. This might in principle be derived from pressure sensitive molecular abundance ratios available in the infrared, but in fact the technique has not been developed, in part because the temperature dependence of the molecular equilibrium tends to be stronger than the pressure dependence.

In a related group of observations, Goorvitch et al (1980), and Tsuji (1982) have noted a potentially serious problem for cool star models. At temperatures below ≈2500K the predominant form of hydrogen in equilibrium should be in molecular H_2. Until recently, molecular H_2 had been reported in the S stars R And and χ Cyg, and nowhere else. The absence of H_2 in cool M stars may be attributed to a number of problems with the models, among which Tsuji favors an underestimate of the temperature of the upper layers of the photosphere. The case of the C stars may be more serious, as in the cooler C stars H_2 should be associated to some depth, and a small error in the model would not appear to explain the discrepancy. If in fact the dominant constituent of the atmosphere is out of equilibrium in an unpredicted way, perhaps due to some combination of small convective timescales and unexpectedly low pressures, the models would be in trouble. Recently however, Johnson et al (1982) have reported the identification of the s(1) line of H_2 at 4712cm^1 in four of the coolest in a group of 15 C stars studied. They suggest that a deep chromosphere may account for the absence of H_2 in the hotter stars studied, but there is still a problem for the stars of intermediate temperature.

3.3 Mass Motion in the Stellar Atmosphere

Spectroscopic measurements will provide the major observational data for understanding convection and other mass motions in stellar photospheres. In the sun, convective motions have been shown to account for both micro- and macroturbulent broadening of spectral lines. In cool supergiants, line widths appear to indicate near supersonic velocities, indicating the importance of distinguishing the nature of mass motions in late type stars. The most promising approach is to proceed with fully resolved line profiles of well understood spectral lines, and to carry out the interpretation building on our understanding of the sun and convection. One study has been reported in the photoelectric region (Gray 1982). A first examination of CO line profiles at 2.3 microns (Ridgway and Friel 1981) has shown that similar observations are possible in the infrared, and that the problems of line blending are smaller, permitting the technique to be extended to cooler stellar types. An example is shown in Figure 11. Here a number of CO lines have been overplotted on a common velocity scale (bottom of the figure). Bisectors of the line profile chords at various depths have been overplotted on an expanded scale (top of the figure) and shifted upward for clarity. The distribution of bisectors gives an idea of the line shape deviations from symmetry. Solar lines, for example, tend to give a bisector curve bent to the right at both top and bottom, for a so-called C shape. Our understanding of this shape is that, crudely speaking, the top of the bisector tends to give a mean velocity, the middle tends to show the effects of rising material (due to convection), and the bottom of the profile tends to show the overlying stationary material. In general, the cool stars tend to exhibit an S shape bisector, as seen here. This may be due to the effects of outward mass flow (mass loss) overlying the convective layers. Detailed discussion of the sun is available in Dravins et al (1981), with a review by Dravins (1982).

It is likely that a great deal of stellar physics will be learned from study of line profiles. To aid in disentangling the phenomena which contribute to the line shapes, it will be useful to correlate changes in line profiles with spectral types, and for a single star, with photometric variations (ubiquitous among the cooler stars), and with other mass loss indicators.

3.4 Stellar Magnetic Fields

In some special cases, careful examination of the line profile may lead to a measure of the magnetic field strength. Worden et al (1982) have recorded infrared spectra of the G8IV star λ And. This is a star

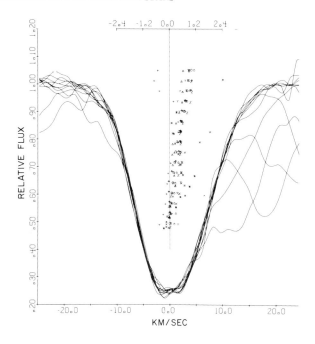

Figure 11. Overplotted line profiles from CO 2-0 band, for the star δ_2Lyrae, $\lambda/\delta\lambda \approx 170000$. The lower velocity scale applies to the profiles, the upper velocity scale applies to the bisectors. The bisectors have been shifted up for clarity. Reported in Ridgway and Friel (1981).

of the RSCVn type, believed to have extensive surface activity in the form of star spots covering a substantial fraction of the stellar surface. By analogy with solar sunspots, a strong magnetic field is expected. The detectability of Zeeman magnetic splitting of sensitive lines should go as the wavelength according to Worden et al, so they worked in the 1.6 micron region. They detect the Zeeman splitting, and partially resolve the individual components. The authors estimate that approximately 20% of the surface area of the star presented a magnetic field of several kologaus at the time of their observation.

3.5 Stellar Pulsation

Proceeding to yet more difficult problems, the infrared spectra provide unique access to the pulsation phenomena prominent in cool stars, and especially in the extreme case of the Mira type stars. Hinkle and collaborators (1979, 1982) have laid out the first reasonably detailed observational basis for understanding the Mira pulsations. As part of the photospheric gas rises at supersonic velocity with a shock wave, part of the material so accelerated in the previous cycle is slowing and falling back. Additional material has been ejected into a cool, low velocity "reservoir" above the photosphere, and at greater distances material leaving the star at greater than escape velocity is found. Each component of gas leaves its trace in the spectrum, and these traces may be followed through the pulsational cycle. An example of the spectroscopic evidence is shown in Figure 12, where the line doubling due to multiple layers in the pulsating part of the photosphere may be readily seen.

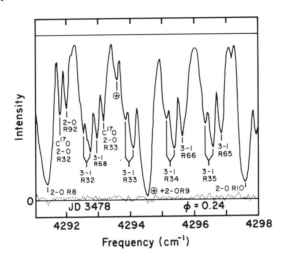

Figure 12. A small section of one spectrum of the mira variable χ Cygni, showing the splitting of the CO line into multiple components due to pulsation in the atmosphere (Hinkle et al 1982).

3.6 Stellar Mass Loss

The discussion of pulsation leads naturally to the issue of mass loss. Here we will concentrate on the aspects of mass loss that relate most directly to knowledge of the stellar photosphere. Mass loss in cool stars is not understood very well. Studies have shown that radiation pressure on dust grains can drive mass loss (Kwok 1975), but a problem remains: it appears that dust may not condense close to the star where the gas density is high, and it is not clear how to obtain a high gas density at a distance from the star where dust formation can occur. The classes of solutions appear to include: raising the density above the static equilibrium values by pulsation (Wood 1979), convection, or other non-radiative means, or by forming dust in cool spots (Salpeter 1974). While the theoretical side still has numerous difficult issues, the observational side is improving steadily, in large part due to improvements in infrared observations.

Work by Boesgaard (1979 and 1981) with Fe emission lines in the blue has contributed substantially to the mapping of gas flows in the very important region inside of about 2 R_*. Infrared studies (Bernat et al 1979, Bernat 1981, Ridgway 1981) have revealed the complexity of gas flows in the region to ≈200R_*, with clumping of the gas into distinct velocity structures of different excitation temperatures. The observations are highly suggestive of episodic mass ejection, with modulation periods of 50 to 200 years, a range which does not match any predictions. However, detailed modeling of one particular case, the extremely thick shell of the C type star IRC+10216 (Keady 1982) shows that continuous flow with several regimes of sudden acceleration can also account for the apparent shell structure. Further understanding of the mass loss phenomenon through infrared spectra can be expected, but it will be achieved painstakingly through detailed modeling of complex physical environments, and the input physics, the boundary conditions, and many of the material properties are among the parameters to be determined.

A somewhat different and dramatic example of mass loss study with infrared spectroscopy is the case of ρ Cas (FIa). Lambert et al (1981) have followed the evolution of a discrete shell which was ejected, presumably with some violence, in 1975. It has slowed and begun to fall back. Pulsations in the underlying photosphere are also observed. Monitoring the pulsations through an ejection episode could be very helpful in understanding the ejection mechanism.

4. Abundances of the Elements and their Isotopes

This, the culminating section of the review, is the appropriate place to recount all that has been learned about the composition of cool stars from infrared spectroscopy. Glancing ahead, the reader will note that it is a short section. This is because the few researchers active in the study of late type stars have thus far found the best investment of their time to concentrate on the somewhat earlier of the late type stars, which present somewhat fewer difficulties, and to push the new techniques in the photoelectric region, with red sensitive, high dynamic range detectors providing observational material of a relatively familiar type but greatly improved quality. This approach has permitted study of stars well into the K spectral region (T⩾4000). Whether it can be extended to the early M spectral region and beyond is doubtful. Results through approximately 1979 have been reviewed (Merrill and Ridgway 1979). The few recently reported uses of infrared spectra in abundance analysis of cool star follow.

Infrared spectra have been used in a study of CNO abundances in barium stars. Smith et al (1980) observed the CO overtone band of ζ Capricorni at 2.3 μm. Their estimate of the carbon abundance was based equally on the infrared CO and photographic spectral region C_2 lines. The star is hot enough that CO is not fully associated, and observing both molecules is significantly more secure than relying on one alone. In any event the choice of model is critical. An extended study of barium stars by Sneden et al (1981) was based on photographic and reticon spectra. An infrared spectrum of the 2.3 μm CO in o Vir was examined as a consistency check.

Ries (1981) and Lambert and Ries (1981) referred to infrared spectra in the 1.4 to 2.5 μm region in a study of CNO abundances in G and K giants. Again, CO was used to confirm the carbon abundance analysis from shorter wavelength data. They note that CO and OH are good temperature indicators. For example, CO and C_2 have opposite dependence on temperature, so requiring that both molecules indicate the same abundance of carbon (via molecular equilibrium) determines the temperature. Similarly with OH and forbidden OI. They note that NH should be relatively insensitive to pressure, and discuss the extension of their analysis to M stars based more completely on infrared spectra.

Tsuji (1979) has come as close as anybody to determining CNO abundances from infrared data. In a study of α Ori, the selection of model was based strongly on the infrared flux distribution. The abundances were determined from fit of synthetic spectra to infrared CO, OH and CN lines. The results, below solar C, above solar N, and possibly no change in O, were generally consistent with several other studies, although considerable uncertainty remains, especially in the oxygen abundance. As a broad lined supergiant, α Ori is not the easiest M type star to work on. No doubt it will be reexamined again from time to time as our understanding of the dynamic phenomena in the photosphere improves.

In 1979 we summarized the early attempts to determine CNO abundances in C stars from infrared spectra. No additional work has been done since, so I will simply repeat my opinion that Thompson's (1977) results are rather convincing that the C enrichment can be attributed to mixing of additional carbon to the surface. The alternative of diluting the surface material with CN cycle processed material from below, which was considered an alternative for many years, simply does not seem to produce the great observed strengths of the CN and C_2 bands. On the other hand, it would be reassuring to see this question reexamined in light of the improved estimates of C star effective temperatures that are now available.

I am not aware of any major new results in the determination of isotopic ratios in late type stars, although this should be a much easier task than elemental abundance determinations. I understand privately, however, that a collection of results based on infrared spectra should be available in the not too distant future.

A thoughtful review of the results for elemental and isotopic abundances of G and K giants and supergiants, including the few results based on infrared spectroscopy, has been given by Lambert (1981). He reported depletion of C and enchancement of N relative to hydrogen, and enhancement of C^{13}/C^{12}, qualitatively as predicted from convective mixing of processed material, and perhaps quantitatively consistent with "dredging up" of additional processed material from the core, also by convective phenomena.

5. Future Prospects

Within the limitations of existing instrumentation and techniques, we can foresee development of infrared spectroscopy in several directions. Low resolution techniques will be used to identify and classify sources found in deeper infrared surveys or by other techniques. High resolution methods are amply developed for extensive study of stellar photospheres, abundances, and mass loss. The major developments will be in the theoretical and computational areas, and the acquisition of laboratory data. The techniques for study of line profiles can be greatly improved with straightforward changes in instrumental technique. Satellite or space shuttle observatories will give access to regions of spectrum scarcely surveyed from the ground. Infrared spectroscopy at all resolutions will gain at least several an order of magnitude in sensitivity with introduction of integrating array type detectors.

6. References

Aitken, D. (1984). This volume.
Augason, G. C., Taylor, B. J., Strecker, D. W., Erickson, E. F. and Witteborn, F. C. (1980). Ap. J. 235, pp. 138-145.
Ayres, T. R. and Testerman, L. (1981). Ap. J. 245, pp. 1124-1140.
Bernat, A. P. (1981). Ap. J. 246, pp. 184-192.
Bernat, A. P., Hall, D. N. B., Hinkle, K. H. and Ridgway, S. T. (1979). Ap. J. 233, L135-L139.
Boesgaard, A. M. (1979). Ap. J. 232, pp. 485-495.
Boesgaard, A. M. (1981). Ap. J. 251, pp. 564-570.
Calvet-Cuni, N. P. (1981). Dissertation, University of California at Berkeley.
Dravins, D., Lindegren, L. and Nordlund, A. (1981). Astron. Astrophys. 96, pp. 345-365.
Dravins, D. (1982). Ann. Rev. Astron. Astrophys. 20, pp. 61-89.
Dyck, H. M., Simon, T. and Zuckerman, B. (1982). Ap. J. 255, pp. L103-L106.
Goebel, J. H., Bregman, J. D. Witteborn, F. C., Taylor, B. J. and Willner, S. P. (1981). Ap. J. 246, pp. 455-463.
Goebel, J. H, Bregman, J. D., Goorvitch, D., Strecker, D. W., Puetter, R. C., Russell, R. W., Soifer, B. T., Willner, S. P., Forrest, W. J., Houck, J. R. and McCarthy, J. F. (1980). Ap. J. 235, 104-113.
Goorvitch, D., Goebel, J. H. and Augason, G. C. (1980). Ap. J. 240, pp. 588-596.
Gray, D. F. (1982). Ap. J. 262, pp. 682-699.
Hall, D. N. B., Klienmann, S. G. and Scoville, N. Z. (1982). Ap. J. 262, L53-L57.
Hall, D. N. B., Kleinmann, S. G., Scoville, N. Z. and Ridgway, S. T. (1981). Ap. J. 248, pp. 898-905.
Heasley, J. N., Ridgway, S. T., Carbon, D. F., Milkey, R. W. and Hall, D. N. B. (1977). Ap. J. 219, pp. 970-978.
Hinkle, K. H. and Barnes, T. G. (1979). Ap. J. 234, pp. 548-555.
Hinkle, K. H., Hall, D. N. B. and Ridgway, S. T. (1982). Ap. J. 252, pp. 697-714.
Johnson, H. R., Goebel, J. H., Goorvitch, D. and Ridgway, S. T. (1983). Ap. J. Lettr. (in prep).
Keady, J. J. (1982). The Circumstellar Envelope of IRC 10216. Dissertation. New Mexico State University.
Kwok, S. (1975). Ap. J. 198, pp. 583-591.
Kwok, S. (1981). "Physical Processes in Red Giants", ed. I. Iben and A. Renzini. D. Reidel, pp. 421-425.
Lambert, D. L. (1981). See Kwok (1981), pp. 115-134.

Lambert, D. L., Hinkle, K. H. and Hall, D. N. B. (1981). Ap. J. 248, pp. 638-650.
Lambert, D. L. and Ries, L. M. (1981). Ap. J. 248, pp. 228-248.
Lebofsky, M. J., Rieke, G. H. and Tokunaga, A. T. (1982). Ap. J. 263, pp. 736-740.
McCarthy, D. W., Low, F. J., Kleinmann, S. G. and Arganbright, O. V. (1982). Preprint.
Manducca, A., Bell, R. A. and Gustafsson, B. (1981). Ap. J. 243, pp. 883-893.
Merrill, K. M. and Stein, W. A. (1976a). PASP 88, pp. 285-293.
Merrill, K. M. and Stein, W. A. (1976b). PASP 88, pp. 294-307.
Merrill, K. M. and Stein, W. A. (1976c). PASP 88, pp. 874-887.
Merrill, K. M. (1977). "The Interaction of Variable Stars with their Environment", IAU Colloquium No. 42, ed. R. Kippenhahn, J. Rahe, and W. Strohmeier. Veroff. Bamberg 11, pp. 446-494.
Merrill, K. M. and Ridgway, S. T. (1979). Ann. Rev. Astron. Astrophys. 17, pp. 9-42.
Neugebauer, G., Becklin, E. E., Beckwith, S., Matthews, K. and Wynn-Williams, C. G. (1976). Ap. J. 205, pp. L139-L141.
Puetter, R. C., Russell, R. W. Soifer, B. T. and Willner, S. P. (1978). Ap. J. 223, pp. L93-L95.
Ridgway, S. T. (1974). IAU Highlights of Astronomy, ed. G. Contopoulos. (Dordrecht: Reidel) pp. 327-339.
Ridgway, S. T. (1981). "Physical Processes in Red Giants", eds. I. Iben and A. Renzini (Dordrecht, Reidel), pp. 305-309.
Ridgway, S. T. et al. (1983). Ap. J. Supp. Ser. (in press).
Ridgway, S. T. and Friel E. D. (1981). "Effects of Mass Loss on Stellar Evolution", ed. C. Chiosi and R. Stalio. D. Reidel, pp. 119-124.
Ries, L. M. (1981). Dissertation, University of Texas at Austin.
Salpeter, E. E. (1974). Ap. J. 193, pp. 585-592.
Scargle, J. D. and Strecker, D. W. (1979). Ap. J. 228, pp. 838-853.
Smith, V. V., Sneden, C. and Pilachowski, C. A. (1980). PASP 92, pp. 809-818.
Sneden, C., Lambert, D. L. and Pilachowski, C. A. (1981). Ap. J. 247, p. 1052.
Strecker, D. W., Erickson, E. F. and Witteborn, F. C. (1978). Ap. J. 83, pp. 26-31.
Strecker, D. W., Erickson, E. F. and Witteborn, F. C. (1979). Ap. J. Suppl. 41, pp. 501-512.
Thompson, R. I. (1977). Ap. J. 212, pp. 754-759.
Thronson, H. A. and Harvey, P. M. (1981). Ap. J. 248, pp. 584-590.
Tsuji, T. (1977). PASJ 29, pp. 497-510.
Tsuji, T. (1978). Astron. Astrophys, 62, p. 29.
Tsuji, T. (1979). New Zealand J. Sci. 22, pp. 415-428.
Tsuji, T. (1981). J. Astrophys. Astron. 2, pp. 95-113.
Tsuji, T. (1982). Preprint.
Wollman, E. R., Smith, H. A. and Larson, H. P. (1982). Ap. J. 258, pp. 506-514.
Wood, P. R. (1979). Ap. J. 227, pp. 220-231.
Worden, S. P., Giampapa, M. S., Boice, D., Robinson, R. D. and Nicholson D. N. (1982). BAAS 14, p. 796-829.

INFRARED SPECTROSCOPY OF EVOLVED OBJECTS

David K. Aitken and Patrick F. Roche

Physics (RAAF) Department, Melbourne University, Australia
and
Department of Physics & Astronomy, University College London

1. Introduction

   In this review, we are concerned with spectroscopic observations of evolved objects made in the wavelength range 1-300µm. Because of absorption by molecular species in the Earth's atmosphere, observations at wavelengths longer than 13µm have sensibly to be made from high altitude, that is from aeroplane, balloon or satellite platforms. It is therefore not surprising that most measurements have been obtained in the atmospheric windows accessible from ground-based telescopes, in the J,H,K,L and N photometric bands, with a few observations made in the M and Q windows. Spectroscopic measurements beyond 20µm of objects such as planetary nebulae have only become available within the last few years.

   Spectroscopic observations can conveniently be divided into studies of narrow lines, bands and broader continua. The vibrational frequencies of molecular groups fall mainly in this spectral region and appear as vibration-rotation bands from the gas phase, and as less structured, but often broader, features from the solid state. Many ionic lines, including recombination lines of abundant species and fine structure lines of astrophysically important ions also appear in this region. The continuum can arise from a number of mechanisms - photospheric emission, radiation from dust, free-free transitions in ionized gas and non-thermal processes.

   In this paper, we will not cover spectral features observed in stellar photospheres. Spectroscopic measurements made before about 1976, mostly with filter wheel spectrometers, are represented in a series of papers by Merrill and Stein (1976a,b) and have been reviewed by Merrill (1977), and Merrill and Ridgway (1979). These papers have effectively covered most aspects of the spectra of late-type stellar photospheres as seen at low or moderate ($\lambda/\Delta\lambda < 100$) resolution. To these we add here a more recent spectrum of the carbon rich mira, V CrB, obtained by Goebel et al (1981) covering the whole 1-13µm region and showing deep photospheric absorption bands with dust emission becoming dominant at longer wavelengths. Instrumentation advances have subsequently led to Fourier transform spectrometers with resolving powers of ~100,000 which produce high quality spectra showing large numbers of atomic and molecular absorption features; this area is covered by Ridgway in this symposium.

Line emission will be discussed in planetary nebulae and Wolf-Rayet stars, two types of object in which the line emission arises from regions of very different physical conditions. In the former, the gas density is generally low, and forbidden lines dominate the spectra, whilst in the latter, collisional de-excitation of fine structure levels is much more important, and recombination lines are more prominent.

To a large extent, infrared spectroscopy has been a study of the properties of dust, and emission from dust grains has been found to be important in a wide variety of astrophysical environments. Dust emission occurs in many evolved objects, including circumstellar shells around giant stars and late-type Wolf-Rayet stars as well as symbiotic systems, novae and planetary nebulae. Although dust emission often dominates the middle infrared, where free-free radiation is generally weak, the latter mechanism gives rise to the near infrared continua in many nebulae. However, in stars undergoing heavy mass loss with high velocity outflows, the free-free emission can be strong, and this mechanism produces the middle infrared excess continuum in some Wolf-Rayet (and other e.g. Be) stars.

We have attempted a brief outline of each broad subject area and then concentrated on recent results, referring the reader to previous reviews and papers for details of prior work. It is hoped that this leads to a nearly complete, if implied, reference network. Lastly, we must apologise if this review is biased towards the 8-13μm region which is where our own interests mostly lie, and to anyone whose work has been ignored or misrepresented.

2 Circumstellar Shells of Giants

2(a) M and C Giant stars.

The most common bright evolved galactic objects are the giant stars. Many of these suffer large mass loss and have developed circumstellar shells containing solid condensates. The Si-O stretching resonance near 10μm, which is taken as the signature of silicate dust, was first observed and identified in some oxygen rich late type stars by Woolf & Ney (1969) in the late sixties using a circular variable filter wheel (CVF) and a bolometer. Since then similar features have been observed in emission and absorption in many HII regions, and the Trapezium region of the Orion nebula is frequently taken as the archtypal astronomical silicate feature. Silicates have a second feature in the 20μm region due to the bending resonance of the O-Si-O bond, and a broad feature peaking near 18μm attributed to this resonance has been observed in oxygen rich stars (Forrest et al, 1979; Treffers and Cohen, 1974) and the Trapezium (Forrest & Soifer, 1976).

It was quickly realised that the details of the astronomical feature were not well represented by the laboratory spectra of terrestrial crystalline silicates, which display much more spectral structure. The laboratory materials which appear most similar to the astronomical feature are hydrated and amorphous silicates (Stephens & Russell, 1979), carbonaceous chondrite meteorites (Penman, 1976) and artificial smokes of silicate produced in the laboratory by Day & Donn (1978). Spectropolarimetric observations of some heavily obscured sources (Capps and Knacke, 1976; Dyck and Lonsdale, 1981) provide additional independent evidence that the interstellar silicates are disordered (Martin, 1975).

The dust which forms in the envelopes of oxygen rich giant stars usually has a distinguishable 10 micron spectral signature from that characterising the interstellar medium and HII regions, as has been pointed out by Willner at this symposium. (However a few circumstellar shells are better fitted by the Trapezium emissivity function than that derived from typical O-rich giants.) The stellar spectra appear to be very nearly photospheric short of 8 microns (eg μ Cep Russell et al, 1975) and it seems that the ISM contains an additional component presumably associated with mass loss from carbon rich objects, such as carbon stars, some planetary nebulae and novae, and possibly including a featureless component from oxygen rich objects. The difference between the ISM and the circumstellar feature may also be due in part to different physical characteristics such as crystalline structure or grain size. The circumstellar feature is quite well represented by disordered olivine and olivine smokes (Kratschner & Huffman, 1979), if the effective shell temperature is in the region 200-300 K. The ISM (Trapezium), circumstellar (μ Cep), and laboratory features are compared in Fig 1. Papoular and Pegourie (1983) have recently discussed the width of the Si-O signature in terms of grain size, and found evidence that grains in giants are larger than in supergiants, and approach a few microns in diameter. Occasionally, a smooth 1000K dust component is seen around oxygen rich stars; this is discussed in 2(c) below.

In carbon stars C/O > 1 and essentially all the oxygen becomes locked up in the very stable CO molecule so that only carbon rich grain materials can condense. The spectra of carbon stars often show a broad emission feature peaking near 11.5μm and attributed to silicon carbide (Treffers & Cohen, 1974; Merrill & Stein, 1976a,b). In many objects, the SiC feature is accompanied by featureless grain emission with colour temperature close to 1000K. All conducting grain species will have essentially indistinguishable, smooth infrared emissivities but in carbon stars, the smooth component is usually taken to be graphite from abundance arguments. Iron based condensates have been suggested by Lewis and Ney (1979) as these will condense at around the same temperature of 1000K from either a carbon rich environment as cohenite (Fe_3C) or as metallic iron from an oxygen rich environment.

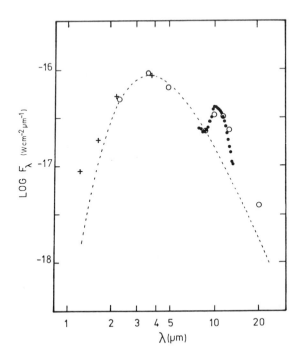

Figure1: Emissivity functions of olivine and silicon carbide obtained in the laboratory compared to those derived from the circumstellar shells of μ Cep and Y Tau and the Trapezium. These curves are taken from the papers referred to in the text.

It is not immediately obvious how to separate the SiC feature from the smooth component in the general case, but fortunately their ratio varies widely from source to source, demonstrating that they are separate constituents, and in some stars the smooth ~1000K component seems to be absent. Fig 1 shows the feature derived from one such star, Y Tau, compared with that of laboratory silicon carbide (Aitken et al, 1979a; Friedman et al, 1981).

A further strong broad emission feature extending from 24-30μm has been observed in four carbon stars (Forrest et al, 1981), all of which have strong SiC features. Forrest et al suggest that this feature may be due to Fe_3C, or to carbyne, a long chain polymer of carbon.

2(b) Giant stars of classification between M and C

The sequence of classification M-MS-S-SC-CS-C is one of increasing C/O ratio, with C/O=1 occuring between SC and CS. About 25 stars in the region S-C have been searched for dust signatures (Clegg et al, 1983), and there does appear to be a transition between silicate and silicon carbide between SC and CS. However the sample of SC and CS stars is very small and two of the S stars also show silicon carbide, while most of them either have silicates or are featureless. It may be that the classification of these in terms of C/O ratio is not secure, or a second parameter may be important. In any case it seems that the transition between silicate and silicon carbide dust features occurs at C/O= 1.00+-0.05. While nearly all the oxygen rich stars with dust shells show the circumstellar silicate, one of them, WX Ser, curiously has a Trapezium excess; WX Ser, however, is an M8e emission line star.

2(c) Symbiotic stars.

Symbiotic stars may be defined as apparently single stars having spectral characteristics of late M giants plus a profusion of high excitation emission lines. They are usually interpreted in terms of a binary system in which a hot unseen star excites the outflow from an M giant. In the near IR about a quarter have cool colour temperatures near 1000 K while the remainder are dominantly photospheric with temperatures around 2500 K, characteristic of the M giant. The cool group are interpreted as showing a large contribution from heated dust. A recent review of the properties of symbiotic stars has been presented by Allen (1984).

The most complete spectral coverage to date appears to be for the object Haro 1-36 (Allen, 1983). An extensive table of emission lines from the UV to the IR is given while the near infrared continuum shows steam bands at 1.4, 1.9 and 2.7 microns, and CO absorption between 2.3 and 2.5 microns, typical of a very late spectral type such as a Mira variable; this spectral region is dominated by the stellar photosphere. At wavelengths longer than $3\mu m$, dust emission dominates, and the $3-4\mu m$ spectrum is featureless with a dust temperature of 700K. In the $10\mu m$ region, there is a silicate emission feature superposed on the smooth continuum, giving further evidence of an oxygen rich mira.

Roche et al (1983a) have obtained spectra of 22 symbiotic stars in the 3-4 and 10 micron regions. The properties of the sample, strongly biased to dusty symbiotics, may be summarised as follows:
   2 are photospheric
   4 show smooth dust emission only
   2 have silicate emission superposed on a photosphere
   13 have silicate emission plus the smooth dust emission
and 1 (HDE 330036) has the narrow unidentified features at 3.28, 8.65, and $11.25\mu m$, (see section on planetary nebulae) and is probably carbon rich.

In most cases silicate emission is best represented by the circumstellar feature, but in four of the objects the feature is more subdued and improved fits are obtained with Trapezium like emissivity. However, equally good fits are obtainable in these cases with the circumstellar feature using large grains of order 1 micron, when the emissivity approaches saturation near the peak.

The appearance of a dust continuum with colour temperature 1000K in such a large proportion of the symbiotics is remarkable since the dust shells of single oxygen rich giants radiate little below 8 microns (Merrill & Stein, 1976a; Russell et al, 1975). A solution in this case is suggested in terms of the presence of the hot star. It is proposed (Roche et al,1983a) that in M giant shells not only silicate but also a minority of small, conducting, iron based grains condense. In the presence of the radiation field of an M star alone the silicate grains dominate the emission, but a hot companion with a substantial UV component will heat the small grains more efficiently so that they reach (colour) temperatures near 1000K (c.f. Dwek et al, 1980).

3. Planetary Nebulae

3(a) Emission lines

Planetary nebulae (pn) are thought to represent the next stage of evolution of some of the red giants as they evolve off the asymptotic giant branch. They show a rich emission line spectrum through the UV, optical and IR. Recombination lines of H, HeI and HeII are seen together with many collisionally excited fine structure lines of a variety of ions, and also emission from vibrational states of molecular hydrogen. The ionic line intensities are related to element abundance and excitation state and, in planetary nebulae where the optical extinction is not large, the infrared lines are essentially unaffected by extinction. They are relatively insensitive to electron temperature, and in those cases where there are optical lines from the same ion, have in general a different dependence on electron density. Dinerstein (1983) has shown how in these cases optical and IR observations lead to better estimates of temperature and density, and hence of ionic abundance. Some ions have observable transitions only in the infrared which then offers the only way to directly measure their abundance.

The first predictions of infrared line intensities in planetary nebulae were made by Delmer, Gould and Ramsay (1967) and later developed and presented by Petrosian (1970), Osterbrock (1974) and Simpson (1975), amongst others. Abundances are referred to hydrogen either through radio observations of the optically thin free-free continuum, or from hydrogen recombination lines. Beam size

Table 1   Ionic IR Lines in Planetary Nebulae.

Ion	λ (μm)	IP Range	Objects	
MgIV	4.49	80.1-109.3	N7027	1,2
MgV	5.61	109.3-141.3	N7027	1,2
NiII	6.62	7.6- 18.2	N7027	3
ArII	6.98	15.8- 27.6	IC418	4
ArIII	8.99	27.6- 40.7	many	4,5,6
SIV	10.52	34.8- 47.3	many	4,5,6,7,8
ClIV	11.76	39.6- 53.5	I2165	9
NeII	12.81	21.6- 41.0	many	4,5,6
SIII	18.71	23.3- 34.8	several	10,11,12
NeV	24.28	97.1-126.2	N7662,N7354,N7027	11,12
OIV	25.87	54.9- 77.4	I2003,N7354,N7662, N7027,N2392	11,12
OIII	51.81	35.1- 54.9	N6543,N6302	13
OIII	88.36	35.1- 54.9	N2440,N6445,N6543 N6826	14
NIII	57.33	29.6- 47.5	N6543	14
OI	63.17	0.0- 13.6	N7027	15

1 Russell et al, 1977.
2 Beckwith et al, 1983.
3 Bregman et al, 1981.
4 Willner et al 1979.
5 Beck et al, 1981.
6 Aitken & Roche, 1982.
7 Dinerstein, 1980.
8 Roche et al, 1983b.
9 Roche & Aitken, 1983.
10 Greenberg et al, 1977.
11 Forrest et al, 1980.
12 Shure et al, 1983.
13 Watson et al, 1981.
14 Dinerstein et al, 1983.
15 Melnick et al, 1981.

corrections frequently have to be made to compare these with the IR observations and corrections for extinction if optical recombination line intensities are used. Ideally comparison with infrared Brackett lines using the same sky aperture would avoid these problems, but these have not yet appeared in the literature. In one case (IC 2165) reference has been made to the Humphries 7-6 line at 12.36μm (Roche & Aitken, 1983) but the line is weak and its errors dominate the abundance determination. Mendoza (1983) has recently presented a critically evaluated compilation of collision strengths calculated by various groups using sophisticated methods. In several cases, and notably for S^{3+}, the values obtained are substantially different from earlier estimates and will affect prior abundance determinations.

Measurements of IR emission lines over the past decade or so have expanded our knowledge of element abundance and the physical conditions within planetary nebulae. Essentially element abundances appear to be within a factor of 2-3 of solar, although as mentioned above, those of sulphur should be treated with caution. This subject is effectively summarised in the recent review by Dinerstein (1983), from which table 1 has been adapted, listing the IR lines so far observed in planetary nebulae. The reader is referred to this review and references therein. Here we will briefly discuss only some recent observations at long wavelength.

Some of the wavelengths are severely affected by atmospheric water vapour and can only be made from an airborne observatory. Many far IR lines have been observed in HII regions (e.g. Lacy, 1981; Emery & Kessler, this volume), but instruments with sensitivities capable of detecting these lines from pn, which are much less luminous, have only recently become available. The [NeV] (24.28μm) line detected in NGC 7027 by Forrest et al (1980) has been confirmed and detected in two other high excitation pn by Shure et al (1983), together with [OIV] (25.87μm) in these and a further two. These results are important because they can provide information on physical conditions in high excitation regions in pn. The interpretations do depend critically on the values adopted for the collision strengths, and a tentative result, for NGC 7662, that the electron temperature in the NeV region is similar to that in the OIII region may be evidence that the NeV collision strength is in error (Shure et al, 1983).

The far IR lines are collisionally de-excited at lower electron densities than those at shorter wavelengths. Where ground state triplets are involved, the two transitions themselves have significantly different critical densities, as for the 51.8 and 88.4μm lines of OIII (see e.g. the listing in Simpson, 1975). Dinerstein and co-workers (1983) have observed [NIII] at 57.33μm together with the [OIII] lines in NGC 6543 and NGC 6302. The observation of both transitions from the triplet ground state of O^{++} allows the determination of both $<n_e>$ and $<n_e^2>$, giving information on the degree of clumping.

Melnick et al (1981) observe an anomalously strong 63.2μm line from neutral oxygen in NGC 7027, which cannot be accounted for on the charge exchange model. The line presumably originates in a neutral shell surrounding the ionized region, and both shock excitation and UV heating have been considered as possible mechanisms.

3(b) Molecular hydrogen emission

Shortly after the detection of emission lines from molecular hydrogen in the Orion nebula (Gautier et al, 1976), some of these were also observed in NGC 7027 (Treffers et al, 1976). Molecular hydrogen has since been observed in several other pn including some objects thought to be protoplanetary. For NGC 6720 (Beckwith et al, 1978) the emission is correlated with [OI] at λ6300 and either arises in neutral clumps within the ionized gas or is external to the ionized region. On the basis of the observed spatial distribution and line intensity ratios, Beckwith et al (1980) conclude that H_2 is shock excited at the ionization front in NGC 7027. Recent multiwavelength mapping in the two micron region demonstrates that this is likely to be the case in two other pn, NGC 3132 and IC 4406 (Storey, 1983). The excitation temperature of the lines is ~2000K in both molecular clouds and pn, an unexpectedly large temperature for neutral regions. The general consensus has developed that in most cases, the emission is consistent with shock excitation where the shock may be associated with ionization fronts or with high velocity outflows. The implications of molecular hydrogen emission have been reviewed by Beckwith (1980).

Observation of molecular hydrogen emission is important because it gives information on the shocked regions of the neutral gas and, if the spectral resolution is high enough, their kinematic properties. Comparative studies of the bipolar nebula CRL 618 (Persson et al, 1982) reveal H_2 velocities which, if attributed to shocks, are much larger than those which would dissociate the hydrogen molecule. CRL 618 is considered to be a protoplanetary nebula in which a B0 star excites a compact HII region obscured from view by a dusty torus. The observations of Persson et al are interpreted in terms of detailed structural models in which the molecular emission arises from shock excitation in dense clumps which are themselves carried by high velocity flows.

3(c) Dust emission

The discovery of warm emitting dust in pn was first made by Gillett et al (1967) in NGC 7027, and dust emission has since been found to be commonplace in dense, compact planetaries. The 8-13μm spectrum of NGC 7027 (Gillett et al, 1973) revealed two prominent, narrow (but resolved) emission features not previously seen in any object. Spectroscopy of NGC 7027 over a wider wavelength range (Russell et al, 1977) showed that these were part of a family of six features at 3.28, 3.4, 6.2, 7.7, 8.65, and 11.25μm. The precise identification of these is still uncertain, but because they are resolved, and do not break up into bands at higher resolution (Tokunaga and Young, 1980), they have been attributed to solid state transitions on dust grains or their mantles. The features occur in roughly a third of pn so far studied in this way (Aitken and Roche, 1982), and in a variety of other sources including some Galactic HII regions (Sellgren, 1981; Aitken et al, 1979b) and almost invariably in galaxy nuclei which contain giant HII regions (Phillips et al, 1983). From the very small sample of sources for which detailed spatial work has been done (Aitken et al, 1979b, Aitken & Roche, 1983; Wynn-Williams et al, 1981), it appears that the features arise from the neutral side of ionization fronts.

There have been three suggestions as to the nature of the excitation mechanisms. Willner, in this volume, has considered the first two, both of which involve UV radiation, in one case (Dwek et al, 1980) to heat small grains, and in the other to excite fluorescent transitions in grain mantles (Allamandola et al, 1979). Barlow (1983) has recently proposed that graphite grains would react with atomic or ionized hydrogen in the vicinity of HII regions, forming surface hydrocarbon complexes. Duley and Williams (1981) have suggested that the features arise from excitation of surface functional groups on grains and can in this way account for the 3.3, 3.4, 7.7 and 11.3µm bands. This model does not require pre-existing mantles on the grains whose existence would be difficult to account for in so many pn, and naturally explains the tendency to be associated with ionization fronts. A fuller description of these features and the proposed excitation mechanisms may be found in Allamandola (this volume).

The remaining pn show either the silicate or silicon carbide emission features in roughly equal number (Aitken and Roche, 1982), and the 24-30µm feature observed in some carbon stars has been seen in two of the pn which show the silicon carbide feature (Forrest et al, 1981). The presence of the dust signatures has been used as an indicator of the C/O ratio (Aitken et al, 1979a; Aitken and Roche, 1982), and where UV data are available and a detailed analysis of the UV and optical line intensities carried out, the C/O ratios obtained confirm those found from the dust grain chemistry (Seaton, 1983).

Observation of dust in planetary nebulae has been the subject of a recent comprehensive review by Barlow (1983). Barlow considers UV and optical observations related to dust, the energetics of the dust emission and the likely long wavelength emissivity law, the dust chemistry and the relationship between the dust emission and the ionized gas, and offers new correlations between the presence of unidentified features and the C/O ratio, as well as on their emission mechanism and origin. The present work will add only a description of some more recent results and refer the reader to Barlow's review and references therein.

Near and mid infrared observations of pn are usually biased towards nearby, compact, high density objects, since these will be brightest to observe in both line and dust continuum with relatively small beam sizes. To try to redress this balance, 8-13µm spectroscopic observations have been obtained of a sample of eight more extended pn (Roche et al, 1983b), with diameters in the range 15-50 arcsec. In these cases the continuum is much weaker than in the compact objects while the [SIV] line is strong, as expected for moderately high excitation objects. In only one case, the high excitation nebula NGC 2392, is there no evidence for the [SIV] line; however sulphur is likely to be dominantly in higher ionization states and [OIV] at 25.87µm is seen (Shure et al, 1983). None of these pn shows convincing evidence for the unidentified features at 8.65 and 11.25µm.

In some of the objects, the [SIV] flux dominates the 10µm region and here spectroscopy or narrow band photometry is necessary to determine the weak continuum which does exist, and even in these cases exceeds the extrapolated free-free flux by factors of 3-10 (cf. 10-100 for compact pn). Four of the sample have also been studied photometrically by Moseley (1980) at longer wavelengths, where he finds a mean temperature of 80 K, assuming a grain emissivity $Q = 1/\lambda^2$. Compared with compact nebulae the sample have much less excess emission at 10µm, and in one of them (NGC 6543) there is little evidence for a separate warm grain population.

It is clear that in NGC 7027 (Becklin et al, 1973), and presumably other compact pn, the warm dust emitting at 10µm is intimately mixed with the ionized gas and that a substantial part of the IR luminosity comes from nebular heating. How much of Moseley's cool 80 K emission comes from the ionized region or from a neutral shell is not so clear, although in NGC 7027 the 8.65 and 11.25µm features have been shown to arise from such a shell, exterior to the warm continuum emitting dust and ionized gas (Aitken and Roche, 1983). It may be that this neutral shell is the source of some of the cool continuum. However, many of the extended nebulae will be density bounded, and some show significant absorption of resonance line photons by dust. In these cases some of the cool continuum must come from the ionized region, but there may be a contribution from neutral inclusions.

4. Wolf-Rayet stars

Population I Wolf-Rayet stars are very massive luminous objects in which evolution has proceeded very rapidly, and represent a particularly young kind of evolved object; most Wolf-Rayets are considered to be severely hydrogen deficient.

Until comparatively recently there had been little IR spectroscopy of Wolf-Rayet stars, although photometry defining the energy distribution over a wide range of wavelengths has been presented (Hackwell et al,1974; Cohen et al, 1975). In most cases there is an infrared excess over the stellar continuum. Except for some of the WC sequence of type WC8 and later, this excess can be attributed to free-free emission from an outflowing stellar wind. Because of optical depth effects in the expanding envelope, a wind expanding so that the density has a $1/r^2$ dependence gives rise to an IR-radio spectrum of form $S_\nu = \nu^{0.6}$ (Wright and Barlow, 1975; and see the discussion by Wynn-Williams in this volume). Observations have shown that this is approximated in Wolf-Rayet atmospheres, with small deviations indicating departures from a strict $1/r^2$ law probably due to accelerating winds. At 10µm, the contribution from the underlying star is only ~10% and 10µm measurements in conjunction with radio observations have been used to determine mass loss rates (Barlow et al, 1981). For the late WC stars an excess typical of dust emission is often seen. The objects are sometimes heavily obscured in the visible, and the silicate feature in absorption has been observed (Aitken et al, 1980; Roche & Aitken, 1984), but apart from this, the dust emission is well represented by emission from featureless materials, presumed to be graphite grains.

Spectroscopy at near infrared wavelengths (Cohen and Vogel, 1978) of WN and earlier WC stars shows helium recombination lines expected in hydrogen deficient objects and for the WC stars additional lines at 2.06 (see below) and 3.3µm, the latter being reminiscent of the unidentified dust feature at 3.28µm. The correct identification of this line was given by Williams, (1982b); it is the 11-10 CIV hydrogenic recombination line. The observation of other hydrogenic alpha lines in CIV have confirmed this. 10µm spectroscopy has only recently become available because the objects are generally faint and the brightest non-dusty WR star by an order of magnitude, γ Vel, is a southern hemisphere star. The infrared spectrum of this source (Aitken et al, 1982) shows, in addition to recombination lines attributable to HeI, HeII and CIV, fine structure lines of [SIV] and [NeII]. Because the density in γ Vel falls off roughly as $1/r^2$ and the fine structure levels are collisionally de-excited at high densities, there are two nearly equal contributions to the fine structure line emission, from within and without a radius where the density equals the critical density for collisional de-excitation. This radius is several orders of magnitude larger than the radius from which the recombination lines originate, and so comparative studies of line profiles could be a useful probe of the velocity structure of WR stars.

Because the recombination transitions in the middle infrared come from high levels, the lines are likely to be optically thin and so easier to interpret than lines at shorter wavelengths. The infrared line strengths have been used to estimate the HeI/HeII and C/He ratios in γ Vel by Hummer et al (1982) who find that C/He > three times solar and $He^+/He^{++} \simeq 5$. The intensities of the [SIV] and [NeII] lines suggest that they too are overabundant with respect to helium. These lines have since been measured in several other WR stars, and analysis should be able to determine relative ion abundances in different spectral types (Williams et al, 1983).

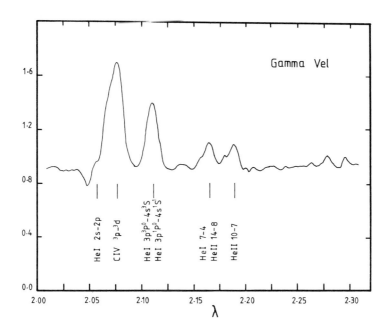

Figure 2: The 2.0-2.3μm spectrum of γ Vel (Hillier & Hyland, 1983); prominent emission lines are identified, and the position of the 2.058μm HeI 2p-2s line is indicated.

Higher resolution spectroscopy of γVel has recently been obtained by Hillier and Hyland (1983) in the 1.5-1.8 and 2-2.3μm region, with resolving power 500-700. The line near 2.06μm is observed to peak at 2.076μm and must now be identified with CIV (3d-3p) rather than the HeI 2p-2s transition, which explains naturally its presence in WC and absence in WN stars (Williams, 1982a). Other uncertainties in the short wavelength range of the earlier spectrum (Aitken et al, 1982) are also clarified; HeI 3p-4s appears at 2.112, HeII 10-7 at 2.189μm and what may be a P Cygni edge to the 2.058μm HeI 2p-2s line (Fig 2). These results demonstrate the need for higher resolution infrared work than possible with a CVF to help solve the problem of line blending in WR stars and especially to separate the helium and carbon lines near 2.07μm.

Hillier (1982) has reported CVF and cooled grating observations of Southern WN stars and high resolution observations of the WN5 star HD 50896 with resolving powers ~700 (Hillier et al, 1983). This star has very little HeI and the high level HeII lines 19-8 and 24-8 have been used to derive upper limits for the H^+/He^{++} ratio <0.3 confirming previous analyses that this star is very hydrogen deficient (Smith, 1973) with H^+/He^{++} <0.05 determined from the Pickering decrement.

Hillier et al's studies also derive a wind temperature appropriate to the HeII lines in the 2 micron region of 33,000 K.

5. Novae

Infrared photometric studies of recent novae have supplied new information on their development. In particular the transition phase has been found to be associated with the development of a large infrared excess, commonly interpreted as due to the formation of a dust shell of sufficient optical depth in the visible to obscure the light from the central source (Ney and Hatfield, 1978). An alternative explanation (Bode and Evans, 1981), has been that the hardening radiation from the nova core shifts the spectral maximum into the UV which heats small pre-existing conducting grains efficiently.

So far there has been little infrared spectroscopy of novae. Spectral studies at short wavelengths (Grasdalen and Joyce, 1975), have followed the development of hydrogen recombination lines in Nova Cyg 1975 (V1500 Cyg), and shown the transient appearance of an unidentified feature near 3.5µm. Narrow band photometry had not revealed the presence of any dust emission features until Nova Aql 1982, and previous energy distributions have been smooth, typical of featureless grains with black body temperatures in the range 800-1100K. Because of this, the appearance of a feature near 5µm which has been attributed to molecular carbon (Clayton & Wickramasinghe, 1976), and the presumed overabundance of carbon in novae, the grains have usually been considered to consist of graphite. Most novae develop a pronounced infrared excess, but the very fast nova, V1500 Cyg 1975, showed no evidence for dust formation.

Nova Aql 1982 is the first nova for which low resolution IR spectroscopy of the dust shell phase has been obtained. The nova was discovered on 27.9 January 1982 (taken as day zero) and its rate of decline classed it as a fast/very fast nova. The outflow velocity was high and two systems were apparent, at 4000 km/s and 10000 km/s; the faster system disappeared after 30 days (Snijders et al, 1983). An infrared excess characteristic of a dust shell with temperature 1100K was observed on day 37 (Williams and Longmore, 1982), itself remarkably early and implying dust formation as early as day 16 according to the grain growth model of Clayton and Wickramasinghe (1976). Unusually, the visible minimum, which in other novae has been coincident with the onset of the dust excess, did not occur until day 60.

CVF spectroscopy in the 2-4µm region obtained by Bode and Evans on 18 April, well into the dusty era, showed Br_γ and possibly Br_α together with a minimum near 4.0µm which may be due to SiO absorption, as seen in many late type stars with C/O < 1 (Rinsland and Wing, 1982). There was no sign of 2.3µm CO absorption.

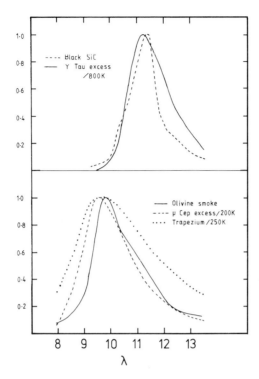

Figure 3: The Infrared flux distribution of nova Aql 1982 on day 156 (Bode et al, 1984); the dashed line is an 800K blackbody.

8-13μm spectra were obtained on days 142, 156, 179 and 275 (Aitken et al, 1984) and all show a pronounced silicate emission feature (Fig 3). The feature peaks close to 10μm and is unlike the SiC feature. It is fit much better by the circumstellar (μCep) emissivity, or by laboratory amorphous olivine than by that of the Trapezium, and requires an additional smooth component; the ratio silicate/smooth emission increases with time. The required olivine temperature is quite low, ~150K, although the fitted temperature depends critically on the form of emissivity function near 8μm.

Warm (~1000K) featureless emission has been seen in previous novae where it was attributed to graphite condensing from a carbon rich atmosphere; such grains are not expected to condense from the same atmosphere that gives rise to silicates. As we have seen before, this situation is not unique to this nova, smooth 1000 K emission is occasionally seen in oxygen rich objects, together with the silicate feature. Analysis of IUE and AAT data (Snijders et al,1983) indicates that in the nova ejecta the heavy elements are dramatically enriched and that C/O <1, which provides further evidence for the formation of a featureless condensate in an oxygen rich environment.

References:

Aitken, D.K., Barlow, M.J., Roche, P.F. & Spenser, P.M., 1980.
    M.N.R.A.S., 192, 679.
Aitken, D.K. & Roche, P.F., 1982. M.N.R.A.S., 200, 217.
-------- ibid --------------- 1983. M.N.R.A.S., 202, 1233.
Aitken, D.K., Roche, P.F. & Allen, D.A., 1982. M.N.R.A.S., 200, 69P.
Aitken, D.K., Roche, P.F., Spenser, P.M. & Jones, B., 1979.
    Ap.J., 233, 925.
------------ ibid ---------------------- 1979. Astr.Ap., 76, 60.
Aitken, D.K., Roche, P.F. & Whitmore, B., 1984. to be published.
Allamandola, L.J., Greenberg, J.M. & Norman, C.A., 1979. Astr.Ap., 77, 66.
Allen, D.A., 1983. M.N.R.A.S., 204, 113.
Allen, D.A., 1984. Astr.Sp.Sci. in press.
Allen, D.A., Beattie, D.H., Lee, T.J., Stewart, J.M. & Williams, P.M.,
    1978. M.N.R.A.S., 182, 57P.
Barlow, M.J., 1983. IAU Symp. 103, Planetary Nebulae, ed D.R. Flower p105.
Barlow, M.J., Smith, L.J. & Willis, A.J., 1981. M.N.R.A.S., 196, 101.
Beck, S.C., Lacy, J.H., Townes, C.H., Aller, L.H., Geballe, T.R. &
    Baas, F., 1981. Ap.J., 249, 592.
Becklin, E.E., Neugebauer, G. and Wynn-Williams, C.G., 1973.
    Ap.Lett., 15, 87.
Beckwith, S., 1980. IAU Symp 96. Infrared Astronomy, ed C.G. Wynn-
    Williams and D.P. Cruikshank, p167.
Beckwith, S., Evans, N.J., Natta, A., Russell, R.W & Wyatt, J., 1983.
    in preparation.
Beckwith, S., Neugebauer, G., Becklin, E.E. & Matthews, K., 1980.
    Astr. J., 85, 886.
Beckwith, S., Persson, S.E. & Gatley, I., 1978. Ap.J., 219, L33.
Bode, M.F. & Evans, A., 1981. M.N.R.A.S., 197, 1055.
Bode, M.F., Evans,A, Whittet, D.C.B., Aitken, D.K., Roche, P.F. &
    Whitmore, B., 1984. M.N.R.A.S., submitted.
Bregman, J.D., Dinerstein, H.L., Goebel, J.H., Lester, D.F., Witteborn,
    F.C. & Rank, D.M., 1981. BAAS, 13, 852.
Capps, R.W. & Knacke, R.F., 1976. Ap.J., 210, 76.
Clayton, D.D. & Wickramasinghe, N.C., 1976. Ap. Sp.Sci., 42, 463.
Clegg, R.E.S., Whitmore, B., Aitken, D.K. & Roche, P.F., 1983.
    in preparation.
Cohen, M., Barlow, M.J. & Kuhi, L.V., 1975. Astr.Ap., 40, 291.
Cohen, M. & Vogel, S.N., 1978. M.N.R.A.S., 185, 47.
Day, K.L. & Donn, B., 1978. Ap.J., 222, L45.
Delmer, T.N., Gould, R.J. & Ramsay, W., 1967. Ap.J., 149, 495.
Dinerstein, H.L., 1980. Ap.J., 237, 486.
Dinerstein, H.L., 1983. IAU Symp 103, Planetary Nebulae, ed D.R. Flower.
Dinerstein, H.L., et al., 1983. in preparation.
Duley, W.W. & Williams, D.A., 1981. M.N.R.A.S., 196, 269.
Dyck, H.M. & Lonsdale, C.J., 1981. IAU 96, Infrared Astronomy, ed
    C.G. Wynn-Williams & D.P. Cruickshank, p223.
Dwek, E., Sellgren, K., Soifer, B.T. & Werner M.W., 1980. Ap.J., 238, 140.
Forrest, W.J., Houck, J.R. & McCarthy, J.F., 1981. Ap.J., 248, 195.
Forrest, W.J., McCarthy, J.F. & Houck, J.R., 1979. Ap.J., 233, 611.

------------------ ibid ------------------- 1980. Ap.J., 240, L37.
Forrest, W.J. & Soifer, B.T., 1976. Ap.J., 208, L129.
Friedmann, C., Gurtler, J., Schmidt, R. & Dorschner, J., 1981.
    Astr.Sp.Sci., 79, 405.
Gautier, T.N., Fink, U., Treffers, R.R. & Larson, H.P., 1976.
    Ap.J., 207, L129.
Gillett, F.C., Forrest, W.J. & Merrill, K.M., 1973. Ap.J., 183, 87.
Gillett, F.C., Low, F.J. & Stein, W.A., 1967. Ap.J., 149, L87.
Grasdalen, G.L. & Joyce, R.R., 1975. Nature, 259, 187.
Goebel, J.H., Bregman, J.D., Witterborn, F.C. & Willner, S.P., 1981.
    Ap.J., 246, 455.
Greenberg, L.T., Dyal, P. & Geballe, T.R., 1977. Ap.J., 213, L71.
Hackwell, J.A., Gehrz, R.D. & Smith, J.R., 1974. Ap.J., 192, 383.
Hillier, D.J., 1982. IAU Symp 99, Wolf-Rayet Stars, ed C. de Loore
    & A.J. Willis.
Hillier, D.J., Jones, T.J. & Hyland, A.R., 1983. Ap.J., in press.
Hillier, D.J. & Hyland, A.R., 1983. in preparation.
Hummer, D.G., Barlow, M.J. & Storey, P.J., 1982. IAU Symp 99,
    Wolf-Rayet Stars, ed C. de Loore & A.J. Willis, p79.
Lacy, J.H., 1981. IAU 96, Infrared Astronomy, ed C.G. Wynn-Williams
    & D.P. Cruickshank, p237.
Lewis, J.S. & Ney, E.P., 1979. Ap.J., 234, 154.
Kratschner, W. & Huffman, D.R., 1979. Ap.Sp.Sci., 61, 195.
Martin, P.G., 1975. Ap.J., 202, 393.
Melnick, G., Russell, R.W., Gull, G.E. & Harwit, M., 1981.
    Ap.J., 243, 170.
Mendoza, C., 1983. IAU Symp 103, Planetary Nebulae, ed D.R. Flower, p143.
Merrill, K.M., 1977. Proc IAU Coll 42, The interaction of Variable Stars
    with their environment.
Merrill, K.M. & Ridgway, S.T., 1979, Ann.Rev.Astr.Ap., 17, 9.
Merrill, K.M. & Stein, W.A., 1976a. P.A.S.P., 88, 285.
---------ibid ------------- 1976b. P.A.S.P., 88, 294.
Moseley, H., 1980. Ap.J., 238, 892.
Ney, E.P. & Hatfield, B.F., 1978. Ap.J., 219, L111.
Osterbrock, D.E., 1974. "Astrophysics of Gaseous Nebulae", Freeman.
Papoular, R & Pegourie, B., 1983. Astr.Ap. in press.
Penman, J.M., 1976. M.N.R.A.S., 175, 149.
Petrosian, V., 1970. Ap.J., 159, 833.
Persson, S.E., McGregor, P.J., Duncan, D.K., Lanning, H., Geballe, T.R.
    & Lonsdale, C.J., 1982. 16th ESLAB Symposium: Galactic and
    Extragalactic Infrared Spectroscopy. Eds. Kessler, M.F.,
    Phillips, J.P. and Guyenne, T.D. ESA SP-192, p. 41-44.
Phillips, M.M., Aitken, D.K. & Roche, P.F., 1983. M.N.R.A.S., in press.
Rinsland, C.P. & Wing, R.F., 1982. Ap.J., 262, 201.
Roche, P.F. & Aitken, D.K., 1983, M.N.R.A.S., 203, 9P.
Roche, P.F. & Aitken, D.K., 1984. M.N.R.A.S., submitted.
Roche, P.F., Allen, D.A. & Aitken, D.K., 1983a. M.N.R.A.S., in press.
Roche, P.F., Aitken, D.K. & Whitmore, B., 1983b. M.N.R.A.S., in press.
Russell, R.W., Soifer, B.T. & Forrest, W.J., 1975. Ap.J., 198, L41.
Russell, R.W., Soifer, B.T. & Willner, S.P., 1977. Ap.J., 217, L149.

Seaton, M.J., 1983. IAU Symp 103, Planetary Nebulae, ed D.R. Flower, p129.
Sellgren, K., 1981. Ap.J., 245, 138.
Shure, M.A., Herter, T., Houck, J.R., Briotta, D.A., Forrest, W.J.,
        Gull, G.E. & McCarthy, J.F., 1983. Ap.J. 270, 645.
Simpson, J.P., 1975. Astr.Ap. 39, 43.
Smith, L.F., 1973, IAU Symp 49, Wolf-Rayet and High Temperature Stars,
        ed M.K.V. Bappu & J. Sahade, p 15.
Snijders, M.A.J., Bath, T., Blades, J.C., Morton, D.C. & Seaton, M.J.,
        1983. in preparation.
Stephens, J.R. & Russell, R.W., 1979. Ap.J., 228, 780.
Storey, J.W.V., 1983. in preparation.
Treffers, R.R. & Cohen , M., 1974. Ap.J., 188, 545.
Treffers, R.R., Fink, U., Larson, H.P. & Gautier, T.N., 1976.
        Ap.J., 209, 793.
Tokunaga, A.T. & Young, E.T., 1980. Ap.J., 237, L93.
Watson, D.M., Storey, J.W.V., Townes, C.H. & Haller, E.E., 1981.
        Ap.J., 250, 605.
Williams, P.M., 1982a. M.N.R.A.S., 199, 93.
----- ibid ---- 1982b. IAU Symp 99, Wolf-Rayet Stars, ed C. de Loore
        & A.J. Willis, p73.
Williams, P.M., Aitken, D.K., Roche, P.F. & Whitmore, B., 1983.
        in preparation.
Williams, P.M. & Longmore, A.J., 1982. IAU circ 3676.
Willner, S.P., Jones, B., Puetter, R.C., Russell, R.W. & Soifer, B.T.,
        1979. Ap.J., 234, 496.
Woolf, N.J. & Ney, E.P., 1969. Ap.J., 155, L181.
Wright, A.E. & Barlow, M.J., 1975. M.N.R.A.S., 170, 41.
Wynn-Williams, C.G., Becklin, E.E., Beichman, C.A., Capps, R.W. &
        Shakeshaft, J.R., 1981. Ap.J., 246, 801.

SECTION IV: GALACTIC CENTRE AND EXTRAGALACTIC SOURCES

THE GALACTIC CENTER

Ian Gatley

United Kingdom Infrared Telescope, Mauna Kea, Hawaii

ABSTRACT

Infrared studies of the interstellar medium in the central ten parsecs of the Milky Way are reviewed. Far infrared (30 - 150μm) photometry gives the density distribution of the dust: the material is confined chiefly to the plane of the Galaxy, and there is a central cavity of two parsecs radius within which the density is extremely low ($A_v \ll 1$). Spectroscopic observations of ionised neon at 12.8μm show that a small fraction of the volume within this central cavity is occupied by compact, high density clumps of plasma; together these clumps constitute the HII region known as Sgr A (West). Spectroscopic observations of molecular hydrogen lines near 2μm show that the higher density gas immediately outside the central cavity is excited by shocks. The distribution of shocked gas and hot dust is symmetric about the position of IRS 16, the near infrared source thought to be the true nucleus of the Galaxy. Spectroscopy of IRS 16 shows that the HeI 2.06μm line is very broad (FWHM ~ 1500 km/sec); this may be indicative of mass loss.

These observations together suggest that an unusual compact object is present at the very nucleus. A possible picture of the interaction of this object with the surrounding medium is described. Mass loss from the central object both creates a void in the interstellar medium — a mass-loss bubble — and shocks the higher density material immediately outside the bubble. The high density plasma clumps exist within the bubble, in the mass-loss flow. The clumps are probably created and excited by the central source, and may be infalling. Most of the radiation from the central object escapes beyond the mass-loss bubble as visible light, only to be absorbed by dust in the surrounding higher density medium.

I. INTRODUCTION

The motivation to study the nucleus of our Galaxy is simply provided by its proximity — our galactic center is one hundred times closer even than Andromeda (M31). It is therefore somewhat surprising that we know, in fact, relatively little about the detailed physical conditions in the Galactic center. There is a simple reason and a straightforward remedy for this state of affairs. The visual extinction to the galactic center is large ($A_V \sim 27$ mag.), and so observations must necessarily be made at infrared and radio wavelengths. In the last few years, with the advent of large telescopes and new instrumentation, it has become possible to make high angular resolution observations which isolate phenomena within the central few parsecs. The improvements in infrared spectroscopic capabilities have been particularly dramatic.

In this review, emphasis will be placed upon infrared observations relating to the condition of the interstellar medium in the central ten parsecs of the Galaxy. Photometric maps reveal both the density distribution (§II) and the total luminosity (§III), while spectroscopy of ionised neon (§IV) and shock-excited molecular hydrogen (§V) give details of the excitation and dynamics of the gas. Symmetries in the distribution of hot dust and shock-excited gas suggest the possibility of an exotic central object at the position of IRS 16, the near-infrared source identified with the very nucleus. Spectroscopy of IRS 16 itself is suggestive of mass loss, and a simple model is presented in which the presence of a luminous central mass-loss object dictates the nearby density distribution, dust heating, gas excitation and gas dynamics.

The particular point of view taken in this review owes its origins in part to earlier work presented by Oort (1977), Gatley and Becklin (1981), and at the Workshop on the Galactic Center, California Institute of Technology, 1982.

II. THE DENSITY DISTRIBUTION

The dusty environment of the Galactic nucleus ensures that most of the power radiated in the form of ultraviolet or visible light is ultimately absorbed by interstellar grains. This power is reradiated by the dust in the thermal infrared, at wavelengths around 100μm. Because this emission is optically thin, it is possible to map both the dust temperature and the column density of radiating grains (see also, Gatley 1982). Figure 1 a,b,c shows maps of the Galactic center region with 30" resolution (1.3 parsecs at a distance of 10kpc) at wavelengths of 30μm, 50μm and 100μm (Becklin, Gatley and Werner, 1982). The position of the Galactic center is indicated by a cross in each frame. At 30μm the source is compact, and coincides with the Galactic nucleus, but at the longer wavelengths the source takes on a double-lobed appearance, with the lobes extended along the plane of the Galaxy. <u>The position of the Galactic center lies between the lobes</u>.

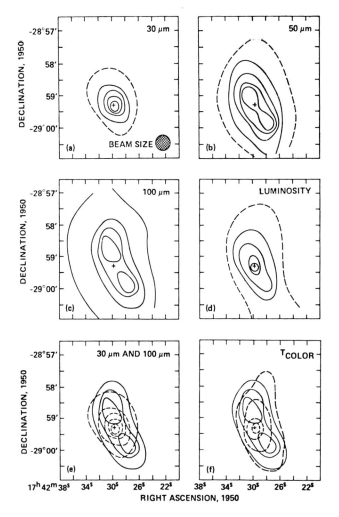

Figure 1. The far infrared surface brightness at 30μm(a), 50μm(b) and 100μm(c) of the Galactic center region measured with 30 arcsec (1.3pc) resolution. The position of the Galactic nucleus is shown by a cross (+). The distribution of luminosity(d) and color temperature(f) are deduced from the data in a,b,c.

The temperature distribution resulting from these data is shown as dashed contours in Figure 1f, superposed on the contours of 100μm surface brightness. The observed grain temperature peaks at the position of the Galactic nucleus. The temperature falls off smoothly in all directions, and is distributed very differently from the 100μm surface brightness. Specifically, there are not temperature maxima coincident with the bright 100μm lobes.

Because the far infrared surface brightness maps the thermal emission from hot dust and is optically thin, it follows that maxima in the surface brightness maps correspond to positions either of enhanced grain temperature or of enhanced dust column density. The 30μm emission peaks at the center, where the grains are hottest, and the 100μm emission peaks in the double lobes, where the dust is densest.

The peaks in color temperature and far infrared luminosity (Figure 1d) seen at the position of the Galactic center suggest that the dust is heated primarily by a centrally concentrated source of luminosity. The double lobed appearance of the source at longer wavelengths arises naturally if the dust density is highest in the Galactic plane: non-ionising radiation streaming out of the central parsec (where the dust density is evidently too low to absorb it) enters a region of higher dust density which is shaped like a ring lying in the Galactic plane. Seen in projection, the emission from hot dust in this ring causes the double lobes. In this picture, therefore, the dust density in the central region is very low; in particular most of the optical and ultraviolet light escapes the central two parsecs, and so $A_V \ll 1$ in this region. The thickness of the surrounding higher density ring in the Galactic plane corresponds to a region of $A_V \sim 1$ in which the luminosity escaping from the center is absorbed by dust; this implies a gas density of about $10^{3.5} cm^{-3}$ in the ring (Gatley et al, 1977; Becklin, Gatley and Werner 1982).

III. THE TOTAL LUMINOSITY

Given the density distribution described above, it is straightforward to calculate the total luminosity of the sources of optical and ultraviolet radiation within the central parsec. The total luminosity within the lowest solid contour in Figure 1d is $5 \times 10^6 L_\odot$. Under the assumption that this power is radiated from a ring of dust which subtends 2π sr as seen from the Galactic center, the central luminosity is $10^7 L_\odot$. An upper limit to the central luminosity and a consistency check on the assumed heating mechanism is provided by comparing the grain temperature observed in the ring with that observed in other sources of comparable luminosity and size, such as in the galactic HII region M17 (Gatley et al 1979). This comparison suggests $L < 3 \times 10^7 L_\odot$ for the Galactic center (Becklin, Gatley and Werner 1982). Thus the central luminosity is found to lie in the range $1 - 3 \times 10^7 L_\odot$.

This large value for the total luminosity far exceeds that deduced from observations of late type stars within the central parsec (Becklin and Neugebauer 1968, Gatley et al 1977). It seems most natural, rather, to associate most of this $\sim 10^7 L_\odot$ with the source which causes the ionisation of the gas in Sgr A (West), as described in §IV (Rieke, Telesco and Harper 1978). Yet to make this association in fact immediately brings to light a remarkable property of the Galactic center, namely, that the source of ionisation has an extremely soft spectrum, as follows: the ultraviolet power required to ionise the plasma in Sgr A (West) is only $\sim 10^6 L_\odot$ (Lacy et al 1980), which is a very small fraction of the $\gtrsim 10^7 L_\odot$ deduced for the total luminosity of the ionising source. This corresponds to an effective spectral type later than O8. The softness of the ionising spectrum is, however, confirmed directly by spectroscopic observations of the ionisation state of the gas, which show $T_e < 35,000K$ (Lacy et al 1980).

The simple fact that 90% of the power of the central source is emitted as non-ionising radiation allows further progress via infrared photometric mapping. For, despite the fact that the dust in the plasma clumps (§IV) present within the low density central parsec will appear hot and bright in the thermal infrared because of resonant trapping of Lα recombination radiation (Gatley 1982), the energy involved in this process is dwarfed by the power in the non-ionising radiation field. This being so, high angular resolution maps of the grain temperature distribution may in fact be used to establish the location of the major source or sources of luminosity, for the grains closest to the heat source(s) will be hottest.

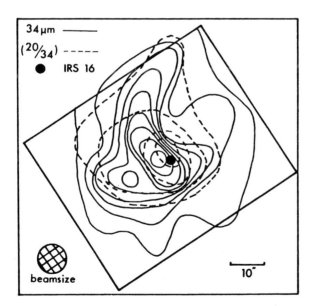

Figure 2. Grain color temperature (dashed contours) and 34μm surface brightness (solid contours) in the central 2pc of the Galaxy. The position of the nuclear source IRS 16 is indicated.

Figure 2 shows the distribution of 34μm surface brightness and of 20μm/34μm color temperature measured with 10" resolution (0.5pc) across the central few parsecs (Gatley, Becklin and Telesco, 1983) — this is the region of very low density within the central cavity (§II) inside which the plasma clumps reside (§IV). Again the distribution of color temperature is quite different from that of surface brightness, and the interpretation proceeds as before. The color temperature has a single central maximum coincident with the near infrared source IRS 16, which is located at the true nucleus (Becklin and Neugebauer, 1975), and the temperature distribution falls off smoothly in all directions. This result suggests that the luminosity of the central parsec of the Galaxy originates from a source or distribution of sources concentrated at the very center of the Galaxy.

As before, the failure of the position of peak surface brightness to coincide with the position of highest grain temperature is indicative of the density distribution: the surface brightness is primarily a map of density distribution. Most of the gas within this region is ionised, and so it is not surprising that the 34μm map appears quite similar to the radio free-free map of the region given below.

Before proceeding with a description of the plasma in Sgr A (West) it is worthwhile to point out again that the analysis of the photometric data alone already leads to the conclusion that it is neither necessary nor desirable to assume that each clump of plasma contains its own source of excitation. Rather, the whole of the central region in which the plasma clumps reside is transparent to optical and ultraviolet radiation ($A_v \ll 1$) and the major source of power is centered on the nucleus, at IRS 16.

IV. THE IONISED GAS

The distribution and dynamics of the ionised gas in the Galactic center were first described in detail from the high resolution observations of the 12.8μm [NeII] line by Lacy et al (1980). This work, and related observations of other ionic species, is described in a recent review (Lacy, 1982) from which figure 3 is taken. Figure 3 shows a 10μm continuum map (Rieke, Telesco and Harper, 1978) of the central three parsecs, with representative spectra of the ionised gas superposed. Detailed comparison has shown that the distribution of [NeII] emission and 10μm continuum emission is closely similar, a result anticipated in §III. Therefore the very complexity of the surface brightness distribution at 10μm is immediately suggestive of the fact that the gas is clumpy. Furthermore each peak in the 10μm map has a different velocity in [NeII]. Fourteen distinct clouds of ionised gas have been identified; they are remarkably similar in their properties. Each cloud has a density $n_e \sim 10^5 cm^{-3}$, a diameter $D \sim 0.2pc$, a mass $M \sim 1 M_\odot$, an internal velocity dispersion $\Delta V \sim 100 kms^{-1}$ and a luminosity $L \sim 10^5 L_\odot$. These high density clumps together fill only a small fraction of the central volume, and the density between the clumps is very low. The internal velocity

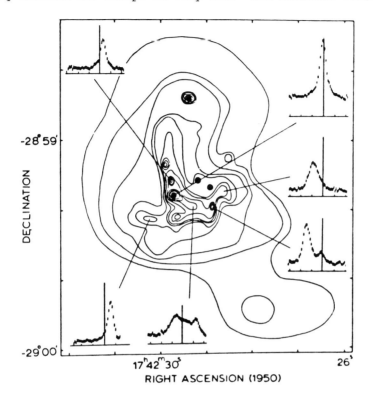

Figure 3. A 10μm map of the Galactic center with representative [NeII] spectra. The central vertical axis of each spectrum is at $V_{LSR} = 0$, and the horizontal axes are marked in 200km/sec intervals — from Lacy (1982).

dispersion of the clumps is high, and a naive lifetime calculation indicates that the clouds will dissipate in ~10^4 years, requiring a creation rate of 10^{-3} clouds/year if the current appearance of the Galactic center is to be maintained. Observations of the ionisation state of the gas from the ratio of fine-structure line fluxes (Lacy 1982) show that, if the plasma is ionised by a single central source, then that source has an effective temperature $T_e < 35,000$K and a bolometric luminosity ~$2 \times 10^7 L_\odot$. This value is in good agreement with other observations (§III). The similarity of the ionisation state of the individual plasma clumps is consistent with a common source for their excitation.

The origins and fate of this clumpy gas, and the nature of the source of its ionisation, are not well understood. A successful model must answer the following questions, which have been widely discussed:

Why is the plasma clumpy? Where does the material in the clumps originate? Why are the properties of the individual clouds so similar? How is the gas ionised? What is the lifetime of the clumps? Where does the material go? Is accretion or expulsion important? Can the clouds be considered as test particles of the gravitational potential? Does star formation occur in the Galactic center? Is the presence or nature of the clumps indicative of the action of an exotic nuclear source?

These questions presently remain unanswered. The observations clearly show that Sgr A (West) is unlike other galactic HII regions, and that the properties of the inner Galaxy are unique. Yet models of widely differing character are currently described: for example, both star formation and tidal disruption of stars near a massive black hole are discussed (see The Workshop on the Galactic Center, California Institute of Technology, 1982). Problems with the star formation model include the absence of high density molecular material (which so characterises that process in the spiral arms) and the curious apparent absence of stars earlier than type O8. Problems with the black hole model include the failure of tidal disruption calculations to predict the production of suitable gas clouds, and of accretion calculations to match the required ionising spectrum. A difficulty for any model involving a single central object is that IRS 16 is rather fainter at 2μm than would be expected for a 35,000K blackbody of luminosity $10^7 L_\odot$. It is worthwhile to remember, however, that these objections may not be insuperable, especially given the uniqueness of physical conditions in the nucleus. Further observations are likely to allow more insights in the near future, of which some examples are given below.

V. THE SHOCKED GAS

Figure 4 shows a radio map of Sgr A (West) from the Very Large Array (Brown, Johnston and Lo, 1981). Comparison with Figures 2 and 3 shows the similarity of radio and infrared surface brightness remarked earlier: this similarity arises naturally from the clumpy distribution of material within a region of generally very low density. Superposed on the radio map are a series of sample positions, shown as square boxes, located on or near the position of the higher density (neutral) ring of material which surrounds the central low density cavity containing the plasma. This ring lies in the Galactic plane, as indicated in Figure 4; the sample positions are labelled according to their location, either NE or SW of the center, followed by an arbitrary alphabetic key A,B,C. . .

THE GALACTIC CENTER   359

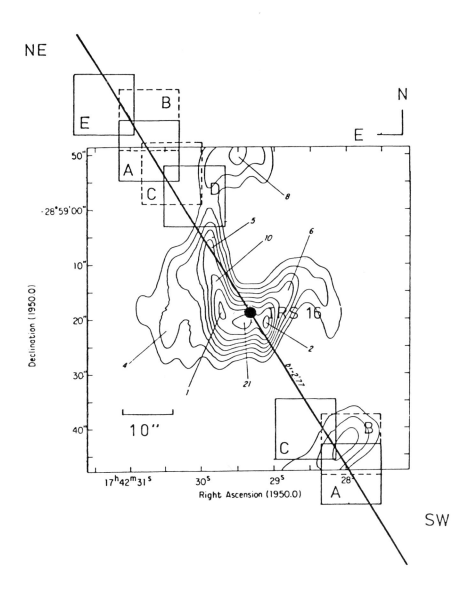

Figure 4. The solid contours show the radio free-free map of Sgr A
           from the VLA (Brown Johnston and Lo, 1981). Superposed are
           the sample positions for the molecular hydrogen spectra
           (Figures 5, 6, 7), shown as square boxes corresponding to
           the beam size employed. The position of the nuclear source
           IRS 16 and the orientation of the Galactic plane are
           indicated.

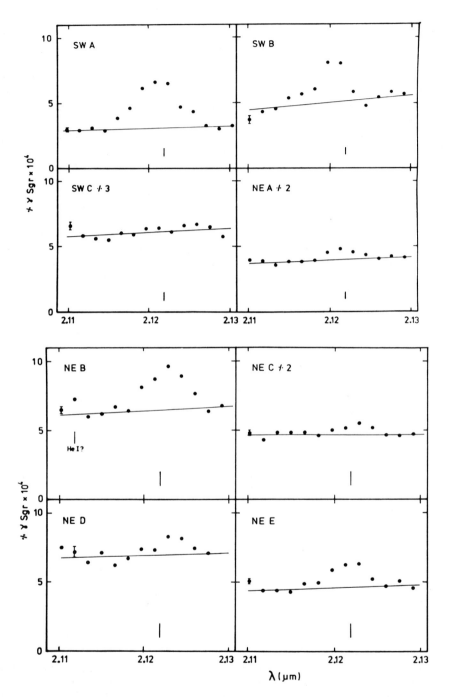

Figure 5. Spectra of the v = 1 → 0 S(1) line of molecular hydrogen (2.122μm) sampled at the eight positions indicated in Figure 4. The lines are unresolved in this experiment.

Figure 5 shows spectra at the wavelength of the $V = 1 \rightarrow 0$ S(1) line of molecular hydrogen (H_2) measured at each of the eight sample positions in Figure 4 (Gatley et al 1983). The molecular line is clearly detected at most of these positions, most strongly at SW-A and NE-B. The spectral resolution of the measurements is 500, and so the observed line widths are unresolved in this experiment. The rest wavelength of the line is shown against each spectrum, and careful comparison of SW with NE positions shows a marginally significant velocity shift of ~260kms^{-1}, consistent with the idea that the ring rotates in the sense of galactic rotation.

The origin of the excitation of the molecular hydrogen is easily deduced by comparison of the strengths of the $V = 1 \rightarrow 0$ and $V = 2 \rightarrow 1$ S(1) lines (e.g. Beckwith, 1981). Figure 6 shows a spectrum of the $V = 2 \rightarrow 1$ line taken at position SW-A. (Also included in this figure is a detection of the $V = 1 \rightarrow 0$ S(0) line at the expected strength, which incidentally confirms the identification of molecular hydrogen emission.) Comparison of the strengths of the $V = 1 \rightarrow 0$ and $V = 2 \rightarrow 1$ S(1) lines shows that the $1 \rightarrow 0$ line is much stronger. This eliminates ultraviolet fluorescent excitation and strongly supports the notion that <u>the gas is shocked</u>.

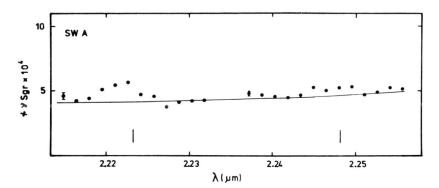

Figure 6. A spectrum taken at the position SW-A of Figure 4, which includes the wavelengths of two molecular hydrogen lines: the $v = 1 \rightarrow 0$ S(0) line (2.223 µm) and the $v = 2 \rightarrow 1$ S(1) line (2.248µm). The relative strength of the $v = 2 \rightarrow 1$ and $v = 1 \rightarrow 0$ S(1) line (Figure 5) is evidence that the molecular gas is shock-excited.

Figure 7 shows the measured line strength as a function of position along the Galactic plane from IRS 16. The striking result is that the strength of the emission from the shocked gas peaks sharply and symmetrically at 40" (~2pc) on either side of IRS 16. This is <u>not</u> an artifact of extinction, for the extinction in the ring itself is only $A_v \sim 1$ (§II), and the foreground extinction is fairly uniform (Lebofsky 1979). Rather, the distribution of shocked gas is symmetric around IRS 16, and forms a thin ring. The observations are therefore entirely consistent with the idea that the central cavity in the interstellar medium is a mass-loss bubble caused by IRS 16, and that the interaction of the mass-loss flow with the ambient material immediately outside the bubble results in shock excitation of the gas. This idea demands that we examine the spectrum of IRS 16 itself for evidence of mass loss. A mass-loss rate of $\geq 10^{-4} M_\odot yr^{-1}$ is required to account for the strength of the observed molecular hydrogen emission.

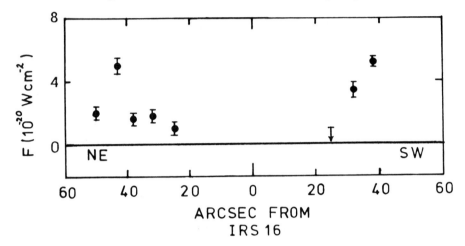

Figure 7. The measured strength of the $v = 1 \to 0$ S(1) line as a function of position along the Galactic plane from the nuclear source IRS 16. The emission peaks symmetrically about the center.

VI. THE SPECTRUM OF IRS 16

An infrared (K band) spectrum of IRS 16 taken with a 3".8 diameter aperture is shown in Figure 8 (Hall, Kleinmann and Scoville, 1982). It shows that the helium emission line at 2.06μm (4857cm^{-1}) is very broad, with a full-width at half-maximum of 1500± 300kms^{-1}. The hydrogen Brackett γ line at 2.17μm (4616cm^{-1}) seen in this same spectrum is not comparably broad, having a FWHM of 240±15kms^{-1}. The strength of the Bγ line is consistent with that expected from the NeII cloud observations at this position (§IV); we conclude that the Bγ line originates in the nearby interstellar medium. The helium line, however, must be more intimately associated with IRS 16, for it is much broader than any emission observed in NeII. This large velocity-width is suggestive of mass loss. Furthermore, comparison with spectra of Wolf-Rayet stars shows that WR stars have very strong emission in the HeI 2.06μm line, and with a velocity width similar to that seen in IRS 16, yet show no Bγ emission (Williams et al 1980).

Therefore IRS 16 shares an important spectral characteristic with a well-known type of mass-loss object.

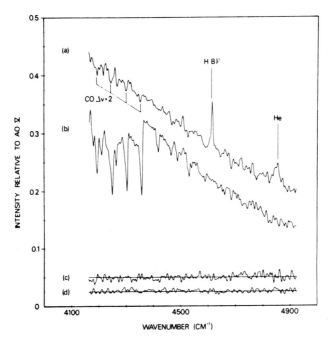

Figure 8. 2μm spectra of the nuclear source IRS 16(a) and the Galactic center M supergiant IRS 7(b) from Hall, Kleinmann and Scoville (1982). The noise levels in these spectra are shown in (c) and (d) respectively.

VII. THE DIVERSITY OF SPECTROSCOPIC CAPABILITIES

The recent growth in spectroscopic capabilities celebrated at this Symposium is perhaps most evident in the study of the Galactic center. We heard in an earlier talk that "Every species detected anywhere is seen in the Galactic center". It would be unwise to dispute this claim, as the recent literature attests. Many of the ideas presented in this review are amenable to immediate observational test, and many more have already been addressed in experiments not fully described here.

For example, as Lacy (1982) reports: "Emission lines of several other ions have also been observed, and provide important information on the atomic abundances and the excitation of the ionised gas. Brackett and Pfund lines of HI have been observed by Soifer et al (1976), Neugebauer et al (1978), Bally et al (1979), Lester et al (1981), and Nadeau et al (1981). [ArII] (7.0μm) has been observed by Willner et al (1979) and Lester et al (1981). Lacy et al (1979, 1980) searched for [ArIII] (9.0μm), [SIV] (10.5μm) and [ArV] (13.1μm), and

McCarthy et al (1980) searched for [SIII] 18.7μm. Of these four lines only [ArIII] was detected, and it only marginally." Bα (4.05μm) and [NeII] (12.8μm) spectra at 80kms^{-1} resolution (Geballe et al 1982) have shown no concentration of moderate velocity gas ($|v|$ <300kms^{-1}) at IRS 16.

Studies of the interstellar dust in Sgr A (West) show a serious difference between this HII region and most other galactic HII regions: namely that the 3.3μm and related "unidentified dust features" (cf Allamandola, these proceedings) are absent from the spectrum (Willner et al 1979). Yet the 3.3μm emission feature is present both in IRS 4 (Willner and Pipher 1982) and in a spectrum of the 2pc ring around the HII region sampled at the position SW-A of Figure 4 (Gatley et al 1983). Therefore local differences in the grain properties exist; this may well be related to the fact that the dust-to-gas ratio decreases towards IRS 16 (Gatley et al 1977; Becklin, Gatley and Werner, 1982).

The energetics of the plasma clouds have been studied by simultaneous line and continuum observations (Aitken, Allen and Roche 1982), which show that the dust in the plasma clumps radiates more power than the trapped Lα alone can provide. This appears to lend support to those who favor _internal_ heat sources for the clumps, but an alternate explanation was given above (§III).

Observations of the distribution and velocity structure of neutral oxygen emission (Genzel et al 1982) show extended emission predominantly from _outside_ the central three parsecs, with motion consistent with galactic rotation, although large scale inhomogeneities and evidence for non-circular motion are also present.

Finally there is spectroscopy of the stars in the Galactic center, a detailed discussion of which is beyond the scope of this review. The presence of late-type giants and supergiants along the line of sight through Sgr A (West) was demonstrated by Neugebauer et al (1976) and Treffers et al (1976). Work in this important area continues (Wollman, Smith and Larson, 1982; Lebofsky, Rieke and Tokunaga, 1982 and references therein). Much of the motivation for the "star formation" models of the Galactic center devolves from these observations and from their comparison with properties of other galaxies (e.g. Rieke and Lebofsky 1982). One obvious complication in studying the very nucleus is that the star formation rate is known to be high at galactocentric distances 10<R<200pc (e.g. Gatley and Becklin 1981), and so placement along the line of sight is always an issue. The spectrum of IRS 16 (Figure 8) corresponds to a spectral type no later than early G (Hall, Kleinmann and Scoville, 1982).

VIII. CONCLUSIONS

A review of the large body of observations of the Galactic center must currently conclude that a detailed, self-consistent, unique model is beyond us: the questions of §IV, and many others, remain unanswered. The observational constraints presently available suggest:

1. There is a central "engine" at the nucleus of the Milky Way.

    (a) This central object radiates $\geq 10^7 L_\odot$ at an effective temperature $T_e < 35,000K$. The ionisation state of the gas demands a soft radiative spectrum, and so implies the high luminosity. Approximately half of this bolometric power is actually observed as thermal emission from dust in the plane. The deduced geometry allows the remainder to escape. The large, directly-observed luminosity is accompanied by a relatively minor amount of ionised gas. Independent of detailed assumptions, there is a great deal of non-ionising power in the Galactic center.
    (b) The distribution of grain temperature is symmetric about, and peaked at, the position of IRS 16. Both the density distribution and the distribution of shocked molecular hydrogen is symmetric about the position of IRS 16.

2. There may be mass loss from the Galactic nucleus.

    (a) There is a void in the interstellar medium centered on IRS 16. This may be a mass-loss bubble.
    (b) The spectrum of IRS 16 shows a very broad (1500kms^{-1}) helium line at 2.06µm.
    (c) Shocked molecular hydrogen occurs symmetrically about IRS 16 at the surface of the mass-loss bubble. The emission can be powered by a mass-loss rate $\gtrsim 10^{-4} M_\odot yr^{-1}$.

3. The presence of the central object dictates the appearance of the interstellar medium in the inner ten parsecs.

    (a) The similarity of the individual plasma clouds is suggestive of a common origin and excitation.
    (b) It is plausible energetically that the plasma clouds are ionised radiatively by the central object.
    (c) Mass loss sweeps material out of the central cavity and shocks the higher density molecular material immediately outside the mass-loss bubble.
    (d) The mass-loss flow impacts the 2pc ring, which is rotating in equilibrium. There is no angular momentum in the flow, and so a braking force is exerted on the ring. This will return material inward, and may contribute to the production of the clumps of material within the central cavity. The

mass-loss rate deduced for IRS 16 is comparable to that required for cloud production.

(e) The plasma clouds are within the mass-loss flow, and so are continuously braked; they may be infalling. An entirely different possibility — that of plasma outflow — has been discussed by Brown (1982) and by Ekers et al (1983)

4. The nature of the central object is unknown.

(a) "Normal" star formation like that seen in the spiral arms is not taking place in the Galactic center.
(b) The deduced luminosity and mass-loss rate of IRS 16 are similar to those of R136, the central object of the 30 Doradus nebula in the Large Magellanic Cloud (e.g. Cassinelli, Mathis and Savage, 1981).
(c) Several further remarkable properties can be attributed to the Galactic nucleus: (1) A non-thermal radio source, unique in the Galaxy, coincides with IRS 16 (e.g. Lo 1982). (2) Time-variable positron annihilation radiation is emitted from the Galactic center (e.g. Lingenfelter and Ramaty 1982). (3) Explosive phenomena centered on the nucleus are observed at various scale sizes (Oort 1977). Therefore,

5. There may be a black hole in the Galactic center.

ACKNOWLEDGEMENTS

Answers to questions raised by Drs. Aitken, Becklin, Geballe, Hall, Joseph, Longmore, McKee, Moorwood, Panagia, Ring and Willner have been incorporated in the text. I want to thank all these people for their lively discussion of this interesting subject, and to thank Mrs. Yolanda Boyce for her careful preparation of the manuscript. This work was supported in part by the Royal Observatory, Edinburgh.

REFERENCES

Aitken, D. K., Allen, M. C., and Roche, P. F.: 1982, in "The Galactic Center," AIP Conference Proceedings 83, G. R. Riegler and R. D. Blandford (eds.), p. 67.
Bally, J., Joyce, R. R., and Scoville, N. Z.: 1979, ApJ 229, 917.
Becklin, E. E., and Neugebauer, G.: 1968, ApJ 151, 145.
Becklin, E. E., and Neugebauer, G.: 1975, ApJ Letters 200, L71.
Becklin, E. E., Gatley, I., and Werner, M. W.: 1982, ApJ 258, 135.
Beckwith, S. V. W.: 1981, in "Infrared Astronomy," IAU Symposium 96, C. G. Wynn-Williams and D. P. Cruikshank (eds.) (Dordrecht:Reidel), p. 167.
Brown, R. L., Johnson, K. J., and Lo, K. Y.: 1981, ApJ 250, 155.
Brown, R. L.: 1982, ApJ 262, 110.
Cassinelli, J. P., Mathis, J. S., and Savage, B. D.: 1982, Science 212, 1497.
Ekers, R. D., van Gorkom, J., Schwartz, U. J., and Goss, W. M.: 1983, Astron. and Astrophys. submitted.
Gatley, I., Becklin, E. E., Werner, M. W., and Wynn-Williams, C. G.: 1977, ApJ 216, 277.
Gatley, I., Becklin, E. E., Werner, M. W., and Sellgren, K. W.: 1979, ApJ 233, 575.
Gatley, I., and Becklin, E. E.: 1981, in "Infrared Astronomy," IAU Symposium 96, C. G. Wynn-Williams and D. P. Cruikshank (eds.), (Dordrecht:Reidel) p. 281.
Gatley, I.: 1982, in "The Galactic Center," AIP Conference Proceedings 83, G. R. Riegler and R. D. Blandford (eds.), p. 25.
Gatley, I., Becklin, E. E., and Telesco, C. M.: 1983, in preparation.
Gatley, I., Hyland, A. R., Jones, T. J., Beattie, D. H., and Lee, T. J.: 1983, in preparation.
Geballe, T. R., Persson, S. E., Lacy, J. H., Neugebauer, G., and Beck, S. C.: 1982, in "The Galactic Center," AIP Conference Proceedings 83, G. R. Riegler and R. D. Blandford (eds.), p. 60.
Genzel, R., Watson, D., Townes, C., Lester, D., Dinerstein, H., Werner, M., and Storey, J.: 1982, in "The Galactic Center," AIP Conference Proceedings 83, G. R. Riegler and R. D. Blandford (eds.), p. 72.
Hall, D. N. B., Kleinmann, S. G., and Scoville, N. Z.: 1982, ApJ 260, L53.
Lacy, J. H., Baas, F., Townes, C. H., and Geballe, T. R.: 1979 ApJ Letters 227, L17.
Lacy, J. H., Townes, C. H., Geballe, T. R., and Hollenbach, D. J.: 1980, ApJ 241, 132.
Lacy, J. H.: 1982, in "The Galactic Center," AIP Conference Proceedings 83, G. R. Riegler and R. D. Blandford (eds.), p. 53.
Lebofsky, M. J.: 1979, Astron. J. 84, 324.
Lebofsky, M. J., Rieke, G. H., and Tokunaga, A. T.: 1982, ApJ 263, 736.
Lester, D. F., Bregman, J. D., Witteborn, F. C., Rank, D. M., and Dinerstein, H. L.: 1981, ApJ 248, 524.

Lingenfelter, R. E., and Ramaty, R.: 1982, in "The Galactic Center," AIP Conference Proceedings 83, G. R. Riegler and R. D. Blandford (eds.), p. 148.
Lo, K. Y.: 1982, in "The Galactic Center," AIP Conference Proceedings 83, G. R. Riegler and R. D. Blandford (eds.), p. 1.
McCarthy, J. F., Forrest, W. J., Briotta, D. A., and Houck, J. R.: 1980, ApJ 242, 965.
Nadeau, D., Neugebauer, G., Matthews, K., and Geballe, T. R.: 1981, Astron. J. 86, 561.
Neugebauer, G., Becklin, E. E., Beckwith, S., Matthews, K., and Wynn-Williams, C. G.: 1976, ApJ Letters 205, L139.
Neugebauer, G., Becklin, E. E., Matthews, K., and Wynn-Williams, C. G.: 1978, ApJ 220, 149.
Oort, J. H.: 1977, Ann. Rev. Astron. Astrophys. 15, 295.
Rieke, G. H., Telesco, C. M., and Harper, D. A.: 1978, ApJ 220, 556.
Rieke, G. H., and Lebofsky, M. J.: 1982, in "The Galactic Center," AIP Conference Proceedings 83, G. R. Riegler and R. D. Blandford (eds.), p. 194.
Soifer, B. T., Russell, R. W., and Merrill, K. M.: 1976, ApJ 207, L83.
Treffers, R. R., Fink, N., Larson, H. P., and Gautier, T. N.: 1976, ApJ 209, L115.
Williams, P. M., Adams, D. J., Arakaki, S., Beattie, D. H., Born, J., Lee, T. J., Robertson, D. J., and Stewart, J. M.: 1980, MNRAS 192, 25P.
Willner, S. P., Russell, R. W., Puetter, R. C., Soifer, B. T., and Harvey, P. M.: 1979, ApJ 229, L65.
Willner, S. P., and Pipher, J. L.: 1982, in "The Galactic Center," AIP Conference Proceedings 83, G. R. Riegler and R. D. Blandford (eds.), p77.
Wollman, E., Smith, H. A., and Larson, H. P.: 1982, ApJ 258, 506.

GROUND-BASED EXTRAGALACTIC INFRARED SPECTROSCOPY AND RELATED STUDIES

Malcolm G. Smith

Royal Observatory, Edinburgh

1. INTRODUCTION.

   1.1 Classification of extragalactic objects in the infrared.
   1.2 The importance of dust as a tracer of star formation in galaxies.
   1.3 Optical and i.r. technology - the need for infrared detector arrays.
   1.4 Spectral signatures and physical effects of dust in galaxies.
   1.5 Infrared spectral signatures of gas in galaxies.
   1.6 Examples of infrared spectra of galaxies.

2. NORMAL GALAXIES.

   2.1 The Milky-Way.
      2.1.1 Dust.
      2.1.2 The central regions.
   2.2 Nearby Galaxies.
      2.2.1 Direct infrared radiation from stars in normal galaxies.
      2.2.2 Dust in normal galaxies.
      2.2.3 Gas in normal galaxies.
      2.2.4 Starbursts.

3. ACTIVE GALAXIES.

   3.1 Definitions and introduction.
   3.2 Narrow-lined x-ray galaxies.
   3.3 Stars in Seyfert galaxies.
   3.4 Atomic hydrogen emission from Seyfert nuclei.
   3.5 Molecules in Seyferts.
   3.6 Dust in Seyferts.

4. QUASARS AND BL Lac OBJECTS.

    4.1 IR and millimetre-wave continuum observations of quasars and blazars.
        4.1.1 The near infrared continuum.
        4.1.2 The shape of the millimetre-wave continuum - radio loud QSOs.
        4.1.3 Radio-quiet QSOs.
        4.1.4 BL Lac objects.
        4.1.5 The continua of the very red radio sources.
    4.2 Dust in quasars?

5. COSMOLOGICAL INVESTIGATIONS.

    5.1 Evolution of elliptical galaxies.
    5.2 The infrared luminosity vs velocity width relation.
    5.3 Infrared observations of supernovae.

1. INTRODUCTION

1.1 Classification of extragalactic objects in the infrared.

Our starting point will be the comprehensive review by Rieke and Lebofsky (1979) of infrared emission from extragalactic sources. They attempted to include all relevant material published before September 1978. They discussed three broad classes of extragalactic objects.

In the first class, the near-infrared output is dominated by direct radiation from stars, even in the nuclei; the spectrum of this radiation shows prominent CO bands near 2.3µm. Such bands are strong in late-type giants, but are not present in the spectra of M dwarfs (see, for example, the paper presented at this conference by Ridgway); more detailed modelling has confirmed the dominance of late type giants in galaxies of this class.

In their second class of objects, the original source of radiated energy is obscured from view. Most of the infrared output is in the form of thermally re-radiated flux; in many cases - the true "infrared galaxies" - this infrared continuum radiation dominates the total observed luminosity output. Indeed, the intrinsic luminosities of the obscured nuclei of many spiral and irregular galaxies are much higher than had at first been realised, and the energy outputs can in some cases be difficult to reconcile with the simplest conventional ideas of star formation.

Finally, Rieke and Lebofsky discuss active galactic nuclei, such as quasars, blazars and at least some Seyferts, in which the bulk of the luminosity in the source is thought to arise in phenomena other than nuclear processing in stars. In these objects, the exact nature and relative importance of possible emission mechanisms is far from certain.

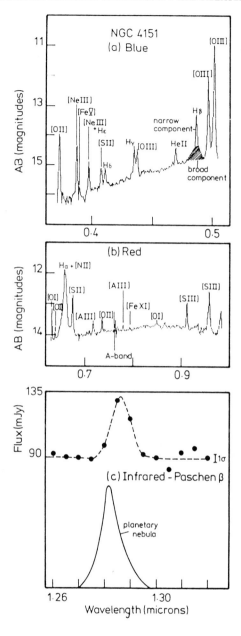

Figure 1: Optical and infrared spectroscopy of the Seyfert galaxy NGC4151. Extragalactic infrared spectroscopy is still a hard struggle. The broad and narrow components to the permitted lines are discussed in Section 3.

1.2 The importance of dust as a tracer of star formation in galaxies.

The next volume of Annual Reviews of Astronomy and Astrophysics will contain a wide-ranging article on extragalactic dust by Stein and Soifer (1983); more detailed discussions have been presented at this conference by Allamandola, Watson, Williams and Willner. Anyone attempting to use infrared astronomy as a tool for studying galaxies and their evolution quickly comes to realise the importance of dust. It is obvious that extinction, reddening, dust heating and re-radiation alter the observed flux of radiation in a systematic, wavelength-dependent way; this alteration has of course to be well understood and corrected for before one can hope to gain further insight into the physical processes occurring in galaxies which contain significant amounts of dust. Of even greater significance however is the apparent link between strong infrared emission by dust, and star formation. Dust appears to be a very sensitive tracer of star formation in galaxies. As Stein and Soifer point out, dust can be used to probe the current level of star formation activity in a wide variety of galaxies and hence to make rather direct tests of galactic evolution models.

1.3 Optical and i.r technology - the need for infrared detector arrays.

Although we shall be concentrating on astrophysics rather than on the equipment used for the observations, it should nevertheless be realised that extragalactic spectroscopy in the infrared is very much in its infancy. We have heard at the beginning of the conference from Hall about the importance of infrared arrays for such spectroscopy. I would like to illustrate his point by demonstrating the large gap in detector array technology which presently exists between state-of-the-art optical and infrared spectrometers. Figure 1 includes an illustration of the optical spectrum of NGC 4151 taken by Oke and myself in March 1982, using the Palomar 5-metre telescope and the twin-CCD double spectrograph (Oke and Gunn 1982). This spectrum was obtained in only 5 minutes. Figure 1(c) is a spectrum of Pβ in NGC 4151 taken by a group of us at UKIRT using Richard Wade's Mk I, single-element cooled grating spectrometer "CGS 1" at UKIRT, also during March 1982. A spectrum of a typical planetary nebula is shown to illustrate the resolution used; the slight shift corresponds to the redshift of NGC1068. This infrared scan, just beyond the long-wavelength limit of the optical work, took almost as many hours as the optical spectrum had taken minutes (admittedly the infrared observing time had been lengthened by poor weather conditions; furthermore, since the conference, we have used a new seven-element InSb array in Richard Wade's MkII spectrometer on UKIRT, with much reduced observing times). Hall's point is clear - we can expect a more rapid rate of progress in infrared astronomy - especially extragalactic infrared spectroscopy - following the more widespread introduction of infrared arrays.

1.4 Spectral signatures and physical effects of dust in galaxies.

From the point of view of infrared spectroscopy of galaxies, the primary signatures of dust are: (i) the stretching and bending modes of the Si-O bond at 9.7μm and 18μm; when these features are compared with laboratory simulations, it appears that interstellar silicates are amorphous in nature; (ii) a series of infrared emission features at 3.3, 3.4, 6.2, 7.7, 8.65 and 11.25μm which, from studies in our own Galaxy, appear to be associated spatially with dust emission at interfaces between ionised and molecular material in HII regions and planetary nebulae; these emission features are believed to arise from solid-state transitions on dust grains (Bregman and Rank 1975, Tokunaga and Young 1980), but which have not yet been attributed successfully to any specific material. As we have heard in the detailed papers by Allamandola, Williams and Willner at this conference, the excitation mechanism of the dust emission features is still highly speculative (see also Aitken and Roche 1983). Because the features have never been resolved, it seems they are not molecular bands, yet the wavelengths of the features are in the range of vibration frequencies of cosmically abundant material such as C-H, O-H, C-O etc. Illustrations of some of these features are provided later in Figures 2 and 4. (iii) A final signature of dust is of course the thermal emission from the heated grains, discussed in the accompanying article by Metzger.

Specific physical effects of dust in galaxies which are discussed by Stein and Soifer (1983) include shielding of gas from ultraviolet radiation, cooling mechanisms involving dust re-radiation, molecule formation on dust grains, radiation pressure instabilities and dust condensation from gas in explosive events.

A useful measure of the presence of dust is the column density of hydrogen per magnitude of reddening; from a model of the size distribution of interstellar dust (derived from the extinction curve), the dust-to-gas ratio can be estimated.

1.5 Infrared spectral signatures of gas in galaxies.

The primary infrared emission lines to look for from gas in extragalactic sources include atomic hydrogen recombination lines (particularly Pβ λ 1.282μm, Bγ λ 2.166μm and Bα λ 4.052μm), molecular hydrogen quadrupole emission (particularly the S(1) v = 1-->0 transition), the (atomic) helium line at 2.06μm and fine-structure lines of [ArII] at 6.98μm, [SIV] at 10.52μm, [NeII] at 12.8μm and [SIII] at 18.7μm. As we have heard at this conference, extragalactic line work is just becoming possible in the far infrared, but the next lines strong enough for detailed extragalactic studies with existing facilities are the hydrogen and molecular lines - especially CO - in the radio region of the spectrum.

Earlier papers at the conference (e.g. Flower, McKee, Wynn-Williams) have demonstated the particular value of infrared emission lines (i) as dynamical probes of highly obscured regions; (ii) as counters of ionising continuum photons when the source of those photons is itself obscured; (iii) as reddening indicators and (iv) as crude abundance estimators. Rieke and Lebofsky (1979) have specifically warned that, as many spiral galaxies suffer heavy extinction, lack of correlation between infrared and optical properties such as emission-line strength may result because different regions are being observed at the two wavelengths. The relatively small levels of obscuration deduced from ultraviolet and visible colours may only reflect inhomogeneities in the dust clouds or that the sources are optically thick at these wavelengths so that one sees only to optical depth unity (corresponding to $A_\lambda \sim 1$). Whenever an extinction $A_\lambda \sim 1$ is encountered in an extragalactic source, one should be aware that these effects may be present.

A rather interesting text-book study of the continuum spectrum of an HII region over six decades in frequency has been provided by Israel et al. (1982). This study, of the giant HII region complex NGC 604 in M33, is based primarily on maps in the radio (6 cm) and near-infrared (1.6 and 2.2µm) continuum as well as Hα along with a measurement of the uv spectrum of the ionising stars. The radio emission is thermal (free-free) and optically thin over the full range observed, from 49 cm to 2.8 cm. The mid- and far-infrared continuum is consistent with the typical hot and warm dust components discussed earlier in this conference by Gatley.

1.6 Examples of infrared spectra of galaxies.

At this point, we collect together references and illustrations which will allow the reader to form an impression of the available infrared spectral data on galaxies, before we move on to a more detailed discussion of the interpretation of these spectra. An illustration of the stellar CO features in M82 near 2.4µm including $C^{13}O$ as well as $C^{12}O$, is given by Rieke (1981). Illustrations of the infrared continua for a range of galaxies have been presented at this conference by Metzger. The 8-13µm spectra of a number of narrow-emission-line galaxies, Seyfert galaxies and broad-lined radio galaxies obtained by Aitken and collaborators are illustrated, later, in Figure 2.

2. NORMAL GALAXIES

2.1 The Milky Way.

Before moving out into the distant reaches of the universe, it is perhaps useful to recap some of the things we have learned already, at this conference and elsewhere, about our own Galaxy. Because we shall be dealing later both with galaxies whose luminous output is dominated by dust re-radiation, and with active galactic nuclei, I shall summarise very briefly the global properties of dust in our Galaxy and recall some of the properties of its central regions.

2.1.1 <u>Dust</u>. In providing this summary, I am once again drawing heavily on the forthcoming review article by Stein and Soifer (1983) and the references contained therein. In our Galaxy, it has recently become clear that the bulk of the diffuse <u>mid</u>-infrared emission (10µm - 30µm) arises from circumstellar shells surrounding late-type stars (Price 1981; Price, Marcotte and Murdock 1982); only a small fraction of the 10µm to 30µm diffuse emission is provided by HII regions. Note that significant emission at these wavelengths implies temperatures of several hundred degrees, and so the dust responsible must be near the heating sources. <u>Far</u> infrared observations of the diffuse emission in our galaxy, which probe right across the galaxy, yield colour temperatures in the range 20-25°K, essentially independent of galactic longitude. We have heard discussions at this conference concerning the somewhat controversial issue of the relative importance of the emission from molecular clouds vis-a-vis extended diffuse HII regions. What does seem reasonably certain is that the far infrared luminosity is associated with sites of star formation activity and represents down-converted energy originally emitted in the form of ultraviolet photons. As we have heard in the other talks, the far infrared luminosity and the blue luminosity probably have similar values - of order $2 \times 10^{10} L_\odot$.

2.1.2 <u>The central regions.</u> For the purposes of comparison with other galaxies, our Galaxy is best described as 'normal', in spite of the activity described at this conference by Gatley (see also Townes et al. 1983). That activity is confined to the central parsec of the Galaxy and involves a total luminosity of $\sim 10^7 L_\odot$ -about two orders of magnitude less than the flux from the central kiloparsec that one would study in typical comparisons of our Galaxy with others placed several Mpc or more away from us; as we have heard in other papers at this conference, it is likely that the bulk of the radiation from this larger central region comes from young, hot stars.

2.2 Nearby galaxies.

In this section we shall be considering objects in the first two classes discussed by Rieke and Lebofsky (1979). First of all we deal primarily with the direct infrared emission spectrum of extragalactic stars, particularly late-type giants as measured through narrow-band filters. Starlight from most external galaxies is too faint for detailed infrared spectroscopy of lines, given the present state of infrared instrumentation; much of section 2.2.1 has therefore to be devoted to infrared filter photometry. Even the excellent narrow-band index work of Frogel, Persson and colleagues is limited to the nearest galaxies of the Local Group. The importance of their work of course lies in investigating essentially the rate of increase of metal abundance with time in different types of stellar system (c.f. the different rates in the disk, nuclear bulge and halo of our galaxy).

We shall later move to galaxies with larger amounts of dust until we reach those spiral and irregular galaxies in which intense bursts of star formation are occurring, over regions often many kiloparsecs in size; the large 10μm excesses in these galaxies are consistent with thermal emission from dust heated by resonance-line radiation from the HII regions and the uv continuum of the young stars which ionise the gas (see, e.g. Rieke et al. 1980).

2.2.1 <u>Direct infrared radiation from stars in normal galaxies.</u>
Aaronson (1978) has shown that spiral and elliptical galaxies form separate sequences in colour-colour plots of U-V versus V-K. Rieke and Lebofsky (1979) summarised the interpretation of these data as follows: the elliptical sequence is probably a metallicity sequence while the spiral sequence is probably a population sequence (see also Griersmith, Hyland and Jones 1982). Rieke and Lebofsky cited as supporting evidence the CO index, which is roughly constant to type Sbc, but decreases in galaxies of later type "indicating a decrease in the amount of radiation from giants". This interpretation and the supporting evidence now seem questionable. For example Cohen et al. (1981) have shown that for <u>carbon</u> stars, CO indices, as measured with the usual narrow-band filters at 2.36μm (line) and 2.20μm (continuum), do <u>not</u> primarily measure CO absorption strengths. In the nuclear bulge of our galaxy, this is unlikely to be a problem as the ratio of luminous C stars to luminous M stars later than M4 is only 0.003 (Blanco, Blanco and McCarthy 1978). However, the deep grism survey of the Magellanic Clouds by Blanco, McCarthy and Blanco (1980) showed that this ratio is 2 in the LMC and 15 in the SMC (Frogel et al. 1982); using intermediate-band CO indices as a measure of the contribution from giants in galaxies such as the Magellanic Clouds is therefore likely to be unreliable (see also Richer and Westerlund 1983).

Persson et al. (1983) have recently completed a study of the integrated light of 84 clusters in the Magellanic Clouds; the clusters span nearly the complete range of cluster ages in the Clouds. This work will obviously be of importance to population syntheses and evolutionary models based on infrared data. They have succeeded in isolating a group of intermediate-age "IR enhanced" clusters. These clusters are separated from the others by their integrated infrared colours alone, apparently because in nearly all cases they contain red, luminous carbon stars. Persson et al. point out that the effects of these luminous carbon stars are strong enough that metal-poor, intermediate-age stellar populations may be detectable in the integrated light of more distant galaxies. This is of particular interest following the discovery of asymptotic giant branch - AGB - carbon stars in dwarf galaxies (e.g. Frogel et al. 1982; Richer and Westerlund 1983; Aaronson, Olszewski and Hodge 1983).

Concentrating on ellipticals and just the bulge regions of early-type spirals, Griersmith, Hyland and Jones (1982) find their UBVJHK colours are very similar in most objects of similar luminosity. Thus, "even though there are dynamical differences between spiral bulges and ellipticals, their stellar populations are similar". However Frogel, Persson and Cohen (1980) have compared data on M31 - a spiral galaxy - with the infrared (stellar) spectra of early-type galaxies. Much redder broad-band colours and strong CO and H_2O indices are found in the early-type galaxies, which seem to require a previously unsuspected population of cool luminous stars. In this case carbon stars cannot be responsible because of the constraints provided by the combination of the CO and H_2O indices. Frogel et al. suggest that (relatively young) upper asympotic giant-branch stars are responsible. If correct, this raises the possibility that star formation has continued over about half the lifetime of the ellipticals. We shall return to the early evolution of elliptical galaxies in section 5.1.

Before we move on to a discussion of dust in normal galaxies, I should conclude our discussion of direct infrared radiation from stars in normal galaxies by mentioning the <u>spectroscopic</u> detection of strong CO absorption bands at 2.3-2.4μm in NGC 253 and M82 (Rieke et al. 1980; see also Willner et al. 1977; Wynn-Williams et al. 1979, 1983; see also the paper presented by Hall at this conference). This CO absorption is consistent with most of the continuum at 2μm arising from a population of late-type luminous stars. There has been some controversy over the relative proportion of supergiants and giants in the mix. A detailed discussion of the issue is given in the paper by Rieke et al. (1980); they contend on the basis of detailed calculations using theoretical stellar evolution tracks, that supergiants may contribute a very significant fraction of the 2μm flux (> 15% for M82). Wynn-Williams et al. (1979) had argued for a much smaller proportion (< 1%) using the ratio of ionising flux to M supergiants in 30 Doradus as a guide. Rieke et al. pointed out that the weakness of this approach lies in the fact that the M supergiants in 30 Doradus were produced in an earlier burst of star formation than the present one (which still produces a very high ionising flux). Other arguments in the debate involving CO indices derived after corrections for extinction seem much less convincing to me, particularly in view of the very complex extinction geometry in NGC 253 (Rieke et al. 1980; Beck, Beckwith and Gatley, 1983).

For M82, Rieke et al. consider a very simple model in which an unresolved nucleus is sited within a plane viewed edge-on and containing stars mixed uniformly with obscuring dust that provides an extinction of A_v = 24 at the depth of the nucleus; a further 1 mag of absorption is added as foreground extinction within, but on the near side of M82. The reddening law is assumed similar to that of the solar neighbourhood. Even such a simple model as this successfully reproduces all the photometric colours and dependences of flux on beam size. The observed Bα/Bγ ratio is observed to be larger than the basic model would predict; this is easily fixed by placing the ionised gas preferentially in regions of higher than average extinction. Rieke et al. make crude estimates of the mass of red stars, assuming all the stellar mass resides purely in K5 giants (which have to produce a near-infrared flux 2 x 10^{10} $L_\odot$); their estimate,

$5 \times 10^8 M_\odot$, is very nearly the total mass interior to a radius of 20" derived from the rotation curve of M82 measured at 12.8μm by Beck et al. (1978). If the 2μm flux were modelled to come mainly from supergiants, the mass of stars required could be ~ 10 times less. Carbon stars similar to the very luminous objects we have discussed earlier probably contribute less than 15% of the flux near 1.8μm [where N-type carbon stars exhibit strong absorption in the C_2 Ballik-Ramsay (0, 0) band, which is not seen in the spectrum of M82]. As we shall see later, constraints set by infrared spectroscopy on the properties of these galaxies become even tighter, because the low mass-to-light ratio and the high IR luminosity of M82 and NGC 253 imply a powerful burst of star formation. We shall come to models of such 'starbursts' in section 2.2.4.

2.2.2 Dust in normal galaxies. In the Magellanic Clouds, the ultraviolet reddening law seems to differ from that in the Galaxy, whereas the infrared reddening law appears quite similar (Morgan and Nandy 1982; Koornneef 1982). These results suggest the existence of different grain components which govern different wavelength regions of the extinction curve. The gas-to-dust ratio is about 4 times larger in the LMC than in our Galaxy and M31. Stein and Soifer (1983) speculate that this may be related to the lower metal abundance in the LMC.

Infrared observations in the Milky Way and other normal galaxies show that dust is a sensitive indicator of sites of star formation (particularly molecular clouds) in spiral galaxies. The large amounts of dust associated with molecular clouds in our Galaxy prompted Telesco, Gatley and Stewart (1982) to build a two-channel photometer (K combined with either J or H) with which they discovered and mapped a massive toroid of dust (and, presumably, molecular gas) surrounding the nuclear bulge in NGC 7331. They conclude from this unusual near-infrared study of dust that the density, mass ($6 \times 10^8 M_\odot$), radius and thickness of this ring are all similar to those of molecular rings in the Galaxy and M81 (see also Young and Scoville 1982).

Dust can be observed in galaxies in absorption near 2μm as described above, or in emission at wavelengths nearer 10μm or beyond; Gatley has advocated observing dust emission nearer 3μm, where ground-based sensitivities are much higher than in the 10μm region. 10μm emission at a flux level well above that expected from stellar photospheres has been found in the nuclei of a number of elliptical galaxies (Rieke and Low 1972; Puschell 1981; Becklin, Tokunaga and Wynn-Williams 1982; Rieke, Lebofsky and Kemp 1982). However, Stein and Soifer (1983) remind us that because mid-infrared emission from ellipticals is almost always associated with their nuclei only, this raises some uncertainties about both the source and the mechanism for 10μm radiation from ellipticals.

One notable exception is the obviously dusty elliptical, NGC 5128. Telesco (1978) observed 10μm emission at 5 extra-nuclear points along the dust lane out to galactocentric radii of nearly 2 kpc. If the 10μm emission is indeed thermal re-radiation from dust, as seems likely, then Telesco demonstrates that the heating sources have to be similarly distributed. Rather than a single central source, the picture of conditions within the dust lane of NGC 5128 is one with an uneven distribution of HII regions in the dark lane.

Strong 10μm radiation is, of course, much more widespread in spiral and irregular galaxies. This has been learned mainly from surveys of late-type galaxies in the mid-infrared (Rieke and Lebofsky 1978; Scoville et al. 1983) and the far infrared (Telesco and Harper 1980). Scoville et al, for example, have recently surveyed 53 spiral galaxies in Virgo at 10μm; they conclude that star formation occurs in the nuclei of spiral galaxies independent of Hubble type, their location in the cluster, or whether or not they have bars. They also conclude that "the rate of star formation in galactic nuclei has little relationship to the size of the nuclear bulge or the gravitational potential near the center".

The mechanism for the 10μm emission in all these cases is assumed to be thermal emission from dust heated (as discussed at this conference by Gatley) in two ways, viz: (i) by the UV continuum of obscured hot young stars - the temperature of the dust heated in this way depends on the stellar luminosity and the distribution of the dust ($T^4 \propto L/r^2$) and (ii) by resonantly trapped recombination radiation in the HII region ionised by these stars. Thus one can expect two components to the dust emission continuum with individual (blended) peaks near 10μm (the hot dust) and 100μm (the warm dust) - see e.g. Israel et al. (1982). The first detections of 10 and 20μm radiation from an extragalactic spiral arm HII region (in M101) were reported by Blitz et al. (1981)).

Gillett et al. (1975b) were the first to make positive identification of dust features in the infrared spectrum of a galaxy. Their spectra of the nuclei of M82 and NGC 253 were similar to those of planetary nebulae in our galaxy (such as NGC 7027), in that they showed the strong dust emission features at 8.6μm and 11.3μm, accompanied by the fine-structure emission line of [NeII] at 12.8μm. Note that in normal low-luminosity HII regions in our galaxy, the 8.6 and 11.3μm features are usually masked by strong continuum emission.

This success was followed by detection of the 3.3-3.4μm dust emission band in the spectrum of NGC 253 by Russell, Soifer and Merrill (1977). Willner et al. (1977) finally hit the jackpot with their spectrum of M82, which contained the entire set of unidentified emission features listed in section 1.4.

The 3.28μm dust emission feature has been clearly detected in the nuclear region of M83 (NGC5236 - Lee et al. 1982) and in NGC3310 it

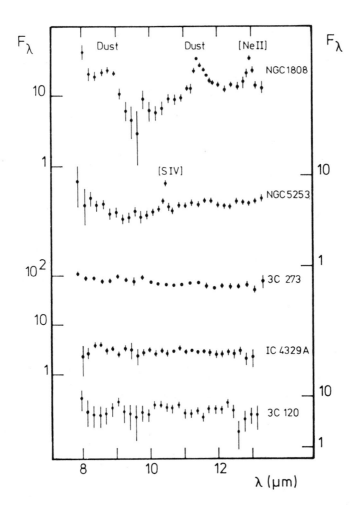

Figure 2: Representative 8-13μm spectra of narrow-lined galaxies (NGC1808, NGC5253) a Seyfert (NGC4329A), a broad-lined radio galaxy (3C120) and a quasar (3C273) as obtained by Aitken, Roche, Phillips and collaborators. See Sections 1.4, 2.2.2., 3.6 and 4.2, for further discussion. (For convenience in plotting, the flux scales have not been normalised). As a rule, most Seyfert galaxies have smooth, featureless 8-13μm spectra, well fitted by power laws, and with no evidence (within this spectral range) for dust emission.

has been mapped out to several kiloparsecs from the nucleus (Brand et al., personal communication). The five brightest galaxies at 10μm in the Scoville et al. (1983) sample (NGC4438, 4536, 4419, 4569, 4303) were also examined by Brand et al.; four showed the 3.28μm feature. The exception, NGC4438, is a Seyfert galaxy; as we shall see later (in section 3.6 - see also Figure 2), Seyfert galaxies often do not show the dust emission features.

Gillett et al. (1975b) also saw the dust in absorption; their identification of the depression in the spectra of M82 and NGC 253 near 10μm was later supported by Houck, Forrest and McCarthy's (1980) detection of silicate absorption near 19μm. Such confirmation was important in view of the warning by Aitken et al. (1979) that the various unidentified dust emission features, including possible emission in the 12μm region, can combine to exaggerate the apparent depth of "silicate absorption" near 10μm. A broad minimum near 10μm is, for example, sometimes seen in sources where the extinction is probably far too small to be compatible with silicate absorption (e.g. Gillett, Forrest and Merrill 1973; Gillett et al. 1975a; Aitken and Roche 1982); silicate absorption features are generally associated with massive extinction - for example along long lines of sight through the plane of the Galaxy to highly obscured regions.

Curiously enough, spectra between 8 and 13μm have been published for only four HII-nucleus galaxies, namely those of M82 and NGC 253 mentioned above, plus spectra of NGC 1614 (Aitken, Roche and Phillips 1981) and NGC 5253 (Aitken et al. 1982). However Phillips, Aitken and Roche (1983, in preparation) have observed five more spiral galaxies with giant HII nuclei. They conclude, from the total sample of nine such objects, that their 8-13μm spectra are remarkably similar. They almost invariably show the [NeII] 12.8μm fine structure line along with the 8.6 and 11.3μm dust emission features; an example, that of NGC1808, is illustrated in Figure 2.

The equivalent widths of the dust emission features in these galaxies are large; this is not the case for galactic HII regions, because it is found that strong continuum emission dominates this part of the spectrum. Phillips et al. argue that this difference in equivalent width can be explained in terms of beam dilution effects. In galactic HII regions, the 10μm continuum emission region normally filling the observing aperture is a compact, high-density core, where intense Lyα resonance heating occurs. A similar aperture centered on the nucleus of another galaxy encompasses a much larger volume, thus giving correspondingly larger weight to the contribution from diffuse, lower-density regions. Evidence is mounting that the unidentified dust emission features arise in ionisation fronts outside the emission-line core (Aitken et al. 1979; Aitken and Roche 1983; Sellgren 1981; Wynn-Williams et al. 1981).

Although the 8-13μm spectra of HII nucleus galaxies are nearly always very similar, the spectrum of NGC 5253 provides a strikingly instructive exception (Aitken et al. 1982; see also Figure 2). Although the dust continuum appears to show silicate absorption, there are no obvious 8.65μm or 11.25μm dust emission features. Furthermore, [SIV] λ 10.52μm fine-structure emission is observed instead of the usual [NeII] line. This unusual level of ionisation, when coupled with the level of the infrared continuum requires an extreme ionising flux of radiation (such as might be produced by several hundred early O stars or a single, massive, highly luminous object) close to the nucleus and obscured in the visible. Aitken et al. (1982) estimate the required flux from this unusually compact infrared core as $\sim 4 \times 10^{52}$ Lyman continuum photons s^{-1}. We discuss star formation in this object in section 2.2.4.

As we shall see later, (see also Figure 2) the 8-13μm spectra of most active galactic nuclei and quasars show no trace of dust-associated emission features.

2.2.3 <u>Gas in Normal Galaxies</u>. Turning first, as usual, to the Magellanic Clouds, we find that astonishingly little infrared spectroscopy (beyond broad-band filter observations) has been done.

The most recent work I am aware of is a study at 2μm (with a circular-variable-filter spectrometer) of the SMC HII region N81 by Israel and Koornneef (1983), in which they report the presence of the expected atomic hydrogen lines and discuss some evidence for the presence of molecular hydrogen.

Moving on to M33 Israel et al.(1982) find that the uv and radio observations of the giant HII complex, NGC604, can be explained in terms of the following stellar mix: (i) 2 or 3 O4 stars, (ii) an associated set of $\sim$ 450 OB stars of types O4 to B0 having a total luminosity of $L = 3 \times 10^7 L_\odot$ and mass $M = 10^4 M_\odot$ (iii) a further mass of $5 \times 10^4 M_\odot$ in the form of later-type stars. These estimates are reasonably consistent with estimates of the ionising flux based on far-infrared observations of the down-converted photons. This stellar mass is a very small fraction of the total mass of gas. The mass of ionised gas is $\sim 10^6 M_\odot$, slightly smaller than the total mass of neutral hydrogen.

The structure of NGC 604 as shown on short-exposure direct photographs is complex; optical "slitless" spectra of this giant HII region in M33, published by Smith and Weedman (1970), also show that excitation conditions in the different condensations are probably rather similar. Reddening varies over the face of the nebula and is greatest in the south-eastern side - the side nearest the edge of one of the brightest HI maxima in M33. At this position, Israel et al. detected infrared emission without a radio or optical counterpart and suggest that this is the most likely site in the NGC 604 complex for molecular line emission and current star formation.

Another somewhat similar study of the 5 brightest HII regions in M101 by Blitz et al. (1981) has located a molecular complex (in the HII region NGC 5461) which is two orders of magnitude more massive than typical giant molecular complexes in the Milky Way. OB star formation activity (and perhaps all star formation) is appreciably higher in NGC 5461 than in Galactic molecular-cloud complexes. NGC 5461 shows the strongest emission of the 5 HII regions at all frequencies, yet is still remarkably similar to Galactic HII regions (though more massive). The 2μm emission observed was found to be consistent with being primarily an extrapolation of the free-free emission measured from the HII region at 6 cm. Direct stellar 2μm emission was not detected.

Telesco's (1978) observations of HII regions in the dust lane of NGC 5128 (mentioned earlier) provide yet another example of considerable extra-nuclear star formation activity. Telesco takes the curious relation between total 10μm flux density and the radio free-free flux density ($\sim$ 10 for Galactic HII regions - Lebofsky et al. 1978; Thronson, Campbell and Harvey 1978 - see also Rieke and Lebofsky 1979 and Rieke et al. 1980) to obtain the level of free-free radio emission from the NGC 5128 HII regions. He then uses Rubin's (1968) relation between radio flux density and the number of ionising photons to obtain a total of more than 4×10^{53} Lyman continuum photons s^{-1} over the region observed (ten times the level in NGC 5253 but in a more extended region).

2.2.4 <u>Starbursts</u>. As stated earlier, a recent 10μm survey of galaxies in Virgo by Scoville et al. (1983) has established that significant star formation is occurring in the nuclei of spiral galaxies of all Hubble types. Gehrz, Sramek and Weedman (1983) have shown that quite formidable 'starbursts' can occur outside the nuclei of some galaxies. Therefore because the extinction at 2μm is an order of magnitude less than at visual wavelengths, one might at first hope to use the infrared to find direct stellar emission from early-type stars in the less dusty galaxies. However this is not possible because, as discussed in 2.2.1, the 2μm continuum in relatively unobscured galaxies is completely dominated by luminous late-type stars.

The most spectacular star-forming episodes in nearby galaxies seem to be found most often in the giant HII region nuclei of dusty spiral and irregular galaxies. We can probe the ionising field by its effect on heating the dust or, to some extent, by the use of the infrared emission lines. As we have seen, the [SIV] λ 10.52μm line in NGC 5253, when coupled with the amount of infrared continuum radiation, indicated a very high ionising flux of radiation (4×10^{52} Lyman continuum photons s^{-1}).

Moorwood and Glass (1982) have recently concluded an interesting study of the current star formation event in NGC 5253. They note that neither the radio, nor the X-ray emission appear to include any significant contribution from supernovae, nor is there any observable enhancement of the 2µm emission which could be attributable to luminous, late-type stars. They therefore argue that the currently observable star formation event is probably the first to have occurred within the last 10^8 years and that the current burst is probably younger than 10^7 years. The star formation appears to be going on in a highly obscured, unusually compact, intense infrared core which is itself embedded in a central HII region complex which occupies about a quarter of the size of the optical galaxy. Moorwood and Glass also follow up van den Bergh's (1980) speculation that an encounter with M83 initiated star formation in the outer regions 10^8 - 10^9 years ago, with a speculation of their own - namely that as HII regions developed around those stars in the outer regions, additional pressure acted on the central regions, causing further cloud collapse and initiating star formation at sites progressively closer to the centre of NGC 5253.

These speculations remind one of some of the processes for production of an active phase in galactic nuclei discussed by Bailey (1982). The initial triggering mechanism is gravitational collapse in his models, whereas here the process may have started through a gravitational interaction with another object. In this context it is of interest to recall Stockton's (1982) recent observations of very closely interacting pairs; as he points out, such interactions could play a major part in accounting for at least some of the manifestations of the QSO phenomena. (See also Condon and Dressel 1978; Gunn 1979; Hummel 1980; Schweizer 1980; Roos 1981; Condon et al. 1982; Bothun and Schommer 1982; Hutchings and Campbell 1983; De Robertis 1983). Gehrz, Sramek and Weedman (1983) have studied the interacting galaxy NGC 3690; its starburst event, involving the formation of > $10^9 M_\odot$ of massive stars, is of particular interest because although it is not in the nucleus of a galaxy, it has a luminosity comparable to the brightest analogous episodes in galactic nuclei. These new data all appear to be consistent with the idea that at least some QSOs may be enormous starbursts in galactic nuclei activated by tidally accreted gas from an interacting galaxy companion.

The most extensive study of massive starbursts published to date is the classic paper by Rieke et al. (1980), "The Nature of the Nuclear Sources in M82 and NGC 253". M82 has been widely adopted as the prototype "infrared galaxy" basically because it contains so much dust to thermalise so much ionising radiation that most of its radiated output eventually emerges in the infrared from a region several hundred parsecs across. We should however remember that most of the evidence that the original source of energy comes from young hot stars is indirect (see, for example, the paper given at this conference by Mezger). The most direct indication is still the presence of emission lines of ionised gas and a corresponding level of free-free radio (continuum) emission at 3 mm.

The models described by Rieke et al. (1980 - see also Rieke 1981) were based on a power-law initial mass function (IMF), $dN/dm \propto m^{-\alpha}$

(α = 2.35 is the Salpeter solar-neighbourhood solution) and an exponentially decaying star formation rate (SFR, the number of stars formed per unit time interval, $\propto e^{-t/t_o}$). The range used for the IMF was assumed to be fixed at the upper end to 31 $M_\odot$, corresponding to a lack of stars hotter than 30,000°K (the Lyman continuum flux $\sim$ 2 x 10^{53} photons s^{-1}). Of the coarse grid models tried, that including a 3.5 $M_\odot$ lower mass limit produced the best fit to the M82 data. Smaller lower limits could not produce enough flux at 2μm within the overall mass limit Rieke et al. allowed for the burst, namely 3 x $10^8 M_\odot$. [The overall mass limit for the burst of star formation itself was set by the requirement to have an earlier population of stars and a total mass of $\sim$ 5 to 8 x $10^8 M_\odot$ - derived primarily from a rotation curve measured at 12.8μm by Beck et al. 1978 but including several other methods which gave consistent estimates, as discussed by Rieke (1981). This incidentally illustrates one of the particularly valuable aspects of mid-infrared spectroscopy, as the longer wavelengths are essential to a representative sampling of the velocity field when dealing with such highly obscured galaxies].

The models by Rieke et al. (1980) were also constrained to try to match the total bolometric luminosity of M82 (4 x $10^{10} L_\odot$), the Lyman continuum flux (2 x 10^{53} photons s^{-1}) measured via the Bα and Bγ line strengths, the absolute magnitude at K(-23.3) and the observed CO index. Rieke et al. (see also Scoville et al 1983) concluded that "Plausible models were found that could meet all of the constraints simultaneously, so long as the formation of low-mass (< 3$M_\odot$) stars is strongly suppressed relative to the initial mass function in the solar neighbourhood." Rieke et al. worked through the observed continuum spectrum as follows:

i) Non-thermal radio flux at 5 GHz: supernova remnants.
ii) Thermal free-free (continuum) flux at 3 mm: ionised gas.
iii) IR excess: dust grains heated by young, hot stars.
iv) Near-infrared flux: direct radiation from luminous red stars formed at an earlier phase of the star burst.
v) Inferred uv flux: the young hot stars.
vi) X-radiation: stellar and supernova remnants.

A period of rapid star formation, lasting 10^7 to 10^8 years can plausibly explain most of the observed characteristics of M82. Possible problems that remain include the exact nature of the ultracompact radio source in M82, and the qualitative nature of the arguments concerning the contribution by supernovae. (See discussions by Biermann 1976; Condon et al 1982; Gehrz, Sramek and Weedman 1983).

We have noted that massive starburst models predict extensive supernova activity - at rates > $1 yr^{-1}$. Although realising that substantial dust extinction may cause problems, M. Phillips (personal communication) is using an optical CCD to make regular checks for supernovae in a number of the 'starburst' nuclei. Laques et al. 1980 have already reported brightening of one of the 'hot spots' in NGC 2903. To overcome the extinction problem, Phillips also intends to make observations with the VLA.

Rieke (1981) posed three major questions for future workers on massive bursts of star formation:
i) How do conditions in the nuclei of galaxies affect the star formation process?
ii) Is there an evolutionary sequence relating the wide range of properties observed in different galaxy nuclei?
iii) What mechanisms trigger and sustain the star formation activity?

A start towards answering these questions has been made; intense bursts of star formation, sometimes over regions kiloparsecs in size and involving luminosities in excess of $10^{10} L_\odot$ have been observed in many galaxies other than M82, viz: NGC 253 (Rieke et al. 1980), NGC 1808, NGC 6946, NGC 7714, NGC 7552, He 2-10 (all five galaxies discussed by Phillips, Aitken and Roche 1983, in preparation), M51 (Telesco and Owensby 1981), NGC 3690 (Gehrz, Sramek and Weedman 1983), NGC 2903, NGC 3310, IC 342 and Maffei 2 [papers in preparation by Beck, Beckwith and Gatley; by Telesco and Gatley (1983); and by Gatley, Sellgren and Werner (1983)].

3. ACTIVE GALAXIES

3.1 Definition and introduction - some pertinent data from optical spectroscopy.

We have already defined an active galaxy (in section 1) as one in which the bulk of the energy in the source is generated by processes other than nuclear processes in stars; one such scenario has been described recently by Bailey (1982). Ward et al. (1982) offer a phenomenological definition roughly equivalent to the theoretical one given here. Among classes of active galaxy that we shall be discussing are (in rather over-simplified terminology):-
(i) **Seyfert galaxies**: Spiral galaxies, (usually), with a bright nucleus. The nuclear light shows emission lines and usually some non-thermal radio emission.
(ii) **N-galaxies**: Elliptical galaxies with a bright nucleus; some are radio sources.
(iii) **Radio Galaxies**: Generally giant and supergiant ellipticals, which usually show double-lobed structure. Frequently the radio structure has a central component of the core-halo type with a compact nuclear source whose radio emission can vary on a time scale of years. Radio jets can show superluminal expansion.
(iv) **Quasi-stellar objects**: Apparently star-like objects with large redshifts. Their optical spectra are reminiscent of Seyfert galaxies. They are usually strong sources of infrared radiation and, particularly at low redshifts, of X-radiation as well. Most (90-95%) are radio quiet, even though quasars were first discovered by radio means. Some are highly variable (OVV's, for Optically Violently Variable).

(v) **BL Lacertae Objects** (BL Lacs, Lacertids, Blazars). Elliptical galaxies (probably) with a very bright nucleus. The nuclear light shows a largely featureless continuum (apart from a few absorption lines in the higher redshift objects and the occasional very weak emission line) along with high, variable polarisation. As a warning of the degree of over simplification involved in these working 'definitions' I refer the reader to the incredible session at the Pittsburgh conference on BL Lac objects in which pages of learned discussion fail to produce a totally acceptable definition of what everyone means by this class.

The most prominent emission lines in optical and ultraviolet spectra of active galaxies involve hydrogen, C^{2+}, C^{3+}, Mg^+, O^{2+} and O^{3+}. The relative intensites can be used to derive electron densities, electron temperatures, abundances and the ionisation equilibrium via a parameter U_H, the ratio of the flux of ionising photons to the electron density. An excellent, very detailed review on the detailed properties and physics of the emission-line regions of active galaxies has been given by Davidson and Netzer (1979) although it is now necessary to read some of the more recent literature - e.g. Baldwin, Phillips and Terlevich (1981), Ferland and Netzer (1983) and Pequignot (preprint) -see also the reviews given at this conference by McKee, Wynn-Williams, Lacy and Flower.

As we shall see, each of the above classes of active galaxy has an enormous number of possible subdivisions, but I shall try to stick to those which appear to have the greatest scientific interest to potential projects in infrared spectroscopy. One of the most important of these is the subdivision of Seyfert galaxies into two basic classes, based on the width of the permitted-line wings. Seyfert galaxies with extremely broad wings to the permitted lines (full width at zero intensity FWZI 10,000 km s^{-1}) are denoted Seyfert class 1. Those with narrow wings, no different from the forbidden lines, are denoted Seyfert class 2. Further subdivisions of these basic classes have since been defined (see, e.g., Osterbrock 1981). Khachikian and Weedman (1971, 1974) based their initial classification on photographic spectra of the (Balmer) Hβ line at 0.4861μm and the pair of [OIII] lines at 0.4959μm and 0.5007μm.

The most obvious significance of the difference between Seyfert 1 and Seyfert 2 galaxies lies in the electron density and the size of the region emitting most of the emission-line flux. In the broad-line region (BLR) responsible for the broad wings on permitted lines in Seyfert 1 galaxies [see, e.g., Figure 1(a)], the electron density is probably in the neighbourhood of 10^{10} cm^{-3}, the size is in the range less than a parsec up to ~ 10 parsecs (for the most luminous quasars) and the mass of ionised gas involved is a few solar masses [Davidson and Netzer 1979; Penston (1982) reviews recent evidence that ionisation conditions in the BLR may be stratified]. The electron density is too high in this region to allow spontaneous forbidden transitions to take place, so the forbidden lines do not have these broad wings. Permitted lines in Seyfert 2 galaxies also fail to show these wings (by definition), indicating the apparent absence of a high-density core. The forbidden lines in both classes of Seyfert are sharp, and are roughly matched in Seyfert 1 galaxies by a similar sharp core to the permitted lines [as shown in Figure 1 (a)]. In the narrow-line region (NLR), the electron density is $\sim 10^4$ cm^{-3}, the size is several kiloparsecs and the mass of gas involved can be several thousand solar masses.

It is perhaps at first surprising that even the spectrum of atomic hydrogen in active galactic nuclei is poorly understood (e.g. Baldwin 1977; see also section 3.4). The observed ratios of the atomic hydrogen lines are steeper than the 0.28: 2.8: 1.0: 0.47 (Pα: Hα: Hβ: Hγ) found in an (unreddened) case B recombination model for the emitting gas (see, e.g., Osterbrock 1974); case B assumes the Lyman lines are optically thick, all other lines are optically thin. Two basic hypotheses involving a variety of complicating effects, could be valid:
(i) The Balmer lines are reddened by dust. The reddening curve may, however, differ from the standard one, and the dust may be internal or external. Scattering by grains with differing albedos in the visible and infrared spectral regions is also possible, but would probably decrease the intensity of the redder lines (see, e.g., Thompson et al. 1978).
(ii) Some mechanism, probably involving high densities (collisional excitation and de-excitation) and large optical depths, selectively populates hydrogen energy levels from the ground up, increasing the intensity of the redder lines. At high optical depths, Lyα trapping maintains a significant population of hydrogen atoms at the n=2 level, so that Balmer and even higher series transitions can become optically thick. An advantage of the infrared is that one can observe lines having lower levels with n > 2, so that one can begin to test these ideas. At moderate optical depths, Hβ has a larger optical depth $\tau_{H\beta}$ and also a smaller escape probability than Pα. Trapped Hβ photons are thus converted into a Pα + Hα cascade. The ratios involving Hβ are correspondingly adjusted (e.g. Pα/Hβ, Hα/Hβ are increased) while the ratio of Pα/Hα is much less affected. At very large optical depths, Hα is so severely trapped that level 3 is selectively populated allowing $\tau_{P\alpha} > \tau_{H\beta}$ and so Pα + Hα photons are converted back into Hβ. A much more detailed account of the physics of ionised gas in Seyfert galaxies and quasars is given by Davidson and Netzer 1979 - see also section 3.4; Lyman-alpha trapping by dust internal to an emission-line cloud can also provide a means whereby the Lyα/Hβ flux can be lowered, while having very little effect on the Lyman continuum or the Balmer lines - but see, e.g., Carswell and Ferland (1980); Smith et al. (1981); Baldwin and Smith (1983).

Lacy et al. (1982) have shown that dust reddening is the only mechanism suggested so far that will enhance hydrogen Paschen alpha (Pα) sufficiently relative to the Balmer lines, so as to correspond to the amounts actually observed. No published models in which collisional excitation and de-excitation and radiative transfer are the most important means of altering the line flux ratios can explain the large values of Pα/Hα (often $\gtrsim 0.2$) that Lacy et al. observed. Collisional excitation tends to decrease Pα/Hα and radiative transfer effects leave it nearly unchanged. The advantage of comparing lines (such as Pα , Hβ ; Bα , Pβ , Hγ ; Pfund β , Bγ) that arise from the same upper level is that the Case B line ratios are determined solely by the branching ratio out of the common upper level, and so should be valid so long as both transitions are optically thin.

3.2 Infrared observations of narrow-line active galaxies.

As we have seen, Seyfert 1 galaxies have Balmer emission lines with broad wings. Most X-ray sources that have been identified with emission- line galaxies have involved the broad-lined Seyfert 1 type. However a small number of them turned out to have narrow optical lines. It has been noticed that they also appear to have heavily absorbed nuclei, judging from direct photographs.

Véron et al. (1980) have proposed that in all these "narrow"-lined x-ray (NLX) galaxies, the nucleus is in fact of Seyfert 1 type, partially hidden behind a dark absorbing cloud. They conclude that although the Hβ line (on which the Seyfert designation was originally based) may be narrow, in several cases a broad component can be detected when accurate profile fitting techniques are used (see also Shuder, 1980; Ward et al. 1980). In particular, NGC 1365 "most likely has a Seyfert 1 nucleus seen through about three magnitudes of visual absorption". NGC 2992 also shows a broad Hα component. MCG-5-23-16 (A0945-30) may have a broad Hα component, but no star-like nucleus could be detected at optical wavelengths. In the few cases where estimates can be attempted, it is seen that the dense BLR suffers several magnitudes of visual obscuration. Lawrence and Elvis (1982) also believe they have found evidence for substantial reddening in NLX galaxies. If we now accept the hypothesis that there exists a continuum of activity from the extreme Seyfert 1's to intermediate Seyfert 1.9's (Osterbrock 1981) extending down to these 'narrow-line' X-ray emitters, it is conceivable that there is a parallel sequence in terms of obscuration of the broad-line (high-density) region of ionised gas. Evidence for high column densities of neutral hydrogen, ($N > 10^{22}$ cm^{-2}) in many of these Seyfert galaxies comes from the X-ray cut-off data presented by Mushotsky (1982), and by Maccacaro, Perola and Elvis (1982).

High obscuration of the emission-line region would manifest itself as a high ratio of Pβ /Hα . [Pα is very difficult to observe reliably from the ground in low-redshift objects because of heavy atmospheric absorption -but see Lacy et al. (1982)]. Clearly, attempts should be made to resolve the Pβ line in some of the 'narrow-lined' objects, thus permitting separate estimates of the reddening in the two distinct regions (NLR and BLR). As mentioned earlier, lines from the same, (preferably high) upper level are best. Hγ however, is usually blended with λ 0.4363μm [OIII] emission, so care has to be taken with deblending; very good optical spectra, with reasonably high resolution and signal-to-noise ratio, are essential. To guard against possible profile variability, observations of the different lines should be as close together in time as possible. In the case of reasonably large Bα equivalent widths, this line might be attempted; however, noise in the high background continuum is likely to swamp the broad wings of Bα unless the reddening of the BLR is very large indeed.

Failure to observe the reddening in the BLR and NLR separately in this manner might introduce greater scatter into the data being fed into models, and may well introduce inconsistencies with the initial assumptions.

Infrared observations which support the concept of significant reddening in "narrow-lined" X-ray galaxies have been reported by a number of authors. NGC 5506 has received particular attention. Its JHKL colours are the reddest of 28 active galaxies measured by Ward et al. (1982). Moorwood and Salinari (1981) report detection of the 3.3 μm dust emission feature (see section 1.4) and 9.7μm (rest frame) silicate absorption in NGC 5506. The 9.7μm feature was detected via a series of broad- and narrow-band filter observations. Its presence, along with the 10.5μm [SIV] emission feature, is confirmed in spectra obtained by Roche et al. (1983, in preparation), using the UCL liquid-helium-cooled grating spectrometer and 25 elements of its 32-element detector array. However, the dust features at 8.65μm and 11.25μm were not detected, which suggested that an attempt should be made to re-observe NGC 5506 at 3.28μm. This has been done very recently at UKIRT by Brand et al. (personal communication); they failed to detect the 3.28μm feature in NGC 5506. NGC 7582 is a strange example of a narrow-lined X-ray galaxy with a flat-spectrum power-law continuum on which a strong 3.28μm dust emission feature is seen - (Moorwood 1982a b,) - this time accompanied by the usual dust features at 8.65μm and 11.25μm (Roche et al. 1983). Multi-aperture data may help to sort out the relative contribution to the continuum spectrum by thermal and non-thermal processes. Estimates of visual absorption to the nucleus of NGC 5506 vary around $A_V \sim 18 \pm 4$ mag; the 10μm and the X-ray extinctions appear to be similar, which led Roche et al to speculate that the sources of radiation at these wavelengths may be the same. It is instructive to note that the Balmer decrement yields $A_V \sim 2$ mag (Wilson et al. 1976)! Because Bγ and [SIV] have very similar absorption optical depths, their essentially reddening-independent line-intensity ratio can be used to estimate the abundance of S^{+++} ions. A value consistent with that found in medium-excitation Galactic planetary nebulae (30% of sulphur in the form S^{+++}) is obtained if one assumes normal total sulphur abundances. All the emission lines appear to suffer similar extinction ($A_V \sim 2-3$ mag), so the [SIV] emission probably arises in a region in front of the source of the i.r. and X-ray continuum. However, in view of some of the doubts raised in the talk at this conference by Flower, these abundance estimates should be regarded with considerable caution.

Glass, Moorwood and Eichendorf (1982) find that the broad-band L-N colour indices of the "narrow-line" X-ray galaxies are on the whole distributed more like the Seyfert 1 galaxies than the Seyfert 2's; the energy distributions of type 1 Seyferts usually rise relatively less steeply in the near infrared than those of type 2 Seyferts (Neugebauer et al. 1976).

3.3 Infrared observations of stars in Seyfert galaxies.

As mentioned earlier, Ward et al. (1982) have presented JHKL photometry for a sample of 28 active galaxies. The two-colour diagram, J-H versus H-K, was used to show that a simple power law was not adequate to explain the 1.2µm to 2.2µm continuum. Instead, one not only has a contribution from a typical QSO, but also from a component whose colour is similar to that of a 'normal' galaxy. It is likely, therefore, that in galaxies whose colours are near J-H = 0.67, H-K = 0.20, we are seeing the direct infrared radiation from stars in these objects, while the light from those objects with colours nearer (J-H) = 0.95, H-K = 1.15 is dominated by a quasar-like non-thermal component.

At this conference, Moorwood (1982b) has reminded us, however, that "a galaxy whose J,H,K colour indices are interpreted as 40% stellar and 60% power law ($\nu^{-1.5}$) could equally well be 65% stellar and 35% blackbody ($\sim$ 1000°K) at 2µm....." He also presents his evidence for CO band absorption from stars in the Seyfert 1 galaxy NGC 7213. The stellar contribution to the 2µm continuum in this galaxy is very large.

A large stellar component in the continuum of NGC 1275 shortward of 3µm has also been reported by Rudy et al. 1982. Earlier observations by Kemp et al. (1977) showed a stronger non-stellar contribution to the 2µm continuum, but their observations appear to have been taken when NGC 1275 was much brighter (presumably associated with a greater fractional contribution by the non-stellar component).

Strong first-overtone absorption bands of CO indicate the presence of late-type luminous stars in the type 2 Seyfert galaxy NGC 1068 (Thompson, Lebofsky and Rieke 1978). The strength of these indicate that the stars contribute about a quarter of the 2.3µm continuum flux within 200 pc of the centre of NGC 1068 (Hall et al 1981) and have a representative spectral type of K5 III; the velocity dispersion of the stars is estimated to be 150 ± 50km s^{-1} (Richstone and Morton 1975). This stellar continuum dominates the spectrum from 0.4 < λ < 1.6µm. The absence of H_2O features indicates that the bulk of the stars responsible for the near infrared emission are of types earlier than M5. The evidence for stars in Seyfert 1.0 - 1.5 galaxies such as NGC 4151 is less direct, being based only on the shape of the 2-4µm continuum (McAlary and McLaren 1981).

Most Seyfert 2 galaxies show a strong stellar continuum which declines beyond $\sim$ 1µm and a dust component which rises steeply around 3µm. This combination produces a characteristic spectral inflection around 2µm which can be detected quite readily with broad-band photometry (see, e.g., Stein and Weedman 1976; Neugebauer et al. 1976; McAlary, McLaren and Crabtree 1979; Balzano and Weedman 1981). More detailed separation of components of the continuum is discussed for NGC 1068 by Hall et al. (1981) and for NGC 4151 by Rieke and Lebofsky (1981).

3.4 Atomic hydrogen emission in Seyfert galaxies.

Baldwin (1977) opened up an intensive area of observational and theoretical research into the emission-line properties of Seyfert galaxies and quasars with a little paper pointing out a discrepancy between the observed and predicted ratios of Lyα and Hβ. From a composite spectrum (necessary in the days preceding suitable ultraviolet satellite data) made up from ground-based optical spectra of high and low-redshift QSOs, Baldwin found a ratio Ly α /Hβ $\sim$ 3, surprisingly small compared with the values $\sim$ 40 produced by the standard theoretical models then in vogue. We have already considered (in section 3.1) the two basic categories of explanation attempted for this effect, namely reddening and selective population of hydrogen energy levels in high-density regions.

An extensive survey of Pα /Hα /Hβ ratios in 18 Seyfert galaxies has been carried out on Mauna Kea by Lacy et al. (1982). Lyα observations were obtained with the IUE satellite for 8 of these galaxies. As stated earlier, the fact that Pα is so much stronger relative to the Balmer lines is suggestive (on the basis of present theoretical models) of the presence of reddening.

There is also dispersion in the Pα /Hα /Hβ ratios orthogonal to the reddening track; although Pα /Hβ is close to the Menzel and Baker case B, radiative-recombination-cascade, line-flux ratio of 0.28 (e.g. Osterbrock 1974), H α is enhanced relative to Pα and Hβ. NGC 1275 is the most extreme object of this type in the sample. The Lyα /Hβ ratios were all less than expected from their case B unreddened values. High densities and large optical depths, in addition to a mixture of internal and external dust reddening, would appear to be required to explain these ratios; even with this amount of freedom to juggle with parameters, existing models do not seem to explain the flux ratios for NGC 1275 (see discussion by Lacy et al. - however McKee has advised me that the models tested by Lacy et al. have generally assumed an isothermal emission-line region). This mixture of dust reddening geometries is needed because (i) the amount of internal dust must not be so severe as to absorb too much ionising continuum whereas (ii) external dust alone cannot explain the Pα /Hβ /Lyα ratios correctly. We shall discuss some of these results further when we come to deal with dust in Seyferts (section 3.8).

Pα has also been detected by Rudy and Tokunaga (1982) in the broad-line-radio galaxy (BLRG) 3C445. The strength of this line, relative to the Balmer series again implies substantial reddening of the emission-line region($E_{B-V} \sim 1.0$).

3.5 Molecules in Seyfert Galaxies.

The first detection of extragalactic molecular hydrogen (in NGC 1068) was reported four years ago by Thompson et al. (1978). Using a resolution of about 500 km s^{-1} and an entrance aperture of 8 arcsec diameter, they found that the strongest molecular hydrogen feature was the v = 1 --> 0 S (1) quadrupole emission line. This line was first detected by Gautier et al. (1976) in the Becklin-Neugebauer source (BN); its rest wavelength is 2.122μm (see Figure 3). Although measurements of remaining features were very approximate, the relative H_2 line intensities were found to be similar to those of the Orion cloud and the planetary nebula NGC 7027. This was taken to indicate that the excitation of the molecular-hydrogen transitions was more likely to be by inelastic collisions in shocked gas than by near-ultraviolet pumping (see, e.g., Hollenbach and Shull 1977). Thompson et al. used the shock model of Hollenbach and Shull to estimate the total amount (600 $M_\odot$) of H_2 needed to produce the observed S (1) flux; they assumed an average temperature of 2000°K in their calculations. Rieke and Lebofsky (1979) concluded that the spectrum of NGC 1068 "shows He, H_2, Brackett γ.... All the lines have a width comparable with those of the lines in the visible ($\sim$ 1000 km s^{-1}).... The detection of molecular hydrogen..... is at least qualitatively consistent with the existence of large-scale mass motions within the nucleus....."

Substantial revision of this overall picture has resulted from more recent observations at higher spectral resolution ($\sim$ 350 km s^{-1}) and sensitivity by Hall et al. (1981). In contrast to Thompson et al. (1978), they found that the H_2 lines are in fact much narrower than the atomic emission lines, in particular those of [OI]. Furthermore, the Bγ to H_2 S(1) v = 1 --> 0 flux ratio is near unity in NGC 1068, whereas this ratio is very large in planetary nebulae (see, e.g., Treffers et al. 1976). These facts led Hall et al. to reject the planetary nebula analogy made by Thompson et al.

Hall et al. (1981) have pointed out that, provided the line width of H_2 in individual clouds is small, the implied line-of-sight velocity dispersion for the H_2 ($\sigma \sim 115 \pm 10$ km s^{-1}) matches the stellar dispersion given by Richstone and Morton (1975) as 125 ± 35 km s^{-1}.

The stellar velocity dispersion was derived by Richstone and Morton (1975) from a line of Fe I at λ0.4046μm (rest wavelength). A more direct, essentially reddening-independent calibration should be possible by means of high-resolution measurements of the CO first-overtone absorption bandhead at a wavelength of 2.3 μm (i.e. much closer to the rest wavelength, 2.122μm of the S (1), v = 1 -> 0 quadrupole emission line of H_2). If the widths of the H_2 and stellar features agree, it might be taken as evidence that they co-exist in regions with similar dynamic properties (unlike ionised hydrogen, which exhibits a velocity width of order 1000 km s^{-1}). Detailed line profiles of the H_2 emission and CO absorption would, however, be needed for tests of even the most elementary models; for example one might expect the H_2 velocity distribution in NGC 1068 to be consistent with a rotating disk or ring, while the stars occur in a central galactic bulge. Unfortunately such observations would be extraordinarily difficult in practice.

It is clearly of interest to attempt observations of the reddening to the molecular and atomic hydrogen in NGC 1068 (and in any other bright galaxies in which H_2 may be detected). We have already seen, (in section 3.1), some of the advantages of using the Bα /Pβ ratio for measuring the reddening to the atomic hydrogen emission-line regions. To measure the reddening to the molecular hydrogen region, one would like, once again, to use transitions from the same upper level (see, e.g., the discussion by Scoville et al. 1982).

Unfortunately, there are a number of practical difficulties which prevent this. The strongest set of emitted lines from the same upper level are the S (1), Q (3) and O (5) v = 1 -> 0 lines (see, e.g., Beckwith et al. 1983). (For those less familiar with the terminology for molecular spectra, Figure 3 is a crude representation of energy levels of interest). Unfortunately, the redshift of NGC 1068 puts the Q (3) and O (5) lines into atmospheric absorption features (see, e.g., Geballe, Russell and Nadeau 1982). Transitions to even - J levels (para-hydrogen, single -degeneracy transitions) are probably too weak to use. The most suitable lines therefore appear to be the S (1) and O (7) v = 1 -> 0 (3.807μm rest wavelength) transitions, even though they do not originate from a common upper energy level; (note that the 'bright' 1 -> 0 S (1) line has a strength of only 2×10^{-20} W cm^{-2}). There is an additional potential problem because Davis et al. (1982) have reported anomalous values for the intensity ratio of these two transitions in Orion. One must therefore measure this ratio in a bright planetary nebula in order to provide an estimate of the intrinsic unreddened ratio in NGC 1068. From the observed line ratio S (1): O (7) in NGC 1068 one can then deduce the reddening to the molecular-hydrogen region.

Hall et al. 'tentatively identify' 4287 cm^{-1} (2.3326μm) emission in NGC 1068 as radiation from overtone, low-temperature CO. Such overtone lines in emission are very unusual in the Galaxy, as most molecules do not stay long at the v = 2 (or higher) levels. Overtone emission band heads of CO have been found by Scoville et al. in BN, a protostar which could have the shocks or very high density ($n_H > 10^{10}$ cm^{-3}) needed to raise a significant number of molecules to the v = 2, 3 and 4 energy levels.

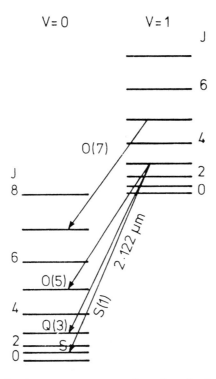

Figure 3: Some of the strongest molecular-hydrogen transitions. The Q(3) and O(5) lines are severely affected by atmospheric absorption, so reddening measurements to the molecular hydrogen in low-redshift galaxies has to depend on the S(1) and O(7) transitions.

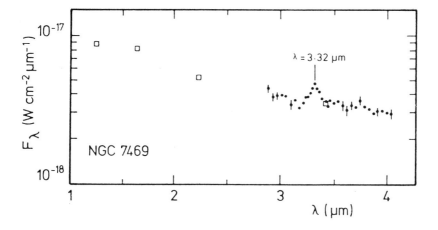

Figure 4: Dust emission features have been convincingly demonstrated in only very few Seyfert galaxies. This observation of NGC7469 by Rudy et al. (1982) is one such case.

3.6 Dust in Seyfert Galaxies

From sections 3.2 and 3.4 we have already seen strong indications from atomic hydrogen lines that dust is likely to be present in both type 1 and type 2 Seyferts. More direct evidence would be provided by observations of a spectral minimum at 9.6μm, normally considered as the product of absorption by silicate particles (associated with a large column density of observing material) in and around the galactic nuclei, and the dust emission features discussed in section 1.4.

The 3.28μm dust emission feature has been reported as strong in a number of Seyfert galaxies, e.g. IC4329A, Moorwood and Salinari 1981, NGC 7469, Rudy et al. 1982; NGC 7582, Moorwood 1982b; see also Lee et al. 1982. From a perusal of this last reference, the reader will quickly become aware of the rather chequered history of reported 3.28μm detections which could not be confirmed by other workers. In some cases, one is probably wise to await confirmation, by spectroscopy in the 8-13μm region, of other strong dust emission features. The 9.6μm silicate absorption was detected in all (except possibly one) of the nine emission- line galaxies (which included 5 with active nuclei) observed by Frogel, Elias and Phillips (1982); they used a Ga:Ge bolometer along with a series of 8 filters to define a spectrum between 7.8μm and 20μm). Lebofsky and Rieke (1979) obtained similar results for a sample of northern galaxies.

Roche et al. (1983) have used the UCL cooled-grating array spectrometer to observe ten Seyfert galaxies and 3C273. Unlike the work described above, they find silicate absorption to be fairly rare, occurring in only two (class 2) Seyferts, NGC 1068 (see also Kleinmann et al. 1976) and NGC 5506. Dust emission features in 8-13μm spectra of Seyferts are also almost always absent. NGC 7582 does, however, show the 8.65 and 11.25μm features (accompanying the 3.28μm feature reported by Moorwood). The 3.28μm feature observed by Rudy et al. (1982) in NGC 7469 (see Figure 4) is also accompanied by the feature at 11.25μm (Aitken, Roche and Phillips 1981). Following the reported detection of 3.28μm emission by Moorwood and Salinari (1981), one might therefore have expected to observe 8-13μm dust emission features in IC4329A, but they are not seen (Roche et al. 1983; see also Figure 2). Brand et al. (personal communication) used the UKIRT recently to check the claimed detection of the 3.28μm feature in IC4329A, but failed to confirm its presence. It is clear that many Seyfert galaxies have smooth, featureless 8-13μm spectra, well fitted by power laws, and with no evidence (within this spectral range) for dust emission. As we have seen, this is in contrast with narrow emission-line galaxies and ionisation fronts in Galactic HII regions which have 8-13μm spectra dominated by dust emission features and which often show emission from Ne^+ (see section 2.2.2).

Roche et al. (1983) also note that Seyfert type 1 galaxies do not have such steep spectral indices in the 8-13μm region as Seyferts of class 2; this appears to be the only difference between the 8-13μm

spectra of the two Seyfert classes, and is the reverse of the situation in the near infrared (where Seyfert 1 galaxies have redder near-infrared colours than Seyfert 2 galaxies). They conclude that if dust is responsible for the observed 10µm continuum emission from Seyfert galaxies, then
(i) it must emit over a large range of temperatures (to produce an approximately power-law flux distribution)
(ii) any silicate dust present must either have its contribution to the 10µm emission swamped by emission from other types of dust or be too cool to emit significantly at 10µm.

A strong argument for the presence of dust in the Seyfert 2 galaxy, NGC 1068 is based on the measured diameter (1" or 90pc) of the 10µm source (Becklin et al. 1973; see also Telesco et al. 1980, who discuss extended 20µm radiation from NGC 1068). While it is easy to see how this 10µm and 20µm source could be dust heated by a central ionising continuum (see, e.g., Lebofsky, Rieke and Kemp 1978), it is not easy to see how to relate it to the non-thermal source observed at optical and ultraviolet wavelengths. Furthermore the continuum slope from 1 to 20µm ($\alpha \sim -3$) is much steeper than that seen over the same wavelength range in any source where non-thermal mechanisms are thought to dominate. Finally, we have the necessary (but not sufficient) condition for dominance of dust emission - NGC 1068 shows no evidence for variability (see discussion by Soifer and Neugebauer 1981). Rieke (1978) has shown that other type 2 Seyfert galaxies are similar to NGC 1068. We should however bear in mind that Gatley (personal communication) has emphasised that the fundamental source of ionisation in NGC 1068 is not clear; the equivalent width of lines such as Bγ decreases sharply as one moves towards the nucleus - ie. the 2.17µm continuum intensity rises, but the line intensity of Bγ does not. Given that the observed wavelengths of line and continuum are similar, we cannot readily appeal to differential extinction as a way out. Thus we cannot be absolutely certain that newly formed stars are the basic source of most of the energy for NGC 1068.

The situation for Seyferts such as NGC 4151 (type 1.5) is even less clear. Its energy distribution is more similar to 3C273 than to NGC 1068. Such a distribution can be fitted by either non-thermal emission (e.g. Stein and Weedman 1976; Neugebauer et al. 1976) or as a sum of (ultraviolet) power-law, stellar and thermal infrared components, i.e. with a significant thermal component (McAlary and McLaren 1981; Rieke and Lebofsky 1981; Penston 1982). Both interpretations are possible.

Given this ambiguity presented by the continuum spectrum of Seyfert galaxies, one has next to appeal to polarisation and spectrum variability. Polarisation in Seyferts is generally weak (Angel and Stockman 1980; Rudy et al 1982; Maza et al. unpublished), but accurate infrared polarisation measurements at polarisation levels $\sim$ 2% or even less are now becoming possible. Axon, Bailey, Hough and Ward have recently performed a new polarisation survey of a small sample of Seyferts. In IC4329A, for example, they find that from ultraviolet polarisation levels in excess of 10%, the level of polarisation falls to less than 2% at wavelengths > 1.2µm. The polarisation level remains approximately constant at least to 2.2µm. (Accuracies of ±0.3% polarisation have been achieved at 1.6µm with their polarimeter). They find, however, that the total energy distribution in IC4329A is rising very steeply from 0.3µm to 2.2µm; the amount of light that is polarised also rises into the infrared. Without further information, the relative balance between dust scattering and non-thermal contributions to the continuum polarisation remains ambiguous. For the grain - size distributions normally assumed, one expects the percentage polarisation induced by scattering from dust to decrease as wavelength increases. If optical-depth effects are important, the total infrared flux can be expected to rise steeply into the infrared; thus the total polarised flux may also increase (less steeply) into the infrared, even if dust scattering is the only significant polarisation mechanism. On the other hand, if optical-depth effects are important, they may also allow a central, polarised, non-thermal component to become increasingly dominant at longer wavelengths; however, in this case, one may be able to observe levels of polarisation substantially higher than expected from dust scattering alone. By combining their spectropolarimetry with spectrophotometry, Axon et al. intend to determine the spectral shape of any polarised component which is present because the polarised flux is independent of dilution. They will use information on the position angle and its rotation with wavelength, to assess the importance of optical depth effects.

Rieke and Lebofsky (1981) find for NGC 4151 that no 10µm variation greater than 8% occurs to accompany changes (by a factor 2 or more) in near infrared and non-thermal flux (see also McAlary and McLaren 1981). They find that thermal re-radiation by dust that lies near or within the BLR (and is heated by the central nuclear source) dominates the continuum at $\lambda \gtrsim 1\mu m$. This dust will, of course, modify the observed non-thermal continuum, as well as the relative emission-line strength. Rieke and Lebofsky require at least 5 components to explain the observed continuum spectrum, including a highly variable non-thermal source whose spectrum deviates substantially from a simple power law; Penston (1982) has reviewed the recent ultraviolet $(0.1\mu m \lesssim \lambda \lesssim 0.2\mu m)$ satellite evidence for a variable, hot, (30,000°K) blackbody component which may represent thermal radiation from an accretion disc (Shields and Wheeler 1978) around a black hole of mass $\sim 10^8 M_\odot$. McAlary and McLaren find that the 2-4µm continuum region of the source spectrum can be adequately fitted by two more black bodies - one is a stellar component, the other is from strong thermal dust emission.

These more complex mixtures of stars, dust and nonthermal components are fairly representative of other Seyfert 1 galaxies (see, e.g. Rudy et al. 1982; Penston 1982).

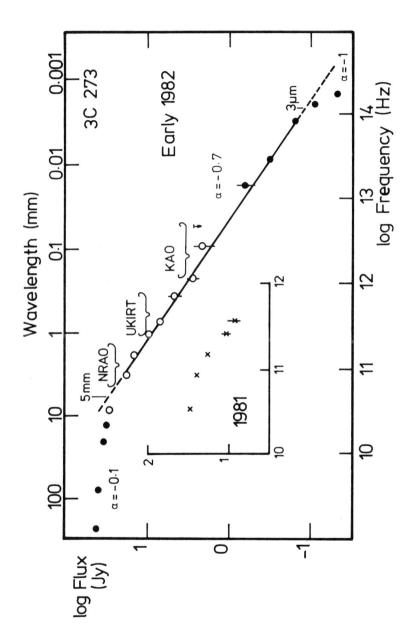

Figure 5: The infrared-radio continuum of 3C273 (adapted from Clegg et al. 1983), the first complete radio-through-optical spectrum obtained for a quasar. The variability of 3C273 and of many BL Lac objects demands that such spectra be obtained within short time intervals.

Rudy and Puetter (1982) discuss models of the emission-line regions (ELR^s) of active galaxies and quasars and show that:
(i) The redder near-infrared colours of Seyfert 1 galaxies can be understood if the broad-line regions harbour a concentration of dust within the central parsec of the nucleus and
(ii) Dust can be confined to the shielded (backside) neutral zones of emission-line clouds; the dust will therefore redden only the components of the lines emitted from those clouds on the side of the ELR nearest to the observer. Thus substantial near-infrared dust emission can occur without producing large amounts of reddening of the emission lines.
(iii) If the broad-line regions of quasars are hotter, dust may not be able to exist in this region.

4. QUASARS AND BLAZARS

4.1 Infrared and millimetre-wave continuum observations of quasars and blazars.

The radiation from most quasars and BL Lac objects is too faint for detailed infrared spectroscopy of lines, and so, as happened in section 2.1.1., we have to relax our definition of spectroscopy to include filter photometry. For BL Lac objects, this is presumably no great loss, as we do not expect strong lines in their spectra.

The non-thermal continuum of optically thin radio sources usually contains a section of slope $\alpha \sim -0.7$ ($S_\nu \alpha \nu^\alpha$ - see, e.g., Figure 5) thought to correspond to synchrotron radiation from electrons having an initial power-law energy distribution $n(E) \alpha E^{-\gamma}$, where $\alpha = -(\gamma -1)/2$ (see, e.g., Moffet 1975; Landau et al. 1983). For $\alpha = -0.7$, $n(E) \alpha E^{-2.4}$, i.e. the power-law electron distribution with exponent -2.4 has given rise to a power-law emission spectrum with exponent -0.7. This optically-thin power-law spectrum is usually modified in most QSOs at the high and low frequency ends. A number of processes may be responsible (see, e.g., Figure 6). As we shall see, recent observational developments in the infrared and millimetre-wave region give grounds for some hope that we shall at last be able to constrain the relative contributions of each. At the high-frequency end, the power-law can be steepened, for example: (i) by a truncation, at high energies, of the initial electron-energy distribution (see e.g., section 4.1.5.) or (ii) by a combination of adiabatic expansion and synchrotron losses (see e.g. Clegg et al. 1983). As we move to the low frequency end, the spectrum often flattens out (see e.g. section 4.1.2), and in most QSOs and BL Lac objects the flux density then decreases once more as the frequency decreases further. (Note that steep-spectrum radio-loud sources usually have an strong extended-lobe component well beyond the core region of most interest to our subsequent discussion). Condon et al. 1981b review the various mechanisms for producing a low-frequency cutoff, which include those illustrated in Figure 6. We begin our discussion at the high-frequency, optical/near-infrared end of the continuum, and move to lower frequencies. We begin to distinguish between radio-loud, radio-quiet and BL Lac objects in the submillimetre-/millimetre-wave region, where their spectra must begin to differ. At the end of section 4.1, we return to the extreme case of the very red radio sources.

4.1.1. <u>The near-infrared continuum</u>. We have seen that the continua of Seyfert galaxies can be classified in terms of the relative importance of the stellar continuum and a non-thermal power-law component in the near infrared. For example, McAlary, McLaren and Crabtree (1979) find that for wavelengths $\lambda < 2\mu m$, type 2 Seyfert galaxies are largely dominated by stellar continua while Seyfert 1 galaxies generally show power-law spectra in this region (see also Neugebauer et al. 1976). The energy distributions of quasars have a variety of continuum shapes which rise steadily with increasing wavelength into the infrared (see e.g., Neugebauer et al. 1979).

Observations of the infrared properties of serendipitous X-ray quasars by Neugebauer et al. (1982) show that the IR characteristics of these quasars do not differ significantly from those of quasars selected by other criteria which had different selection biases. <u>We may therefore regard the known IR properties of quasars as genuinely typical</u>. Glass (1981) and Hyland and Allen (1982) have shown that the colours of non-variable QSOs as a function of redshift can be understood in terms of the shifting of spectral lines through the infrared filter passbands. Some BL Lac objects have colours which clearly reveal the mixture of a non-thermal source with an underlying galaxy. Such a mixture is particularly striking in the case of NGC 1052 where the non-thermal source is of exceptionally low luminosity (Rieke, Lebofsky and Kemp 1982; see also the case of IC5063 discussed by Axon, Bailey and Hough 1982).

Soifer et al (1983) have combined JHK measures with visual spectrophotometry for 21 quasars with redshifts $z > 2.66$. They conclude that the rest frame visual/ultraviolet continua can be described by the sum of a power law continuum of slope -0.4 and a .30μm bump. This bump in QSO spectra was first recognised by Baldwin (1975) and since discussed by many authors (e.g. Richstone and Schmidt 1980). Puetter et al. (1981) discuss difficulties encountered by most of the current explanations (12000°K thermal emission from an accretion disk, synchotron radiation from mono-energetic particles, starlight from a superposed galaxy, Balmer continuum etc). They conclude that Baldwin's original explanation for at least the bulk of this feature (namely in terms of Balmer continuum emission) is the most likely (see also Peutter et al. 1982; Puetter and LeVan 1982). The original objections to the strength and shape of the feature can be met if it also includes a component of blended FeII emission. Gaskell (personal communication) plans further tests of these ideas using large samples of X-ray emitting quasars; Kwan and Krolik (1981) have predicted that both the optical FeII emission and the ratio of Balmer continuum emission to Hβ should be directly correlated with the X-ray luminosity.

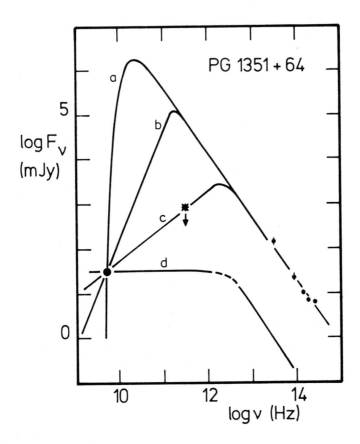

Figure 6: Even an upper limit at millimetre wavelengths (the asterisk) can be of crucial importance in distinguishing between various models which fit optical, infrared and radio data, in this case :-
    a: Free free.
    b: Homogeneous synchrotron.
    c: Inhomogeneous synchrotron.
    d: Relativistic jet.
Illustration adapted from Ennis, Neugebauer and Werner (1982b).

4.1.2. The shape of the millimetre-wave continuum - radio loud QSOs.
There is no systematic difference between the observed infrared energy distributions of radio-quiet and radio-loud quasars so, as Soifer and Neugebauer (1981) point out, the difference between radio-quiet and radio-loud quasars must show up at about 100μm. According to many of the models for the radio emission, the flat radio spectrum observed in many sources arises from the superposition of many compact components, each of which has a long-wavelength cutoff; the shorter the radio wavelength of observation, the smaller the angular size of the quasar turns out to be. Only by going to the millimetre/ submillimetre region (from the radio) do we reach a region of the spectrum where we have peeled off enough layers of the "onion skin" (Marscher 1983) to view the radio core directly. The longest wavelength at which the core can still be seen is an essential parameter in most synchrotron emission models. Figure 5 is adapted from a preprint by Clegg et al. (1983) and shows spectra of 3C273 formed from observations taken within a few months of each other [however one should make a cautionary note that Sherwood et al. (1983) have recently reported very substantial and rapid variations in 3C273 at 300 GHz during late July 1981, including a decrease by a factor of about two in a single day; similar variations at 1mm (Ennis, Neugebauer and Werner 1982a) and at somewhat longer wavelengths (Epstein et al. 1982) had been reported earlier - though the millimetre-wave data was of lower signal-to-noise ratio than that of Sherwood et al.]. 3C273 is the first quasar for which the entire radio through optical spectrum has been measured. It shows that the quasar's radio-emitting region is transparent at submillimetre wavelengths; Metzger has suggested at this conference that VLBI at millimetre wavelengths should be developed to attempt to map out the angular structure of the core region.

4.1.3 Radio-quiet QSOs. The situation concerning the millimetre-wave properties of radio- quiet QSOs remains quite controversial. The significance of attempts to detect radio-quiet QSOs is illustrated in Figure 6, taken from the recent paper by Ennis, Neugebauer and Werner (1982b). Even reliable upper limits at millimetre wavelengths can provide significant constraints on at least some of the simpler single-component models of the continuum energy generation mechanism. The controversy centres on the fact that the only group reporting successsful detections of radio-quiet quasars at 300 GHz is the team at Bonn (Sherwood, Kreysa and Schultz 1981; Sherwood, Schultz and Kreysa 1981; Sherwood et al 1982; Sherwood 1982; see also the paper at this conference by Metzger), while other efforts from the Northern hemisphere taken to similar sensitivity levels, have proven singularly unsuccessful. Ennis, Neugebauer and Werner (1982b) failed to detect any radio-quiet quasar at 300 GHz from a sample of 8 objects; 3σ upper limits to the 1mm flux density were set at 1Jy. At UKIRT, we have encountered a similar lack of success, even though integrations have been taken to the point where 3σ upper limits less than ½ Jy have been achieved (see, e.g., Robson 1982).

Sherwood et al.'s most dramatic result (Sherwood Shultz and Kreysa 1981) has been the detection of Q0420-388, a highly luminous, optically selected QSO (Osmer and Smith 1977, 1980) which has also been detected at a wide range of other wavelengths (Smith and Wright 1980; Condon et al. 1981a; Zamorani et al. 1981; Wright and Kleinmann 1978). Sherwood et al.'s results reveal a continuum which appears to be sharply peaked in the 300 GHz region, such that the luminosity of the quasar is dominated by the millimetre-wave emission. Indeed, from the radio observations by Condon et al. (1981b) of optically selected QSOs, we must expect some objects to have continuum peaks in the millimetre-submillimetre region (see also Capps, Sitko and Stein 1982); using the VLA, Condon et al. were able to establish that in some cases the level of radio emission was lower than that of the near infrared.

4.1.4. <u>BL Lac objects.</u> The continua of BL Lac objects are much smoother than those of either Seyfert galaxies or quasars. As mentioned earlier, BL Lac objects have the following fairly general properties:-
(i) Lack of emission lines.
(ii) Situation at the nucleus of an elliptical galaxy.
(iii) Rapid variability at all observed wavelengths.
(iv) Flat or inverted radio spectrum.
(v) Substantial visible and IR polarisation.
The continuous spectrum rises steeply from the optical through the infrared, becoming considerably flatter in the radio. The frequencies at which changes of spectral slope occur may be used to indicate parameters of simple relativistic beaming models (see, e.g. Marscher 1980a,b; Clegg et al. 1983; Marscher 1983, Landau et al. 1983). The high degree of observed polarisation supports synchrotron emission models, but the detailed physical basis for the various polarisation properties (e.g. Rieke et al. 1977; Puschell and Stein 1980; Impey et al. 1982; Angel and Stockman 1980) is still not fully understood.

Ennis, Neugebauer and Werner (1982a) have recently published their broad-band continuum observations of 3C273, 3C279, BL Lac, 3C84, OJ287 and 3C345 made between 1977 March and 1981 January. With the possible exception of 3C279, all showed evidence for 1mm variability on a time scale of a few months. For blazars and OVV quasars, flux variations at wavelengths of 1mm and 2cm are well correlated; emission outbursts occur simultaneously and have similar amplitudes at the two wavelengths. This behaviour is inconsistent with the canonical expanding - source model of radio variability or any of its variations. According to the basic model, the cloud of synchrotron-emitting electrons expands and the observed flux density from the cloud will increase. However, the optical depth to synchrotron radiation at a given wavelength decreases with time; this optical depth is frequency dependent, and so after the flare the flux will begin to decrease at different times for different wavelengths. The amplitude of fluctuation should be much larger at the shorter wavelengths, and the timescale for flux increases should also be much shorter at the shorter wavelengths. In order to fit the variability data at 1mm and 2cm, Ennis, Neugebauer and Werner require models in which the number of synchrotron-radiating electrons is time variable (as a result, for example, of particle injection or acceleration behind a relativistic shock). More detailed discussion is given by Pauliny-Toth and Kellerman 1966, while Ennis et al. give interesting numerical illustrations.

Landau et al. (1983) found that the millimetre - visual spectral index α_{mv} was close to -0.7 for the 9 objects they observed, typical of optically-thin synchrotron emission. Interpreting this spectral index as a reflection of the intrinsic energy distribution of accelerated electrons, they concluded that synchrotron and self-Compton radiative losses must be neglible over that portion of the spectrum because radiative losses would steepen the electron energy distribution beyond the injected form (c.f. Clegg et al. 1983 and Figure 5). In order to avoid severe losses, Landau et al. concluded that the emitting region must be beamed in our direction and require relativistic bulk motion with Doppler factors $\sim$ 10. A direct check on the presence of radiative losses following an injection of electrons at outburst is to look for changes in spectral shape as the flare subsides. Changes in radio spectral shape seem to be rare, but have been recorded for 1156 + 295 (Glassgold et al.), 2251 + 15 (Landau et al. 1983) and possibly also for 1921-293 (Gear et al. 1983) and 0235 + 164 (Landau et al.).

Bregman and collaborators (Bregman et al. 1981, 1982, 1983; Glassgold et al. 1983) have also begun a major research programame of simultaneous multifrequency observations of BL Lac objects and OVV quasars. Their first four papers in a series have dealt with the red quasar 1413 + 135 (discussed later in this section), the X-ray bright BL Lac object IZw 187, the BL Lac object PKS0735 + 178 and the flaring (OVV) quasar 1156 + 295. Their data on PKS0735 + 178 show that "rapid and dramatic variations evident at infrared and optical wavelengths are absent at radio and X-ray frequencies. These observations support a picture where the IR-UV flux emanates from a small region, while the X-rays are produced by the inverse Compton process in the radio emitting region (partially optically thick synchrotron emission gives rise to the radio flux)". In the more compact core region, they find that particles, photons and magnetic field may not be far from equipartition, whereas the energy density of the particles appears to be dominant in the larger region responsible for most of the X-ray and radio flux. By comparing the VLBI size to the temporal flux variability size, Bregman et al. also derive, for the first time, an upper limit for the degree of relativistic motion; the Lorentz factor in PKS0735 + 178 is less than ten.

Observations of the spectral energy distributions for BL Lac objects have proved difficult until recently, because of their rapid variability and the consequent need to organise assignments of several telescopes at once to cover each waveband. [For example, Landau et al. (1983) report observations of 9 QSOs and BL Lacertids over a wavelength region 4000Å to 20 cm using 5 telescopes over a ten-day period; a consortium of 22 authors from ten different institutions was involved in the multifrequency observations of 1156 + 295 described above and a similar group of 20 authors studied 0735 + 178]. To ease this problem we have recently found the rotatable dichroic system at the UKIRT to be of particular value; it has allowed us to obtain spectral points, on a single night with the one telescope, at J,H,K,L,10µm, 20µm, 800µm and 1100µm. Very recently (February 1983) we have made additional 9-12 point "snapshots" of the continua of 3C273, OJ287, 3C345 and 3C84 (NGC1275).

4.1.5. <u>The continua of the very red radio sources</u>. We turn finally to the continuum shape for the very red radio sources - the empty-field objects first discussed by Rieke, Lebofsky and Kinman (1979). In their original survey, they found six objects (at 2.2µm) at the positions of flat-spectrum radio sources, that had either no optical identification at all or had been identified with very faint red sources. The infrared-to-optical spectra of these sources have power-law spectra of index -3 ($S \alpha \nu^{\alpha}$), and this extreme redness suggested that a distinct class of active galaxy had been discovered. Though this slope is much steeper than the spectra of previously known QSOs, a follow-up study by Impey and Brand (1981) showed that such sources are in fact just the tail of the normal distribution of quasar colours, that they are relatively common, and that a considerable fraction of these sources will have properties similar to the BL Lac objects (see also Beichman et al. 1981a,b).

Rieke and Lebofsky (1980) do however show that <u>optical</u> identification programmes for radio-source samples produce samples biased against very red sources, even if optical colour was not used as a criterion in making the identifications. Furthermore Lebofsky, Rieke and Walsh (1983) have shown that compact <u>steep</u>-spectrum sources are as detectable at 2µm as are compact flat-spectrum sources, although such a sample of steep-spectrum sources contains a large fraction of high-redshift galaxies (see section 5.1). Axon, Bailey and Hough (1982) have reported an extremely red nucleus in the radio elliptical IC 5063 which could be either an $\alpha = -4.5$ non-thermal source, or the Wien tail of a 650°K blackbody. They point out that steep-spectrum, low luminosity sources like this would have escaped detection in existing surveys.

As pointed out explicitly by Beichman et al. (1981a), "the importance of these sources is that their extreme nature forces a close examination of the emission mechanism thought to be operating in the whole class". This fact, and the very red colour of these sources naturally led to searches for these objects at longer wavelengths. Beichman and collaborators (1981a, b) reported detections of 1413 + 135 first at 10µm and then at 1.0 millimetre. Summaries of the multifrequency spectrum of 1413 + 135 were given by Beichman et al. 1981b and by Bregman et al. 1981. Both papers agree that the cause of the steep infrared continuum is an extremely sharp high-energy cutoff in the distribution of synchrotron-emitting electrons (see also Rieke, Lebofsky and Wisniewski 1982, who point out that the cutoff is nearly as sharp as is theoretically possible for synchrotron emission). Bregman et al. also showed that this cutoff is unlikely to be caused by internal reddening. The other properties of 1413 + 135 are generally similar to the BL Lac objects. If the emission at 1mm is due to incoherent synchrotron radiation, then the source is very compact and a large magnetic field ($\sim$ 10G) is required. The X-radiation cannot then be primary radiation, but rather inverse Compton upscattering with Lorentz factor $\gamma \sim 300$. The shape of the spectrum is attributed to synchrotron radiation from a relativistic jet, whose axis makes a small angle to the line of sight (see, e.g. Ennis, Neugebauer and Werner 1982b). Rieke, Lebofsky and Wisniewski stress that the abruptness of the inflection in the infrared spectrum means that if normal synchrotron emission is responsible, then the sources must have remarkably uniform magnetic fields, and the electron acceleration process in some of these sources may produce a nearly mono-energetic electron spectrum.

4.2 Dust in Quasars?

In sections 3.1 and 3.4 we have discussed some of the difficulties in interpreting the spectrum of atomic hydrogen in active galaxies. Davidsen (1980) has given a brief review of the observational situation, while Davidson and Netzer (1979) summarised the theoretical picture up until the date of their review. Of particular recent interest is the series of papers by Puetter and collaborators [Canfield and Puetter 1981; Canfield, Puetter and Ricchiazzi 1981; Puetter 1981; Puetter and Le Van 1982; Puetter et al. 1982; Rudy and Puetter (1982)] and the work of Kwan and Krolik (1981). As we have seen from the observational work on Seyfert galaxies by Lacy et al (1982), it is essential to postulate some dust reddening in Seyfert galaxies. Is dust present in quasars in quantities sufficient to affect their infrared spectral properties?

What seemed to be a crucial result was the apparent detection, in 3C273, of the 3.28μm dust feature by Allen (1980); Hyland and Allen (1982) also use broad-band measures of quasars to demonstrate the presence of excess continuum emission in the 1-3μm rest-frame region, above the continuum component present at shorter and longer wavelengths. They interpret this excess in terms of dust emission at $\sim$ 1200°K, originating in the same region responsible for the broad emission lines. The 3.28μm dust feature was not however seen in spectra of 3C273 taken at higher signal-to-noise ratio by Lee et al. (1982), nor were any dust features seen in 8-13μm spectra by Roche et al. (1983) - see Figure 2. Puetter et al. (1981) have made an attempt to synthesise the intrinsic hydrogen-line spectrum of QSOs and to compare the results with observations of 14 quasars. They find that the resulting spectrum is inconsistent with optically thin recombination models reddened by dust, but agrees (at least qualitatively) with models in which the BLR clouds have very large optical depths in both the Lyman and Balmer lines ($\tau_{Lyc} \sim 10^6$) and in the resonance lines of abundant heavy elements; electron temperatures in such models exceed 1.5×10^{4}°K in the line-emitting region. We have already discussed in sections 3.4 and 3.1 the observations of Pα, Hα and Hβ in Seyfert galaxies by Lacy et al (1982), in which they found evidence for a variety of effects, including (dust) reddening; the Seyfert galaxies were distributed either along a reddening line or along a line of constant Pα/Hβ. Soifer et al. (1981) earlier reported similar observations for a sample of 16 quasars. The quasar data were distributed about the Pα/Hβ = constant line only, and no evidence for significant amounts of dust reddening was present; Pα/Hα weakened gradually as the Balmer decrement Hα/Hβ increased. It seems likely that hydrogen-line ratios in most quasars have to be explained exclusively by radiative transfer and collisional excitation effects.

The spectrum of 3C273 shown in Figure 5, shows no evidence for any significant excess over a power-law spectrum over the entire spectrum from 5mm to 3μm. This is a strong argument in favour of the incoherent synchrotron mechanism rather than multi-component dust re-radiation as the predominant emission mechanism in this spectral region. The same argument against the presence of cool dust holds for the spectra of BL Lac objects.

5. COSMOLOGICAL INVESTIGATIONS

5.1 Evolution of Elliptical Galaxies

Bruzual's (1981, 1983) models of the stellar evolution in elliptical galaxies predict that <u>the intrinsic spectral energy distribution in the near-infrared region should retain an essentially constant shape to look-back times of</u> $z \sim 1$. This is borne out by observations (Lilly and Longair 1982 a, b, c; Ellis and Allen 1983). The maximum of the observed energy distribution at low redshifts occurs near a wavelength of $1.3(1 + z)$μm; it is therefore accessible within the (low-background) near-infrared region over a cosmologically interesting range of redshifts.

Puschell, Owen and Laing (1982a,b) have suggested that at the higher redshifts infrared colours show anomalous behaviour, being generally considerably bluer than Bruzual's non-evolutionary predictions. Ellis and Allen (1983) draw a parallel with QSOs in which Hyland and Allen (1982) noticed a significant drop in J-K for $z \gtrsim 0.65$ - as the Hα emission line enters the J band; Ellis and Allen show a plot (reproduced here as Figure 7) in which the quasar data (filled circles) are superimposed on all the radio-galaxy data, regardless of quality; they emphasise in their caption that the low-redshift QSOs "reveal a similar trend at high redshifts to that delineated by Puschell's data".

Lilly has kindly given me a plot of his most recent data which is reproduced in Figure 8. At UKIRT, J is about as easy to measure as K

GROUND-BASED EXTRAGALACTIC INFRARED SPECTROSCOPY AND RELATED STUDIES

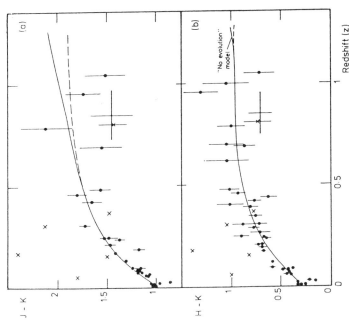

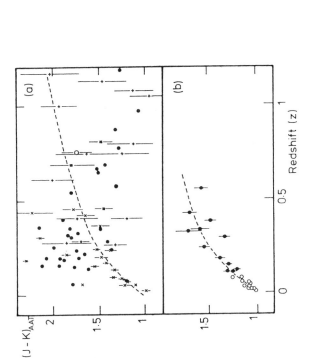

Figure 7: (a) A pot-pourri of infrared data on quasars and radio galaxies presented by Ellis and Allen. Filled circles are data on quasars. See text for discussion.
(b) J-K vs redshift for a sample of optically-selected galaxies adapted from Ellis and Allen (1983).

Figure 8: Infrared colours for a sample of 3C radio galaxies observed by Lilly and Longair. They concluded:
 (i) that the near-infrared continua of narrow-lined radio galaxies (filled circles) are indistinguishable from starlight.
 (ii) that the infrared colours do not evolve with redshift up to $z \sim 1$, but,
 (iii) infrared <u>luminosity</u> evolution must occur.
The solid line corresponds to the predictions of Bruzual (1981) for a luminous elliptical galaxy in the absence of evolution.

for J-K ∼ 1. For higher-redshift objects, which have redder J-K colours (and weaker J), it turns out to be much easier to measure H than J, so the colour-redshift diagram is most complete for H-K, particularly at high redshifts. There is no evidence on this plot for the anomalies reported by Puschell et al., and the points are distributed fairly evenly about Bruzual's "no colour evolution" prediction. The scatter on the points is smaller and the number of objects surveyed is larger than the earlier study by Lebofsky (1981).

There is some conflict about the use of radio galaxies for tests of stellar evolution models. Lilly and Longair (1982a) will continue to favour radio galaxies until optical techniques have been devised for systematic sampling of reasonably large numbers of representative stellar systems, to redshifts $z \gtrsim 1$. They see the use of a radio criterion allows them to work systematically with a large well-defined sample out to much higher redshifts than most purely optical samples have so far managed to reach (see also Lebofsky, Rieke and Walsh 1983). [Persson - (personal communication) - has however observed ∼ 40 galaxies in the redshift range 0.2 to 0.9 from the Hoessel-Oke-Gunn optical survey of first ranked galaxies in distant clusters]. The radio galaxies "comprise the most distant set of readily recognisable objects whose light is dominated by starlight". Ellis and Allen (1983) on the other hand argue that their use of optically selected galaxies is a better approach, because the evolutionary behaviour of the stars in radio galaxies "is hardly likely to be typical of the early types on which the models are based". Ellis and Allen go on to suggest that the stellar populations in supergiant (radio) ellipticals will be distorted by galaxy mergers. Finally they draw on evidence from field counts of galaxies to suggest that the optical properties of radio galaxies evolve differently from those of normal ellipticals.

The optical sample by Ellis and Allen (1983) is able to overlap only with the lower-redshift end of the radio sample, but this is nevertheless a very reasonable test of whether anything about the radio emission upsets the overall optical properties; in a constant 10".8 observing aperture, the contrast in the ratio of nuclear to envelope light is highest at low redshift, and so it is here that one may expect to see the greatest problems in using radio galaxies. In spite of all this speculation, everyone now seems to agree that in fact all the available observational checks show no observeable differences between the near infrared colours of radio-loud and radio-quiet ellipticals at a given redshift, (provided one excludes broad-line radio galaxies -BLRGs - which are found to have very red IR colours and show excesses of 1-2 magnitudes at K over the narrow-lined radio galaxies-NLRGs; the colours and magnitudes of the BLRGs are consistent with a power-law addition to the stellar component). The near infrared continua of NLRGs are indistinguishable from starlight. The fact that the continuum at rest-frame wavelengths > 0.6μm is still dominated by old stars at z=1 is not surprising, as most models for the changes in stellar populations over this timescale predict very small effects in the red and infrared.

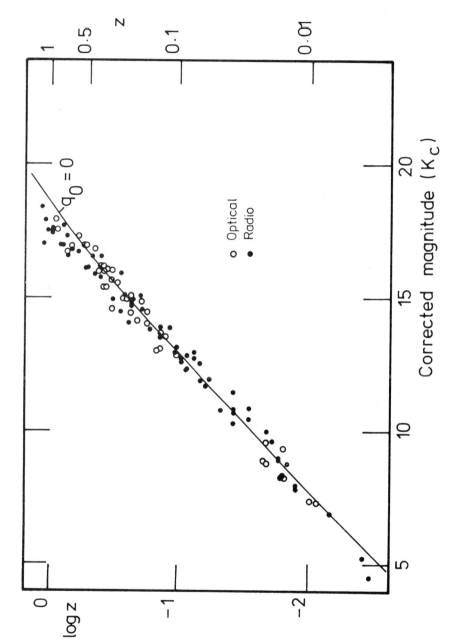

Figure 9: An infrared Hubble diagram compiled and supplied by Rieke, and based on an updated version of data discussed by Lebofsky and Eisenhardt (1983). The solid line represents the predictions of non-evolutionary models for $q_o = 0$.

We have seen that the intrinsic infrared colours do not evolve significantly with redshift up to $z \sim 1$, but what about the infrared luminosities? An appeal must be made to the infrared Hubble diagram. People are now sufficiently realistic (cynical?) about Hubble diagrams to realise that they tell you more about evolution than about the deceleration parameter q_0. However, if we consider the updated version of Figure 2 in the article by Lebofsky and Eisenhardt (1983) - the updated figure was kindly supplied by George Rieke and is reproduced here as Figure 9 - and imagine that the infrared luminosity does not evolve, then apparent values of q_0 well in excess of unity are obtained. [From their more restricted, but perhaps potentially more systematic data set, Lilly and Longair find q_0(apparent) = 3 or 4 under similar constraints of no luminosity evolution]. If, instead, values of $q_0 < 1$ are assumed, then luminosity evolution of the galaxies must occur. Rieke (personal communication) points out that even for $q_0 = 1$, the galaxies at $z \geq 0.6$ tend to be too bright. From Lilly's data set, if $q_0 \sim \frac{1}{2}$, the observed infrared luminosities are about 1 magnitude brighter at $z = 1$ than the predictions of non-evolutionary models. This higher luminosity is expected, for as we look back in time, we expect to find more red giants (the rate of change of turnoff mass with epoch is greater in the past, i.e. stars were peeling off the zero-age main sequence and moving over to the giant branch at a greater rate than now; this effect outweighs that of the increasing IMF, the numbers of stars available per unit time for peeling off down the main sequence). The luminosity function is in fact relatively insensitive to the SFR history, provided that the SFR is a steeply falling function of time (so that one is dominated by the mass of stars formed at the beginning).

The optical-infrared colours, unlike the purely infrared colours, do evolve. V-K is roughly a magnitude bluer by the time one has looked back to $z \sim 1$. Because K was itself about a magnitude brighter then (there were more red giants) this is equivalent to rest-frame U being about 2 magnitudes brighter then than now. Lilly, Longair and McLean (1983) were able to demonstrate directly, at least in the case of 3C352, that this excess of short wavelength light is present throughout the galaxy, and is not concentrated at the nucleus. They conclude that the excess is therefore not a manifestation of some unusual, central, non-thermal phenomenon but is simply the effect of young stars in an evolving stellar population; futhermore, it indicates that the regions of star formation are also not concentrated at the nucleus. These results, if generally applicable to distant elliptical galaxies, are of considerable importance to models of galactic evolution, since they suggest that star formation and interstellar gas were more plentiful in the past - the SFR has decreased.

I have been quite qualitative in my reports of the evolutionary studies of elliptical galaxies, and particularly in estimates of a corrected value for q_0. This is because I believe we still probably do not understand the full extent of evolutionary corrections sufficiently well - including corrections for the effect of environment.

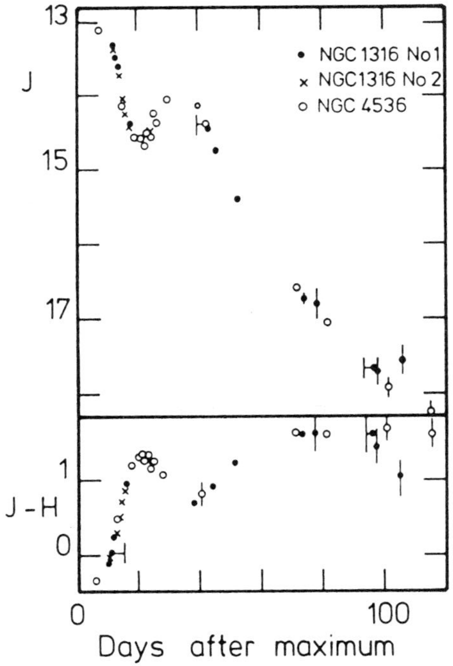

GROUND-BASED EXTRAGALACTIC INFRARED SPECTROSCOPY AND RELATED STUDIES 413

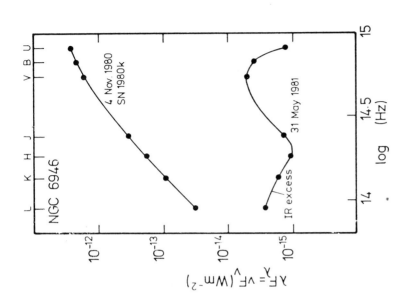

Figure 11: The onset of an infrared excess in the spectrum of SN 1980k. The excess is thought to be caused either by the formation of dust in the SN envelope or by a light - echo effect on interstellar material already in the vicinity of the Supernova explosion. (Dwek et al. 1981).

Figure 10: The infrared light curves of type 1 supernovae may offer a remarkably stable means for measuring distances well beyond the influence of our own Supercluster. Diagram adapted from Elias et al (1981).

Persson, Frogel and Aaronson (1979) first showed that the H-K colour for low-redshift ellipticals does not show any tendency for field ellipticals to differ from cluster ellipticals. However, Aaronson, Persson and Frogel (1981) were then able to show that in fact early-type galaxies do not follow a universal colour-magnitude relation. To use such a relation requires V-K or U-V colours. The infrared energy distributions of galaxies in Virgo and Coma are (perhaps surprisingly) quite different. Persson and his colleagues concluded that many Coma galaxies lost their gas content prior to the epoch of formation of a cool stellar component in Virgo (perhaps consisting, for example, of intermediate-age giants or very metal-rich stars).

5.2 <u>The infrared-luminosity vs HI velocity-width relation</u>. Much less sensitive to variations in the numbers of cool stars is the infrared-luminosity vs HI velocity-width relation studied in late-type galaxies by Aaronson and collaborators (1979, 1980, 1982). A tighter correlation with 21cm line width has been found for H magnitudes than for optical magnitudes, even when the H magnitudes are left uncorrected for inclination or internal absorption. (One should of course realise that substantial absorption can still affect infrared colours - see e.g. Telesco, Gatley and Stewart 1982). Aaronson and collaborators find a Hubble ratio of $\sim$ 60 km s^{-1} Mpc^{-1} for Virgo and Ursa Major, but $\sim$ 95 km s^{-1} Mpc^{-1} for more distant clusters. They attribute the difference to a component of Local-Group motion towards the Virgo cluster, which compares fairly well with observations of a dipole term in the microwave background (although some differences still remain). They map out the Virgo/Local-Group velocity field in some detail.

5.3 <u>Infrared Observations of Supernovae</u>. Elias et al. (1981) obtained the light curves of three type 1 supernovae which went off within the space of a few weeks - two of them in the same E galaxy, Fornax A (NGC 1316). The other was in the spiral galaxy NGC 4536 in the Virgo cluster. In Figure 10, one can see the remarkable similarity of their relative infrared light curves at all phases as well as the double maximum not seen at visual wavelengths. For the two SN I in NGC 1316, <u>no</u> shift has been applied to the intensity axis, as the objects were expected to be at the same distance.

The energy distributions of the three supernovae are also similar. The bottom panel of the figure shows the time evolution of the J-H colour through the outbursts. These data show that (i) the physical mechanisms operating in the expanding supernova envelopes to produce light and colour variations appear to be the same (note the onset of an infrared excess in SN 1980k illustrated in Figure 11) and (ii) subject to verification of their stability, these curves provide a very powerful method of measuring the distances of galaxies out to $\sim$ 50 Mpc, i.e. well beyond any perturbations within our own super-cluster.

Acknowledgements:- I should like to thank numerous colleagues all over the world who responded so well to my request for preprints, reprints and advice. I am particularly grateful to Russell Cannon, Mark Phillips, Tom Geballe, Richard Wade, Martin Ward, Ian Gatley, Dave Aitken, Pat Roche, Simon Lilly, Gerry Gilmore, Walter Gear, Ian Robson, Peter Ade, Adrian Webster and my colleagues on the UKIRT staff for extensive discussions. I should also like to thank the conference organisers for providing this stimulating opportunity; conversations with many of the conference participants assisted me greatly in the preparation of this article. Finally I wish to acknowledge the patience and encouragement of my wife Anamaria and children Paulina and Carolita who, in common with the families of astronomers everywhere, continue to live their lives around the frequent long absences and late nights associated with astronomical research.

REFERENCES

Aaronson, M.: 1978, Ap.J. (Letters), 221, L103.
Aaronson, M., Huchra, J., and Mould, J.: 1979, Ap.J, 229, p.1.
Aaronson, M., Huchra, J., Mould, J., Schechter, P.L., and Tully, R.B.: 1982, Ap.J., 258, p.64
Aaronson, M., Mould, J., Huchra, J., Sullivan III, W.T., Schommer, R.A., and Bothun, G.D.: 1980, Ap.J., 239, p.12.
Aaronson, M., Olszewski, E.W. and Hodge, P.W.: 1983 preprint "Carbon Stars and the Seven Dwarfs".
Aaronson, M., Persson, S.E., and Frogel, J.A.: 1981, Ap.J., 245, p.18.
Aitken, D.K., and Roche, P.F.: 1982, M.N.R.A.S., 200, p.217.
Aitken, D.K., and Roche, P.F.: 1983, M.N.R.A.S., 202, p.1233.
Aitken, D.K., Roche, P.F., Allen, M.C., and Phillips, M.M.: 1982, M.N.R.A.S., 199, 31P.
Aitken, D.K., Roche, P.F., and Phillips, M.M.: 1981, M.N.R.A.S., 196, 101P.
Aitken, D.K., Roche, P.F., Spenser, P.M., and Jones, B.: 1979, Astron. and Ap., 76, p.60.
Allen, D.A.: 1980, Nature, 284, p.323.
Angel, J.R.P., and Stockman, H.S.: 1980, Ann. Rev. Astron. Ap., 18, p.321.
Axon, D.J., Bailey, J., and Hough, J.H.: 1982, Nature, 299, p.234.
Bailey, M.E.: 1982, M.N.R.A.S., 200, p.247.
Baldwin, J.A.: 1975, Ap.J., 201, p.26.
Baldwin, J.A.: 1977, M.N.R.A.S., 178, 67P.
Baldwin, J.A., Phillips, M.M., and Terlevich, R.: 1981, P.A.S.P., 93, p.5.
Baldwin, J.A., and Smith, M.G.: 1983, M.N.R.A.S., 204, in press. "The Location of Material Producing Lyman-Limit Discontinuities in QSO Spectra".
Balzano, V.A., and Weedman, D.W.: 1981, Ap.J., 243, p.756.

Beck, S., Beckwith, S., and Gatley, I. : 1983, preprint.
   "Observations of Brackett α and γ Emission from Galaxies".
Beck, S.C., Lacy, J.H., Baas, F., and Townes, C.H.: 1978, Ap.J., 226, p.545.
Becklin, E.E., Matthews, K., Neugebauer, G., and Wynn-Williams, C.G.:
   1973, Ap.J., 186, L69.
Becklin, E.E., Tokunaga, A.T., and Wynn-Williams, C.G.: 1982, Ap.J., 263, p.624.
Beckwith, S., Evans, N.J., II, Gatley, I., Gull, G., and Russell, R.W.: 1983, Ap.J., 264, p.152.
Beichman, C.A., Neugebauer, G., Soifer, B.T., Wooten, H.A., Roellig, T., and Harvey, P.M.: 1981b, Nature, 293, p.711.
Beichman, C.A., Pravdo, S.H., Neugebauer, G., Soifer, B.T., Matthews, K., and Wootten, H.A.: 1981a, Ap.J., 247, p.780.
Biermann, P. : 1976, Astron. and Ap., 53, p.295.
Blanco, B.M., Blanco, V.M., and McCarthy, M.F: 1978, Nature, 271, p. 638.
Blanco, V.M., McCarthy, M.F., and Blanco, B.M.: 1980, Ap.J., 242, p.938.
Blitz, L., Israel, F.P., Neugebauer, G., Gatley, I, Lee, T.J., and Beattie, D.H. : 1981, Ap.J., 249, p. 76.
Bothun, G.D., and Schommer, E.R.: 1982, Astron.J., 87, p.1368.
Bregman, J.N., Glassgold, A.E., Huggins, P.E., Aller, H.D., Aller, M.F., Hodge, P.E., Rieke, G.H., Lebofsky, M.J., Pollock, J.T., Pica, A.J., Leacock, R.J., Smith, A.G., Webb, J., Balonek, T.J., Dent, W.A., Kn, W.H.-M., Schwartz, D.A., Miller, J.S., Rudy, R.J., and LeVan, P.D.: 1983, Ap.J., submitted. "Multifrequency Observations of the BL Lac Object 0735 + 178".
Bregman, J.N., Glassgold, A.E., Huggins, P.J., Pollock, J.T., Pica, A.J., Smith, A.G., Webb, J.R., Ku, W.H.-M., Rudy, R.J., LeVan, P.D., Williams, P.D., Brand, P.W.J.L., Neugebauer, G., Balonek, T.J., Dent, W.A., Aller, H.D., Aller, M.F., and Hodge, P.E.: 1982, Ap.J., 253, p.19.
Bregman, J.N., Lebofsky, M.J., Aller, M.F., Rieke, G.H., Aller, H.D., Hodge, P.E., Glassgold, A.E., and Huggins, P.J.: 1981, Nature, 293, p.714.
Bregman, J.D. and Rank, D.M.: 1975, Ap.J (Letters), 195, L125.
Bruzual, G.: 1981, Ph.D. Thesis, University of California, Berkeley.
Bruzual, G.; 1983, Ap.J., submitted. "Spectral Evolution of Galaxies. I. Early Type Systems".
Canfield, R.C., and Puetter, R.C.: 1981, Ap.J., 243, p.390.
Canfield, R.C., Puetter, R.C., and Ricchiazzi, P.J.: 1981, Ap.J., 248, p.82.
Capps, R.W., Sitko, M.L., and Stein, W.A.: 1982, Ap.J, 255, p.413.
Carswell, R.F., and Ferland, G.J.: 1980, M.N.R.A.S., 191, p.55.
Clegg, P.E., Gear, W.K., Ade, P.A.R., Robson, E.I., Smith, M.G., Nolt, I.G., Radostitz, J.V., Glaccum, W., Harper, D.A., and Low, F.J.: 1983, Ap.J., in press. "Millimeter and Submillimeter Observations of 3C273".
Cohen, J.G., Frogel, J.A., Persson, S.E., and Elias, J.H. : 1981, Ap.J., 249, p. 481.

Condon, J.J., Condon, M.A., Jauncey, D.L., Smith, M.G., Turtle, A.J., and Wright, A.E.: 1981a, Ap.J., 244, p.5.
Condon, J.J., Condon, M.A., Gisler, G., and Puschell, J.J.: 1982, Ap.J., 252, p.102.
Condon, J.J., and Dressel, L.L.: 1978, Ap.J., 221, p.456.
Condon, J.J., O'Dell, S.L., Puschell, J.J., and Stein, W.A: 1981b, Ap.J., 246, p.624.
Davidsen, A.F.: 1980, Proc IAU Symp., #92, "Objects of High Redshift", eds G.O. Abell and P.J.E. Peebles (Reidel:Holland) p.235.
Davidson, K., and Netzer, H. : 1979, Rev. Mod. Phys., 51, p.715.
Davis, D.S., Larson, H.P., and Smith, H.A. : 1982, Ap.J., 259, p.166.
De Robertis, M.: 1983, Nature, submitted. "QSO Evolution in the Interaction Model".
Dwek, E., A'Hearn, M.F., Becklin, E.E., Capps, R.W., Telesco, C.M., Tokunga, A.T., Wynn-Williams, G., Dinerstein, H.L., Werner, M.W., and Gatley, I.: 1981, Bull. Am. Astr. Soc., 13, p 795.
Elias, J.H., Frogel, J.A., Hackwell, J.A., and Persson, S.E.: 1981, Ap.J., 251, L13.
Ellis, R. and Allen, D.A.: 1983, preprint.
Ennis, D.J., Neugebauer, G. and Werner, M.: 1982a. Ap.J., 262, p.451.
Ennis, D.J., Neugebauer, G. and Werner, M: 1982b, Ap.J., 262, p.460.
Epstein, E.E., Fogarty, W.G., Mottman, J., and Schneider, E.: 1982, Astron. J., 87, p.449.
Ferland, G.J. and Netzer, H.: 1983, Ap.J., 264, p.105.
Frogel, J.A., Blanco, V.M., McCarthy, M.F., and Cohen, J.G.: 1982, Ap.J., 252, p.133.
Frogel, J.A., Elias, J.H., and Phillips, M.M.: 1982, Ap.J., 260, p.70.
Frogel, J.A., Persson, S.E., and Cohen, J.G.: 1980, Ap.J., 240, p. 785.
Gautier, T.N., III, Fink, U., Treffers, R.R., and Larson, H.P.: 1976, Ap.J. (Letters), 207, L129.
Gear, W.K., Robson, E.I., Ade, P.A.R., Griffin, M.G., Smith, M.G., and Nolt, I.G.: 1983, Nature, in press. "Multifrequency Observations of OV236 (1921-293): a case of an unusual spectrum?"
Geballe, T.R., Russell, R.W., and Nadeau, D.: 1982, Ap.J., 259, L47.
Gehrz, R.D., Sramek, R.A., and Weedman, D.W.: 1983, Ap.J., in press (April 15). "Star Bursts and the Extraordinary Galaxy NGC 3690".
Gillett, F.C., Forrest, W.J., and Merrill, K.M.: 1973, Ap.J., 183, p.87.
Gillett, F.C., Jones, T.W., Merrill, K.M., and Stein, W.A.: 1975a, Astron. and Ap., 45, p.77.
Gillett, F.C., Kleinmann, D.E., Wright, E.L., and Capps, R.W.: 1975b, Ap.J. (Letters), 198, L65.
Glass, I.S.: 1981, M.N.R.A.S., 194, p.795.
Glass, I.S., Moorwood, A.F.M., and Eichendorf, W. : 1982, Astron. and Ap., 107, p276.
Glassgold, A.E., Bregman, J.N., Huggins, P.J., Kinney, A.L., Pica, A.J., Pollock, J.T., Leacock, R.J., Smith, A.G., Webb, J.R., Wisniewski, W.Z., Jeske, N., Spinrad, H., Henry, R.B.C., Miller, J.S., Impey, C., Neugebauer, G., Aller, M.F., Aller, H.D., Hodge, P.E., Balonek, T.J., Dent, W.A., and O'Dea, C.P.: 1983, Ap.J.,

submitted. "Multifrequency Observations of the Flaring Quasar 1156 + 295".
Griersmith, D., Hyland, A.R., and Jones T.J.: 1982, Astron. J., 87, p.1106.
Gunn, J.E.: 1979, in "Active Galactic Nuclei", eds C. Hazard and S. Mitton, (Cambridge), p.213.
Hall, D.N.B., Kleinmann, S.G., Scoville, N.Z., and Ridgway, S.T.: 1981, Ap.J., 248, p.898.
Hollenbach, D.J., and Shull, J.M.: 1977, Ap.J., 216, p.419.
Houck, J.R., Forrest, W.J., and McCarthy, J.F.: 1980, Ap.J. (Letters), 242, L65.
Hummel, E.: 1980, Astron. and Ap. 89, L1.
Hutchings, J.B. and Campbell, B. 1983: Nature, submitted. "Are QSO's Activated by Interactions Between Galaxies"?
Hyland, A.R., and Allen, D.A.: 1982, M.N.R.A.S., 199, p.943.
Impey, C.D., and Brand, P.W.J.L.: 1981, Nature, 292, p.814.
Impey, C.D., Brand, P.W.J.L., Wolstencroft, R.D., and Williams, P.M.: 1982, M.N.R.A.S., 200, p.19.
Israel, F.P., Gatley, I., Matthews, K., and Neugebauer, G.: 1982, Astron. and Ap., 105, p.229.
Israel, F.P., and Koornneef, J.: 1983, Ap.J. (Letters), submitted "Two-Micron Spectrophotometry of the SMC HII Region N81".
Kemp, J.C., Rieke, G.H., Lebofsky, M.J., and Coyne, G.V., S.J.: 1977, Ap.J. (Letters), 215, L107.
Khachikian, E.Y., and Weedman, D.W.: 1971, Astrofizika, 7, p.389.
Khachikian, E.Y., and Weedman, D.W.: 1974, Ap.J., 192, p.581.
Kleinmann, D.E., Gillett, F.C., and Wright, E.L.: 1976, Ap.J., 208, p.42.
Koorneef, J.: 1982, Astron and Ap., 107, p.247.
Kwan, J., and Krolik, J.H.: 1981, Ap.J., 250, p.478.
Lacy, J.H., Soifer, B.T., Neugebauer, G., Matthews, K., Malkan, M., Becklin, E.E., Wu, C.-C., Boggess, A., and Gull, T.R. : 1982, Ap.J., 256, p.75.
Landau, R., Jones T.W., Epstein, E.E., Neugebauer, G., Soifer, B.T., Werner, M.W., Puschell, J.J., and Balonek, T.T.: 1983, preprint. "Extragalactic 1-mm Sources: Simultaneous Observations at cm, mm, and Visual Wavelengths."
Laques, P., Nieto, J.-L., Vidal, J.-L., Augè, A., and Despiau, R.: 1980, Nature, 288, p.145.
Lawrence, A. and Elvis, M.: 1982, Ap.J., 256, p.410.
Lebofsky, M.J.: 1981, Ap.J. (Letters), 245, L59.
Lebofsky, M.J. and Eisenhardt, : 1983, preprint.
Lebofsky, M.J., and Rieke, G.H.: 1979, Ap.J., 229, p.111.
Lebofsky, M.J., Rieke, G.H., and Kemp, J.C.: 1978, Ap.J, 222, p.95.
Lebofsky, M.J., Rieke, G.H., and Walsh, D.: 1983, M.N.R.A.S., in press. "Infrared Counterparts to 'Empty Field' Steep Spectrum Radio Sources".
Lebofsky, M.J., Sargent, D.G., Kleinmann, S.G., and Rieke, G.H.: 1978, Ap.J., 219, p.487.
Lee, T.J., Beattie, D.H., Gatley, I., Brand, P.W.J.L., Jones T., and Hyland, A.R.: 1982, Nature, 295, p.214.

Lilly, S.J., and Longair, M.S.:   1982 a, M.N.R.A.S., 199, p.1053.
Lilly, S.J., and Longair, M.S.:   1982 b, in "Astrophysical Cosmology", eds. H.A. Brück, G.V. Coyne, and M.S. Longair, (Pontificia Academia Scientiarum), p.269.
Lilly, S.J., and Longair, M.S.:   1982 c, Proc IAU Symp. #97. "Extragalactic Radio Sources", eds D.S. Heeschen and C.M. Wade, (Reidel:Holland), p.413.
Lilly, S.J., Longair, M.S., and McLean, I.S.:   1983, Nature, 301, p.488.
Maccacaro, T., Perola, G.C., and Elvis, M.   :   1982, Ap.J., 257, p.47.
Marscher, A.P.:   1980 a, Nature, 286, p.12.
Marscher, A.P.:   1980 b, Ap.J., 235, p.386.
Marscher, A.P.:   1983, Nature, "Millimetre and Submillimetre Wavelength Observations of Quasars: Bridging the Gap" article in preparation for "News and Views".
McAlary, C.W., and McLaren, R.A.:   1981, Ap.J., 250, p.98.
McAlary, C.W., McLaren, R.A., and Crabtree, D.R.:   1979, Ap.J., 234, p.471.
Moffet, A.T. in "Galaxies and the Universe", Stars and Stellar Systems Vol. IX, eds A. Sandage, M. Sandage and J. Kristian (U. Chicago press), p.244.
Moorwood, A.F.M.:   1982a, The Messenger (ESO), 27, p.11.
Moorwood, A.F.M. :   1982b, ProcXVIth ESLAB Symposium, "Galactic and Extragalactic Infrared Spectroscopy", eds. J.P. Phillips, M.F. Kessler and T.D. Guyenne (ESA SP-192 ESTEC : Holland), p.47.
Moorwood, A.F.M., and Glass, I.S. :   1982, Astron. and Ap., 115, p.84.
Moorwood, A.F.M., and Salinari, P. :   1981, Astron. and Ap., 100, L16.
Morgan, D.H., and Nandy, K.:   1982, M.N.R.A.S., 199, p.979.
Mould, J.R., Cannon, R.D., Aaronson, M., and Frogel, J.A.:   1982, Ap.J., 254, p.500.
Mushotsky, R.F.   :   1982, Ap.J., 256, p.92.
Neugebauer, G., Becklin, E.E., Oke, J.B., and Searle, L.:   1976, Ap.J., 205, p.29.
Neugebauer, G, Oke, J.B., Becklin, E.E., and Matthews, K.:   1979, Ap.J., 230, p.79.
Neugebauer, G., Soifer, B.T., Matthews, K., Margon, B., and Chanan, G.A. :   1982, Astron. J., 87, p.1639.
Oke, J.B. and Gunn, J.E.:   1982, P.A.S.P., 94, p.586.
Osmer, P.S., and Smith, M.G.:   1977, Ap.J. (Letters), 215, L47.
Osmer, P.S., and Smith, M.G.:   1980, Ap.J. (Suppl.), 42, p.333.
Osterbrock, D.E.:   1974 "Astrophysics of Gaseous Nebulae" (Freeman) p.64.
Osterbrock, D.E.:   1981, Ap.J., 249, p.462.
Pauliny-Toth, I.I.K., and Kellerman, K.I.:   1966, Ap.J., 146, p.634.
Penston, M.V.:   1982, Proc. 3rd European IUE conf., p.69.
Persson, S.E., Aaronson, M., Cohen, J.G., Frogel, J.A., and Matthews, K: 1983, Ap.J., in press (March 1) "Photometric Studies of Composite Stellar Systems V. Infrared Photometry of Star Clusters in the Magellanic Clouds".
Persson, S.E., Cohen, J.G., Sellgren K., Mould, J., and Frogel, J.A.:   1980, Ap.J., 240 p. 779.

Persson, S.E., Frogel, J.A., and Aaronson, M.: 1979, Ap.J. (Suppl.), 39, p.61.
Phillips, M.M., Aitken, D.K., and Roche, P.F.: 1983, preprint "8-13μm Spectrophotometry of Galaxies. 1. Galaxies with Giant HII Region Nuclei".
Price, S.D.: 1981, Astron. J., 86, p.193.
Price, S.D., Marcotte, L.P., and Murdock, T.L.: 1982, Astron. J.,87, p.131.
Puetter, R.C.: 1981, Ap.J., 251, p.446.
Puetter, R.C., Burbidge, E.M., Smith, H.E., and Stein, W.A.: 1982, Ap.J., 257, p.487.
Puetter, R.C., Hubbard, E.N., Ricchiazzi, P.J., and Canfield, R.C.: 1982, Ap.J., 258, p.46.
Puetter, R.C., and Le Van, P.D.: 1982, Ap.J., 260, p.44.
Puetter, R.C., Smith, H.E., Willner, S.P., and Pipher, J.L.: 1981, Ap.J., 243, p.345.
Puschell, J.J.: 1981, Ap.J., 247, p.28.
Puschell, J.J., Owen, F.N., and Laing, R.N.: 1982a. Proc. IAU Symp. #97. "Extragalactic Radio Sources". eds. D.S. Heeschen and C.M. Wade (Reidel:Holland), p.423.
Puschell, J.J., Owen, F.N., and Laing, R.N.: 1982 $\underline{b}$, Ap.J., 257, L57.
Puschell, J.J., and Stein, W.A.: 1980, Ap.J., 237, p.331.
Richer, H.B., and Westerlund, B.E.: 1983, Ap.J, 264, p.114.
Richstone, D.O., and Morton, D.C.: 1975, Ap.J., 201, p.289.
Richstone, D.O., and Schmidt, M.: 1980, Ap.J., 235, p.361.
Rieke, G.H.: 1978, Ap.J., 226, p.550.
Rieke, G.H. : 1981, Proc. IAU Symp. #96, "Infrared Astronomy", eds. C.G. Wynn-Williams and D.P. Cruikshank (Reidel:Holland) p.317.
Rieke, G.H., and Lebofsky, M.J.: 1978, Ap.J. (Letters), 220, L37.
Rieke, G.H., and Lebofsky, M.J. : 1979, Ann. Rev. Astr. and Ap. 17, p.477.
Rieke, G.H., and Lebofsky, M.J.: 1980, Proc. IAU Symp. #92, "Objects of High Redshift", eds. G.O. Abell and P.J.E. Peebles, p.263.
Rieke, G.H., and Lebofsky, M.J.: 1981, Ap.J., 250, p.87.
Rieke, G.H., and Lebofsky, M.J.: 1982, Proc. AIP Conference #83, "The Galactic Center", eds. G.R. Riegler and R.D. Blandford (American Inst. Physics: New York), p.194.
Rieke, G.H., Lebofsky, M.J., and Kemp, J.C.: 1982, Ap.J. (Letters), 252, L53.
Rieke, G.H., Lebofsky, M.J., Kemp, J.C., Coyne, G.V., and Tapia, S.: 1977, Ap.J., 218, L37.
Rieke, G.H., Lebofsky, M.J., and Kinman, T.D.: 1979, Ap.J., 232, L151.
Rieke, G.H., Lebofsky, M.J., Thompson, R.I., Low, F.J., and Tokunaga, A.T.: 1980, Ap.J, 238, p.24.
Rieke, G.H., Lebofsky, M.J., and Wistiewski, W.Z.: 1982, Ap.J., 263, p.73.
Rieke, G.H., and Low, F.J.: 1972, Ap.J. (Letters), 176, L95.
Robson, E.I.: 1982, Proc. ESA Workshop, "Scientific Importance of Submillimetre Observations", (ESA SP-189 Noordwijkerhout:Holland) p.147.

Roche, P.F., Aitken, D.K., Phillips, M.M. and Whitmore, B.: 1983, preprint, "8-13μm Spectrophotometry of Galaxies. 2. Ten Seyferts and 3C273."
Roos, N.: 1981, Astron. and Ap., 95, p.49.
Rubin, R.H.: 1968, Ap.J., 154, 391.
Rudy, R.J., Jones, B., Le Van, P.D., Puetter, R.C., Smith, H.E., Willner, S.P., and Tokunaga, A.T.: 1982, Ap.J., 257, p.570.
Rudy, R.J. and Puetter, R.C.: 1982. Ap.J., 263, p.43.
Rudy, R.J., and Tokunaga, A.T.: 1982, Ap.J., 256, L1.
Russell, R.W., Soifer, B.T., and Merrill, K.M.: 1977, Ap.J., 213, p.66.
Schweizer, F.: 1980, Ap.J., 237, p.303.
Scoville, N.Z., Becklin, E.E, Young, J.S., and Capps, R.W.: 1983, Ap.J., in press. "A 10μm Survey of Star Formation in Galactic Nuclei: Virgo Spirals".
Scoville, N.Z., Hall, D.N.B., Kleinmann, S.G., and Ridgway, S.T.: 1979, Ap.J. (Letters), 232, L121.
Scoville, N.Z., Hall, D.N.B., Kleinmann, S.G., and Ridgway, S.T.: 1982, Ap.J., 253, p.136.
Sellgren, K.: 1981, Ap.J., 245, p.138.
Sherwood, W.A.: 1982, Proc. IAU Symp. #104, "Early Evolution of the Universe and its Present Structure" in press. "QSO Luminosities at 1MM".
Sherwood, W.A., Kreysa, E., Gemünd, H.-P., and Biermann, P.: 1983, Astron. and Ap., 117, L5.
Sherwood, W.A., Kreysa, E., and Schultz, G.V.: 1981, Astr. Ges. Mitt., 52, p.138.
Sherwood, W.A., Shultz, G.V., and Kreysa, E.: 1981, Nature, 291, p.301.
Sherwood, W.A., Schultz, G.V., Kreysa, E., and Gemünd, H.-P.: 1982, Proc. IAU Symp. #92, "Extragalactic Radio Sources", eds. D.S. Heeschen and C.M. Wade (Reidel:Holland), p.305.
Shields, G.A., and Wheeler, J.C.: 1978, Ap.J., 222, p.667.
Shuder, J.M.: 1980, Ap.J., 240, p.32.
Smith, M.G., Carswell, R.F., Whelan, J.A.J., Wilkes, B.J. Boksenberg, A., Clowes, R.G., Savage, A., Cannon, R.D., and Wall, J.V.: 1981, M.N.R.A.S., 195, p.437.
Smith, M.G. and Weedman, D.W.: 1970, Ap.J., 161, p.33.
Smith, M.G., and Wright, A.E.: 1980, M.N.R.A.S., 191, p.871.
Soifer, B.T., and Neugebauer, G.: 1981, Proc. IAU Symp. # 96, "Infrared Astronomy", eds. C.G. Wynn-Williams and D.P. Cruickshank (Reidel-Holland) p.329.
Soifer, B.T., Neugebauer, G., Oke, J.B., and Matthews, K.: 1981, Ap.J., 243, p.369.
Soifer, B.T., Neugebauer, G., Oke, J.B., Matthews, K., and Lacy, J.H.: 1983, Ap.J., in press (February 1). "Infrared/Optical Energy Distributions of High Redshift Quasars".
Stein, W.A. and Soifer, B.T.: 1983, Ann. Rev. Astr. and Ap., 21, in press. "Dust in Galaxies".
Stein, W.A., and Weedman D.W.: 1976, Ap.J., 205, p.44.

Stockton, A.: 1982, Ap.J., 257, p.33.
Telesco, C.M. : 1978, Ap.J., 226, L125.
Telesco, C.M., Becklin, E.E., and Wynn-Williams, C.G.: 1980, Ap.J., 241, L69.
Telesco, C.M. and Gatley, I.: 1981, Ap.J., 247, L11.
Telesco, C.M., Gatley, I., and Stewart, J.M.: 1982, Ap.J. (Letters), 263, L13.
Telesco, C.M., and Harper, D.A.: 1980, Ap.J., 235, p392.
Telesco, C.M., and Owensby, P.D.: 1981, contributed paper at IAU Symp. #96, "Infrared Astronomy", eds. C.G. Wynn-Williams and D.P. Cruickshank Reidel:Holland).
Thompson, R.I., Lebofsky, M.J., and Rieke, G.H.: 1978, Ap.J, 222, L49.
Thronson, H.A., Campbell, M.F. and Harvey, P.M.: 1978, Astron. J., 83 p1581.
Tokunaga, A.T. and Young, E.T.: 1980, Ap.J. (Letters), 237, L93.
Townes, C.H., Lacy, J.H., Geballe, T.R., and Hollenbach, D.J.: 1983, Nature, 301, p.661.
Treffers, R.R., Fink, U., Larson, H.P., and Gautier, T.M.: 1976, Ap.J., 209, p.793.
Van den Bergh, S.: 1980, P.A.S.P., 92, p 122.
Véron, P., Lindblad, P.O., Zuiderwijk, E.J., Véron, M.P., and Adam, G.: 1980, Astron. and Ap., 87, p.245.
Ward, M., Allen, D.A., Wilson, A.S., Smith, M.G. and Wright, A.E.: 1982, M.N.R.A.S., 199, 953.
Ward, M.J., Penston, M.V., Blades, J.C., and Turtle, A.J.: 1980, M.N.R.A.S., 193, 563.
Willner, S.P., Soifer, B.T., Russell, R.W., Joyce, R.R., and Gillett, F.C.: 1977, Ap.J., 217, L121.
Wilson, A.S., Penston, M.V., Fosbury, R.A.E., and Boksenberg, A.: 1976, M.N.R.A.S., 177, p.673.
Wright, E.L., and Kleinmann, D.E.: 1978, Nature, 275, p.298.
Wynn-Williams, C.G., Becklin, E.E., Beichman, C.A., Capps, R.W., and Shakeshaft, J.R.: 1981, Ap.J., 246, p.801.
Wynn-Williams, C.G., Becklin, E.E., Matthews, K., and Neugebauer, G.: 1979, M.N.R.A.S., 189, p.163.
Wynn-Williams, C.G., Becklin, E.E., Matthews, K., and Neugebauer, G.: 1983, Ap.J., in press "Two-Micron Spectrophotometry of the Galaxy NGC 253".
Young, J.S., and Scoville, N.: 1982, Ap.J. (Letters), 260, L41.
Zamorani, G., Henry, J.P., Maccacaro, T., Tananbaun,H., Soltan, A., Avni, Y., Leibert, J., Stocke, J., Strittmatter, P.A., Weymann, R.J., Smith, M.G., and Condon, J.J.: 1981, Ap.J., 245, p.357.

A COMPARISON OF THE FIR LINE AND CONTINUUM EMISSION FROM EXTERNAL GALAXIES WITH THE EMISSION FROM OUR GALAXY

P.G. Mezger

Max-Planck-Institut für Radioastronomie, Bonn, F.R.G.

ABSTRACT

Luminosities of the FIR continuum of dust and of four atomic and ionic lines in our Galaxy are estimated and their origin is discussed. The corresponding emission from external galaxies observed today is discussed and compared to our Galaxy. So far FIR observations of external galaxies relate mostly to pathological cases with central bursts of star formation (e.g. M82, NGC253) or with luminous central sources (as e.g. the Seyfert 2 galaxy NGC1068) or with very compact central synchrotron sources (such as the Seyfert 1 galaxy NGC1275, many QSO's and BL Lac objects). Only one normal Sc Galaxy, M51, has been mapped to date in the FIR continuum. And only four FIR lines of two elements (S and O) have been detected to date in two galaxies (M82 and NGC1068). It is to be expected, however, that IRAS will detect the FIR dust continuum of most of the nearby spiral galaxies, while the observation of compact, self-absorbed synchrotron sources will become the domain of the new generation of large mm/submm telescopes. The need for a European astroplane is stressed if Europe wants to reach and stay at the forefront of FIR astronomy. The potential for future space applications is stressed of new technologies such as carbon fiber reinforced epoxy as lightweight material for precision telescopes or heterodyne radiometers for spectroscopy longward of ~ 100 μm.

I. INTRODUCTION

The topic of this symposium is IR spectroscopy. Broad minded astronomers will accept the quasi Planck spectrum of a dusty galaxy (as e.g. the spectra shown in Fig. 7) or the sharp emission peak of a self-absorbed compact synchrotron source, (as, for example, observed in the QSO 0420-388 and shown in Fig. 6) as spectral features, too. The organizers of this symposium requested me to include results of FIR line spectroscopy of external galaxies in my review. I complied with this request drawing the line between airborne FIR spectroscopy (the results of which will be included) and groundbases MIR and NIR spectroscopy (the results of which will be reviewed elsewhere in this symposium, Smith).

Although observations of compact extragalactic sources are a major research object of some of my colleagues at the MPIfR I will discuss only two examples of this class of object. The main part of my review will relate to the interpretation of the dust emission from galaxies. In two recent papers (Mezger, Mathis and Panagia, 1982; hereafter referred to as Paper I. Mathis, Mezger and Panagia (in prep.), Paper II) a self-consistent model of the emission of our Galaxy from the mm range to the Lyman continuum limit has been obtained. Input parameters to this model are: i) The interstellar radiation field (ISRF) in the solar vicinity. ii) The radial distribution of the four principal stellar components. iii) The radial distribution of interstellar gas and dust. The predicted integrated spectrum of the Galaxy (Fig. 1) shows most of the spectral features and the radial distribution of the galactic FIR emission exhibits the essential features of the spatial structure observed in external galaxies. However, both the relative strength and the luminosity of the different spectral and spatial features varies from one galaxy to another by orders of magnitude.

In the following, the FIR line and continuum emission from external galaxies will be compared with the FIR emission from our Galaxy. In such a comparison, the low angular resolution of airborne FIR telescopes, of typically ∼one arcminute, should be borne in mind. In our Galaxy compact objects such as the Orion molecular clouds or giant HII regions are the most conspicuous FIR sources. But their integrated emission accounts for only 10 ∼20% of the total galactic IR luminosity, L_{IR}. Diffuse emission, which originates inside the solar circle and which is observed in low-resolution balloon surveys (see the review by Okuda, 1980, and references therein) contributes most to the galactic FIR emission. Observed from a large distance only the galactic center region with a luminosity ∼0.1 L_{IR} and a diameter of ∼0.6 kpc would be recognized as a prominent compact source. Therefore, comparing the IR emission of our Galaxy with that of external galaxies, we have to relate it to the diffuse galactic emission rather than to the compact sources, which are the preferred objects of galactic FIR observations.

II. THE CASE OF OUR GALAXY

The Galaxy is a spiral galaxy of type Sbc. The total mass of the galactic disk as inferred from the rotation curve is $\sim 1.7\ 10^{11}\ m_\odot$. It is still not yet clear if the spiral structure is a two-armed or rather a four-armed spiral. Its nuclear activity appears to be average. Most of the galactic emission comes from galactocentric distances $D_G \lesssim 13$ kpc, although stars and atomic hydrogen extend out to $D_G \sim 20$ kpc.

In Table 1 are listed those constituents of the galactic disk which are relevant for the interpretation of its diffuse FIR emission. Stars form the bulk of the mass; their superimposed emission creates the interstellar radiation field (ISRF). About equal fractions of the interstellar matter (ISM) appear to be contained in dense and massive molecular clouds (H_2) and in the more diffuse clouds and intercloud gas, respectively, where hydrogen is atomic (HI). About 1/3-1/2 of the mass fraction Z of elements heavier than ^{4}He are condensed into dust grains, which are well mixed with the gas.

Table 1: Masses and luminosities of the main constituents of the Galactic Disk within galactocentric distances $D_G \lesssim 13$ Kpc.

Component	Mass/m_θ	FIR continuum	L_{IR}/L_θ	FIR Lines	L_{IR}/L_θ	Ref.
Stars	1 E 11	$\lambda \lesssim 8$ μm $\lambda \gtrsim 8$ μm	5 E 10 negligible			1)
Molecular Clouds (H$_2$)	2 E 9	Dust	$\lesssim$ 1 E 8	CO rot. trans. C^o 610 μm	5 E 5 $\lesssim$ 1 E 5	2)
Diffuse Clouds (HI)	1 E 9	Dust	} 4 E 9	C$^+$ 157 μm	5 E 8	3)
Intercloud gas (HI)	1 E 9	Dust		O^o 63.2 μm	3 E 6	4)
Ionized gas (H$^+$) in ELD HII regions	1 E 7	Free-free Dust	$N'_{Lyc} \sim 1.5$ E 53 s^{-1} 6 E 9	O^{++} 88.4 μm (and other ionic lines of Ne,Ar,S)	1 E 7	5)
Relativistic electrons		Synchrotron emission				6)

1) Paper II.
2) L_{IR} (Dust) from Papers I, II. Luminosity of all CO rotational lines as estimated by Güsten (priv. comm.). Luminosity of C^o 610 μm line estimated from Goldsmith and Langer (1978).
3) L_{IR} (Dust) from Paper II. L_{IR} (C$^+$ 157 μm) from Stacey et al. (1982).
4) L_{IR} (Dust) from Paper II, L_{IR} (O^o 63 μm) estimated by Harwit (priv. comm.),
5) S_{ff} is obtained from N'_{Lyc} (Mezger, 1978), the number of Lyc photons absorbed in the ELD HII regions. L_{IR} (Dust) from Paper II. L_{IR} (O^{++} 88.4 μm) estimated by Harwit (priv. comm.).
6) $S_{6\,cm}$ (synchr.) $\sim 10\, S_{6\,cm}$ (free-free) (see text).

Ionized gas (H$^+$) with electron temperatures $\sim$8000 K is found surrounding O stars. HII regions expand from compact sources into extended low-density (ELD) HII regions. The production rate of Lyman continuum (Lyc) photons by all O stars in our Galaxy is estimated to be $N_{Lyc} \sim 3\ 10^{53}$ s^{-1}. About 85-90% of these O stars are associated with ELD HII regions of $\langle n_e \rangle \sim 10$ cm^{-3} (Mezger, 1978). Both molecular and atomic clouds fill only a few percent of the interstellar space, while the intercloud gas (which accounts for $\sim$50% of the mass of HI) may fill up to 50% of the interstellar space. But the largest fraction of interstellar space appears to be filled by an ionized, hot (10^5-10^6 K) and tenuous 10^{-2}-10^{-3} cm^{-3}) "coronal" gas, whose original dust content was probably destroyed if this gas is heated by SN explosions.

In Paper I and II we published results of model calculations of the Galaxy relating to the ISRF from .09 to 1000 μm, which take into account recent observations as well as the distribution of both stellar populations and dust. The composite spectrum of the Galaxy (Fig. 1) and the luminosities given in Table 1 are based on these two papers. The number of Lyc photons absorbed per sec by the gas as given in Table 1 is taken from Mezger (1978). It corresponds to a Lyc photon production rate of

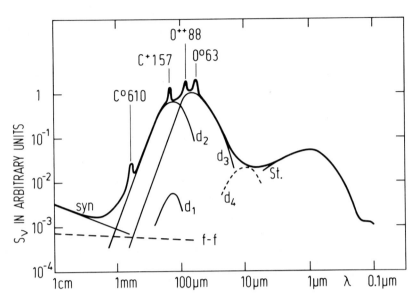

Fig. 1: The composite continuum spectrum of our Galaxy from the radio to the optical region. The position of the four strongest atomic and ionic lines is indicateed. "syn" refers to synchrotron emission, emitted by relativistic electrons gyrating in the galactic magnetic field, "f-f" refers to the free-free emission from galactic ELD HII regions. The dust continuum is composed of contributions from cold dust associated with quiescent molecular clouds (d_1) which probably also emit most of the $C^0 610$ μm line; medium warm dust associated with diffuse atomic hydrogen (d_2), which emits the $C^+ 157$ μm line and possibly (i.e. the hot intercloud gas) the $0^0 63$ μm line; warm dust associated with primarily the ionized gas in ELD HII regions (d_3), which is expected to emit the $O^{++} 88$ μm line, too; and hot dust associated with circumstellar shells (d_4). At wavelengths $\lesssim 8$ μm the stellar radiation dominates the spectrum.

∼30 000 O 6.5 stars. The spectrum of the free-free emission shown in Fig. 1 corresponds to this Lyc photon production rate and $S_{ff} \propto \nu^{-0.1}$ (see also eqs. 3a,b). The spectrum of the diffuse synchrotron emission, produced by relativistic electrons gyrating in the galactic magnetic field, is estimated from the observation that at λ6 cm the brightness temperatures of free-free and synchrotron emission at $\ell \lesssim 40°$, $b \sim 0°$ are about equal, but that the scale height of the synchrotron emission is about ten times greater.

From 8 to 0.1 μm it is found that the galactic emission is dominated by stars. A surprisingly large population of M giants, whose existence was detected through NIR continuum surveys of the galactic plane (see e.g. the review by Okuda, 1980), causes the stellar emission to peak at wavelengths longward of 1 μm. Between 1000 and 8 μm the galactic emission spectrum is dominated by thermal emission from dust which is heated by starlight. About 1/5 of the intrinsic stellar emission (or ∼1/4 of the observable stellar emission) is absorbed by dust and reradiated in the FIR. The model computations show that (as indicated in Fig. 1) three dust components contribute to the galactic FIR emission:

d1: The emission from cold ($\lesssim$13-20 K) dust associated with quiescent molecular clouds with no internal sources of heating but embedded old disk population stars. Its contribution to the total FIR/submm luminosity is negligible.

d2: Is the emission from medium warm (13-26 K) dust associated with the diffuse ISM and heated by the general ISRF.

d3: The emission from warm (29-40 K) dust associated with ELD HII regions and heated by O stars.

d4: Although observations of the diffuse galactic emission in the wavelength range from 4 to $\sim$80 μm are lacking or, at best, fragmented, the galactic flux density appears to be low in this spectral region. This is in contrast with the spectrum of the Seyfert 2 galaxy NGC1068 (Fig. 7), where νS_ν attains its maximum at $\sim\lambda$30 μm. Therefore, a component d4 of hot (200-500 K) dust must be added to explain the dust emission from some external galaxies. Such high dust temperatures can only be attained in shells surrounding very luminous compact objects.

Fig. 2 shows the radial distribution of the volume emissivity of the diffuse galactic FIR emission as derived by Boissé et al. (1981) from their survey of the Galaxy. For a constant scale height of the dust $\varepsilon(D_G)$ would be proportional to the surface brightness of the dust emission of the Galaxy seen face-on, which hence would appear as a compact central source surrounded by a ring of $\sim$10 kpc diameter. Although the galactic center has a FIR surface brightness which is about hundred times higher than that of the ring source, it contains only $\sim$10% of the total FIR luminosity.

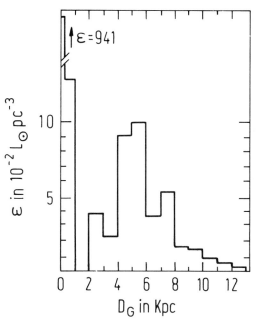

The neutral gas is heated through starlight (ionization of carbon; collision with dust particles), through ionization by cosmic rays and possibly (in dense clouds) through interaction with stellar winds. The major cooling lines in the FIR are: The O° 63 μm line for **neutral gas** with kinetic temperatures of several 1000 K; the C+ 157 μm line for the diffuse atomic gas; and both the C° 610 μm line and the rotational lines of the CO molecule for molecular clouds. The C° 610 μm was originally thought to be the major cooling

Fig. 2: Radial distribution of the diffuse galactic FIR emission as derived by Boissé et al. 1981.

line but observations indicate that it requires a quite specific temperature range for its excitation around $T_K \sim 20$ K (Phillips and Keene, 1982). Therefore, the integrated luminosity of this line is probably considerably lower than that of the CO rotational lines which appear to provide the major cooling of molecular clouds (Goldsmith and Langer, 1978). The O^{++} 88 μm line is an important cooling line of HII regions, which are heated by ionization by Lyc photons. All these lines have been detected in the ISM of our Galaxy either in compact radio HII regions or in molecular clouds with active star formation. But to date the diffuse extended galactic emission of only one line, the C^+ 157 μm line, has been detected in a brilliant experiment by Stacey et al. (1982). The expected galactic luminosity of this line together with the estimated galactic luminosities of the other three lines are given in Table 1.

III. RADIATION MECHANISMS

i) Atomic and ionic lines

I refer to the review by Harwit (this symposium).

ii) Synchrotron emission

Synchrotron radiation is the dominant component in the total radio emission of galaxies (see, e.g. Fig. 1). If relativistic electrons with an energy spectrum $N(E)dE \propto E^{-\gamma} dE$ are gyrating in a magnetic field they emit a radio radiation with the spectrum $S_\nu \propto \nu^\alpha$, and $\alpha = (1-\gamma)/2$. Observed spectral indices of extragalactic objects cluster around $\alpha \sim -0.75$ and ~ -0.25 to 0.0, respectively. Sources with steep spectral indices are extended and optically thin so that $\gamma \sim -2.5$ appears to be related to the mechanism, which accelerates the electrons. However, there are also some more compact sources, such as the Crab Nebula, with spectral indices ~ -0.25 and hence $\gamma \sim -1.5$. Then there are the sources with flat ($\alpha \sim 0.0$) or inverted ($\alpha > 0$) spectra, which are compact with a low-frequency cutoff above 100 MHz due to synchrotron self-absorption. At frequencies below the cutoff, the sources are opaque and, for a homogeneous source, should have a spectral index $\alpha = 2.5$. The spectrum attains its maximum value, S_m (in Jy) at the frequency ν_m (in GHz)

$$\nu_m \sim f(\gamma) B^{1/5} S_m^{2/5} \theta^{-4/5} (1+z)^{1/5} \tag{1a}$$

with $f(\gamma)$ a function, which varies only slowly with γ and which is ~ 8 for $\gamma = 2$ (see e.g. Kellermann and Pauliny-Toth, 1981). Here, B is the magnetic field in gauss, θ the angular size of the source in milli arc sec (= mas) and z the redshift of the source. There is an upper limit to the observable brightness temperature of synchrotron sources, $\sim 10^{12}$ K, (Kellermann and Pauliny-Toth, 1969) due to inverse Compton radiation yielding X-rays rather than radio emission.

Only compact extragalactic synchrotron sources have been observed in the submm/FIR region. At frequencies $\nu > \nu_m$ their flux density should decrease with $\alpha \sim -0.25$ to -0.75. Actually, in many cases the spectra of compact sources continue to be flat. This is interpreted as the superposition of a number of compact synchrotron source components with different cutoff frequencies. For typical source parameters $T_b \sim 5 \cdot 10^{11}$ K and with the relation $T_b \sim 1.7 \cdot 10^{12} \, S \, \theta^{-2} \, \nu^{-2}$, we have

$$\theta \sim 2 \, S^{1/2} \, \nu_m^{-1} \tag{1b}$$

which, for the observed values of $S \sim 1\text{-}10$ Jy in the frequency range, $0.3 \lesssim \nu_m/\text{GHz} \lesssim 300$, yields $\theta \sim 20\text{-}0.01$ mas. Thus the hypothesis of compact self-absorbed synchrotron sources can be experimentally tested, since the highest angular resolution attained with VLBI observations at $\lambda \sim 1$ cm is ~ 0.1 mas.

iii) Thermal emission from dust

Balance between the energy absorbed and emitted by a dust particle requires

$$\int_0^\infty 4\pi J_\lambda \sigma_\lambda \, d\lambda = 4\pi \int_0^\infty B_\lambda(T) \sigma_\lambda \, d\lambda \tag{2}$$

with $4\pi J_\lambda$ the mean intensity of the ISRF at wavelength λ, $B_\lambda(T)$ the Planck function and σ_λ the absorption cross section of the dust particle. For the interpretation of thermal dust emission one needs to know dust absorption cross sections from the UV through the optical to the FIR/submm region. It is shown in Paper I that the absorption cross section of dust observed as a function of wavelength from the Lyc region till ~ 1 mm can be described by the MRN (Mathis, Rumpl and Nordsiek, 1977) model, which consists of a mixture of graphite and silicate grains with size distribution (a = grain radius) $f(a) \propto a^{-3.5}$ and $a_{min} \sim 0.01$ μm, $a_{max} \sim 0.25$ μm. Absorption cross sections per H-atom are shown in Fig. 3. They can be related to extinction cross sections through $\sigma_{ext} = \sigma_a (1-\gamma)^{-1}$, with the albedo γ given in Paper II. Especially, in the visual ($\lambda = 0.55$ μm), $\gamma_V = 0.63$.

Throughout this symposium we have heard about observations related to mantles of interstellar molecules which are condensed on cores which may be identical with the MRN composite grain mixture. One should bear in mind, however, that most of these observations relate to very dense condensations of molecular clouds but that the characteristics of the grain cores (and hence the absorption cross sections given in Fig. 3) probably relate to the bulk of dust in the galactic ISM.

For $\lambda \gtrsim 40$ μm, σ_λ can be approximated by power laws $\sigma_\lambda \propto \lambda^{-m}$. With $\int \lambda^{-m} B_\lambda \, d\lambda \propto T^{4+m}$, and T the temperature of the dust grain we can rewrite the above relation in the form

$$T^{4+m} \propto \frac{\langle \sigma_{opt} \rangle}{\langle \sigma_{FIR} \rangle} \int 4\pi J_\lambda \, d\lambda$$

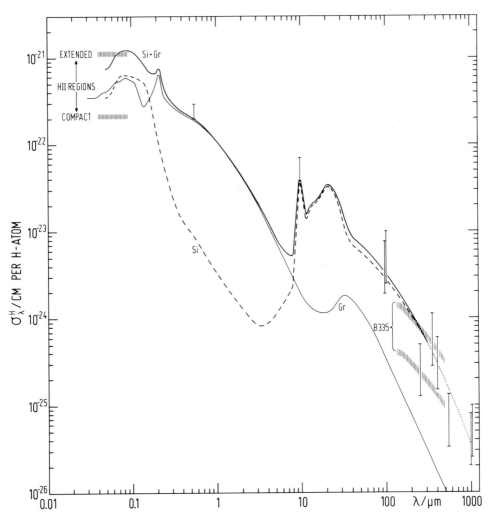

Fig. 3: Dust absorption cross sections from Paper I. The model fit consists of a mixture of graphite (gr) and silicate (si) grains with size distribution $f(a) \propto a^{-3.5}$, $a_{min} = 0.01$ μm and $a_{max} = 0.25$ μm.

which is well suited for a discussion of the characteristics of the thermal dust emission. With $\langle\sigma_{opt}\rangle \propto a^2$, $\langle\sigma_{FIR}\rangle \propto a^3$ we see that $T^{4+m} \propto a^{-1}$, i.e. in a given radiation field smaller particles get hotter. It is shown in Paper I that this does not lead to a serious distortion of the Planck emission spectrum of particles with the MRN size distribution. However, the very different wavelength dependence of σ_λ (Fig. 3) yields quite different temperatures for graphite and silicate dust grains exposed to the same ISRF, $T_{gr} \sim 2\, T_{si}$. It should be stressed that the Planck spectrum of dust particles exposed to the radiation of the (known) ISRF can be used to put constraints on the dust material (see, e.g. the fit to the emission spectrum of the globule B335 using the MRN dust model (Paper I)).

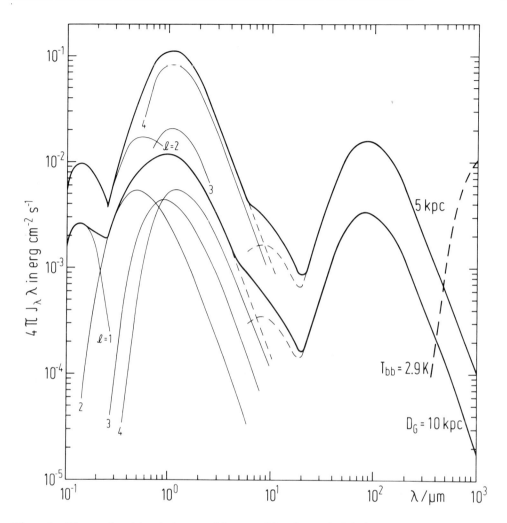

Fig. 4: The galactic interstellar radiation field (ISRF) at galactocentric distances $D_G = 5$ and 10 kpc, respectively. For wavelengths $\lesssim 8$ μm the ISRF is dominated by direct stellar radiation, $\ell = 1$ through 4 refers to four discernable stellar disk populations which contribute to the ISRF with different dependences on galactocentric distance. For $\lambda \gtrsim 8$ μm the ISRF is dominated by reradiation from dust grains.

Fig. 4 shows the galactic ISRF outside molecular clouds and ELD HII regions for two galactocentric distances $D_G = 10$ and 5 kpc (Paper II). Components 1, 2 and 3 are in essence identical with the ISRF of Werner and Salpeter (1969). Component 4 had to be added to explain the diffuse galactic NIR emission observed by different Japanese groups (see, e.g. the review by Okuda, 1980, and references therein). It is attributed to M type giants. The NIR part of the ISRF can penetrate deep inside molecular clouds and there is in fact the main source of heating for graphite grains, while the FIR part of the ISRF is the main heating source of silicate grains, which maintains dust temperatures of 5-7 K even at the center of clouds with $A_V \sim 200$ mag.

Evaluating eq. (2) for the absorption cross sections (Fig. 3) and the (interpolated) radiation field (Fig. 4) yields the emission spectrum of dust grains associated with the diffuse atomic hydrogen in the galactic disk and shown as component d2 in Fig. 1 (Paper II). It is found that the emission from this dust can account for only 40% of the total galactic FIR/submm luminosity and that additional dust with temperatures of ~ 40 K is needed to explain the observed spectrum of the diffuse galactic FIR emission.

The spectrum of this missing radiation is shown as component d3 in Fig. 1. Since we know both temperature and absorption cross section of the emitting dust of this component we can estimate the mean intensity of the radiation field required to heat the dust grains. Evaluating eq. (2) in this way it is found that the integral $\int 4\pi J_\lambda \sigma_\lambda d\lambda$ has to attain values which are about hundred times the corresponding values obtained for the ISRF in the vicinity of the sun. (This, of course, does not imply, that the intensity of the radiation which heats the dust is hundred times that of the solar vicinity, since UV radiation is absorbed by dust grains much more readily than radiation in the visual).

In fact, there is now general agreement that the radiation from OB stars is the most likely source of heating for the "warm dust" (component d3 in Fig. 1). Since O stars spend most of their lifetime in association with ELD HII regions it is probably dust located in these HII regions or just outside of them which is primarily responsible for the emission of component d3. (Contrary to some statements made during this symposium the bulk of the O star emission is probably absorbed by dust inside the HII region. Note that Lyman alpha photons, which are practically all absorbed by dust inside an HII region, already account for 20 ~ 25% of the total luminosity of the ionizing O stars).

In ELD HII regions the radiation field of the ionizing O star, including the Lyman continuum, adds to the ISRF. Dust absorption cross sections attain their maximum in the UV (see Fig. 3). The radiation from O and B stars is therefore very effectively absorbed by dust. This accounts for the considerably higher temperatures of dust in the vicinity of OB stars. Therefore it is probably correct to interpret extremely high IR luminosities of galaxies such as M82, or NGC253, whose emission peaks at wavelengths $\lambda < 100$ μm, to a recent burst of OB star formation. In this case we can connect the FIR spectrum with the free-free radio spectrum. The infrared excess is the total IR luminosity of an HII region expressed in units of the energy available in form of Lyman alpha photons, (IRE) = $L_{IR}/N'_{Lyc} h\nu_\alpha$, with $N'_{Lyc} \sim N_\alpha$ the number of Lyc photons absorbed per sec by the gas. In compact HII regions (IRE) ~ 7; for the ELD HII regions in our Galaxy one obtains (IRE) = 4.6. For this latter value one derives a relation between free-free flux density at the frequency ν and integrated dust emission, viz.

$$\nu^{0.1} S_\nu(ff) = 1.1 \ 10^{-16} \int S_\nu(IR) d\nu \qquad (3a)$$

with ν in GHz, dν in Hz and S_ν in Jy. Integration of the normalized spectrum d3 of Fig. 1 yields $\int (S_\nu/S_{max}) d\nu \sim 5 \ 10^{12}$ Hz and

$$\nu^{0.1} S_\nu(ff) = 5.5 \ 10^{-4} S_{max}(IR) \qquad (3b)$$

This is the relation between curves f-f and d3 in Fig. 1. The maximum flux density of the thermal emission of dust associated with ionized gas lies about a factor of 2000 above the free-free emission of the ionized gas.

In the central regions of galaxies, where the intensity of the general ISRF is high, dust associated with neutral gas can attain temperatures which are similar to that of dust in ELD HII regions. This probably accounts for the high IR luminosities of the giant molecular clouds located close to the galactic center, whose submm emission was mapped by Hildebrand et al. (1978).

In summary, a high IR luminosity of a galaxy is primarily an indication of its high dust content, especially if the ratio of IR luminosity of (visible) star light, $L_{IR}/L_* > 0.2$, the value derived for the Galaxy. In Table 2 we approximate this quantity by $L_{IR}/L_* \sim [S(IR)_{max}/\lambda_{max}]/[S_{1.25\mu m}/1.25]$. One must have additional evidence to tie an observed high IR luminosity to a correspondingly high formation rate of OB stars, such as for example dust temperatures well above 30 K or a high Lyc photon production rate. Even then the conversion of IR luminosities into star formation rates usually yields much less reliable results than the corresponding conversion of Lyc photon production rates. On the other hand, while the determination of Lyc photon production rates from radio free-free emission is straightforward the separation of thermal and nonthermal radio emission usually introduces large uncertainties, since the integrated nonthermal radiation always dominates the total radio spectrum of galaxies (see, e.g. Fig. 1).

IV. OBSERVATIONS

(i) Atomic and ionic lines $\lambda > 10 \ \mu m$

The first two rotational transitions of the CO molecule and the Ne^+ 12.8 μm line have been observed in external galaxies. These results will be summarized elsewhere in this symposium.

To date, extragalactic FIR spectroscopy has yielded positive results only for two galaxies. One is M82, an irregular type II galaxy, which is also a strong FIR continuum source ($L_{IR} \sim 3 \ E10 \ L_\odot$), which has been resolved ($\sim 30"$ or ~ 500 pc) in the direction of the galactic plane (Telesco and Harper, 1980). Ionized gas appears to be smoothly distributed over this region, while the distribution of atomic and molecular hydrogen is considerably more extended. The activity in the central region of M82 may be the result of a recent burst of star formation. A further peculiarity of this galaxy is its association with a dusty intergalactic cloud, which may be connected with the nearby large spiral galaxy M81.

Houck et al. (1980) have detected the S^{++} 18.7 µm line. They estimate a S^{++}/H^+ abundance which is about half of the solar system abundance S/H. They also observe a broad absorption dip near 19 µm, which together with a similar, but much deeper absorption feature near 10 µm (Gillett et al., 1975) is interpreted as absorption of the continuum emission from hot dust by cold silicate grains. This result and similar results in other galaxies (see e.g. Lebofsky and Rieke, 1979) as well as in compact HII regions in our Galaxy indicate, that the dust composition is similar.

Recently Watson et al. detected the fine structure lines O^o 63.2 µm and O^{++} 88.4 µm in the direction of the nucleus of M82 with a telescope beam of 44" FHPBW. The luminosities of both lines are about equal and $\sim 7\ 10^7\ L_\odot$, which is about 0.2% of the total continuum luminosity. From a comparison with the Ne^+ 12.8 µm line it is concluded, that $O^{++}/O \sim 0.2$, and from this ratio an effective temperature of the ionizing radiation of ~ 35000 K is inferred. The radial velocity, V_{LSR} = +220 km s^{-1}, and line width ~ 300 km s^{-1} FWHM, are in good agreement with the rotation curve derived from Ne^+ 12.8 µm line observations.

While the O^{++} 88.4 µm line appears to arise from the extended low-excitation HII region in the central region of M82 the origin of the O^o 63.2 µm line is less clear. Its central velocity, V_{LSR} = +130 km s^{-1}, and its FWHM, $\lesssim 180$ km, are quite different from that of the 88.4 µm line. Two alternative explanations are discussed. The line could be emitted by a galactic source or it could be emitted from the extragalactic cloud of neutral gas, with which M82 apparently is associated.

The second galaxy with a detected FIR line is NGC1068, a Seyfert 2 galaxy, whose FIR continuum emission will yet be discussed. In the proceedings of the 180th AAS meeting Forrest and Houck (1982) report the detection of strong O^{+++} 25.87 µm emission in this galaxy. They estimate an abundance of $O^{+++}/H^+ \sim 1.2$-$2.3 \cdot 10^{-3}$. Since the solar system abundance is $\sim 7\ 10^{-4}$ these observations imply a surprisingly high abundance of oxygen in the central region of NGC1068 with much of it in the triply ionized state. The authors also mention a search for the Ne^{++++} 24.28 µm line without giving details.

ii) Compact sources of synchrotron emission

Figure 5 shows as an example the spectrum of the BL Lac object 0212+73 (Biermann, 1982). It was detected in the 5 GHz survey with the MPIfR 100-m telescope (Kühr et al., 1981). It has a flat spectrum ($\alpha \sim 0$) throughout the cm- and mm-wavelength range and decreases with $\alpha \sim -1$ from the FIR through the optical and X-ray range. VLBI observations yield a size of the emitting region of 1.2x0.3 mas, corresponding to a linear

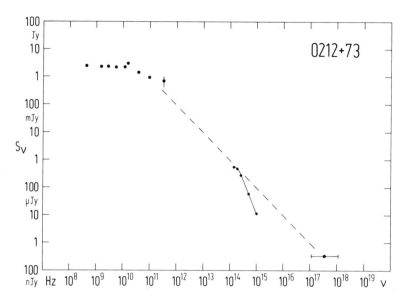

Fig. 5: Spectrum of the BL Lac object 0212+73. Biermann (1982) interprets this spectrum as synchrotron-self-Compton emission with either i) synchrotron emission accounting for the mm radiation and inverse Compton emission for the IR and optical radiation in first order and X-ray emission in second order; or ii) synchrotron emission accounting for the mm through optical radiation and the X-ray radiation being first order Compton emission.

size of ~13x3 pc (for H_o = 50, q_o = 0.05, z = 1), (Eckart et al. 1982). Even more interesting in the context of this symposium is the class of radioquiet quasars which have been previously observed at optical wavelengths only. Fig. 6 shows the spectrum of the QSO 0420-388 (Sherwood, Schultz and Kreysa, 1981). At radio wavelengths it was first detected at 1 mm and subsequently at longer wavelengths at flux levels ~0.1 Jy. At wavelengths <1 mm the spectrum again decreases with $\alpha \sim -1$. Most of these sources are too weak to be observable with present-day FIR telescope facilities but will certainly become a major area of research for the new generation of mm/

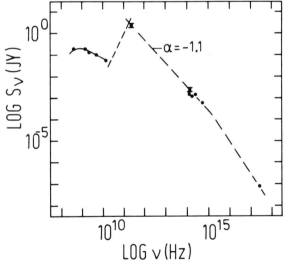

Fig. 6: Spectrum of the radioquiet QSO 0420-388 (Sherwood, Schultz and Kreysa, 1981).

submm telescopes such as the IRAM/MPIfR 30-m MRT or dedicated submm telescopes of $\sim$10 m aperture diameter which are now under construction at CalTech, and the MPIfR/U. of Arizona, respectively.

iii) Galaxies with active nuclear regions

The best investigated members of this class of objects are M82, NGC253 and NGC1068. For a detailed discussion of these and related objects I refer to Hildebrand et al. (1977), Telesco and Harper (1980), Rieke et al. (1980) and the review of Rieke and Lebofsky (1979). The FIR characteristics of the observed galaxies are: i) Emission comes from the nuclear region, comparable to the size of the FIR source located at the center of our Galaxy ($D_G \lesssim 500$ pc). The FIR continuum is thermal emission from dust. The heating appears to be provided mostly by distributed early type stars (in the case of the spiral and irregular galaxies) and, in addition, by a central source in the case of the Seyfert 2 galaxy NGC1068. iii) The luminosity is $\sim$30 times the luminosity of the central part of the Galaxy (M82, NGC253) and another order of magnitude higher for NGC1068. iv) Spectra of M82, NGC253 and NGC1068 between 1 mm and 3 μm from Telesco and Harper (1980) are shown in Fig. 7. The full curves are modified Planck functions $\nu^m B_\nu(T_d)$. Their radio spectra are similar to that of the Galaxy (Fig. 1) and always dominated by synchrotron emission. See also Fig. 1 of Rieke and Lebofsky (1979).

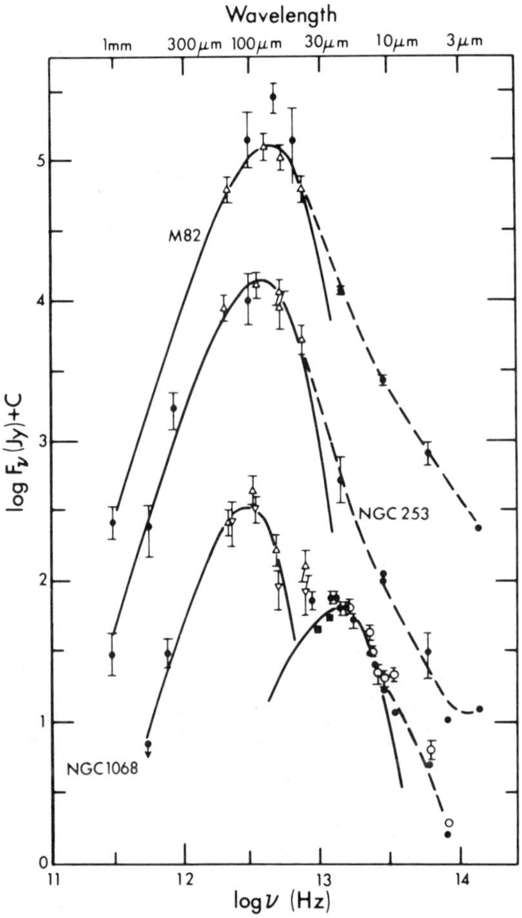

Fig. 7: FIR dust continuum of the galaxies M82, NGC253 and NGC1068. The constant is C=0 (NGC1068), =1 (NGC253) and =2 (M82). From Telesco and Harper, 1980.

The salient characteristics related to the FIR continuum emission from these and other galaxies included in this review are compiled in Table 2. Emission from hot dust becomes negligible at wavelengths $\lambda \lesssim 2$ μm indicating, that dust grains are destroyed at temperatures somewhat above 1500 K.

Temperatures of "warm dust" in NGC253 and M82 are similar to that of component d3 in our Galaxy (Fig. 1) indicating, that the heating of dust in the central regions is dominated by O stars. This led to the suggestion of bursts of star formation. It is of interest to note that Rieke et al. (1980) find that the IMF of the star burst in the central part of M82 has to be truncated at $\lesssim 3.5$ $m_\odot$ in order to match both the total IR luminosity, the Lyc photon production rate and the absolute magnitude at 2.2 μm. It appears thus that the star formation mechanism in star bursts in the central regions of spiral galaxies is similar to that in spiral arms in our Galaxy, where circumstantial evidence (Mezger and Smith, 1977) and the explanation of observed abundance gradients (Güsten, 1981) also suggest a truncation of the low-mass end of the IMF. It must be borne in mind, however, that practically all stellar emission from the central part of these galaxies is absorbed by dust (see Table 2) and that lower mass stars still contribute to the heating of the dust and hence to the observed IR luminosity. This makes a straight forward estimate of the star formation rate associated with the star bursts difficult.

Table 2: Characteristics of galaxies mentioned in this review

| Object | Type | $L_{IR}/L_\odot$ | | S_{max}/λ_{max} | S_{max}/S_{ff} | T_d/K | | | m | Ref. |
		centr.reg.	disk	$S_{1.25}/1.25$		med.warm	warm	hot	5)	
Galaxy	Sbc	1E9	1E10	2E-1	2E3	25(13)	40(30)		$\sim 2(\sim 1.5)$	1)
M51	Sc	6E9	2E10	4E-1	2E3	20			~ 2	2)
NGC253	Sc	3E10		29	6E3		$\lesssim 40$		$\gtrsim 1.5$	3)
M82	IrrII	3E10		14	8E2		$\lesssim 45$		$\gtrsim 1.5$	4)
NGC1068	Seyf.2	3E11		18			~ 30	~ 170	~ 2 ; ~ 1	

1) FIR and dust characteristics and $S_{max}/1.25$ from papers I and II. S_{max}/S_{3mm} this paper. T_d values for graphite grains and in brackets for silicate grains.

2) Smith, 1982. As in the case of the Galaxy $(S_{max}/\lambda_{max})/(S_{1.25}/1.25)$ $\sim L_{IR}/(L_{IR}+L_*)$ with L_* the observable stellar luminosity and L_{IR} the IR luminosity related to the total galaxy. S_{ff} for M51, NGC253 and M82 from Wielebinski (priv. comm.).

3) FIR and dust characteristics for NGC253, M82 and NGC1068 from Telesco and Harper (1980), $S_{1.25}$ from Glass (1973).

4) $S_{1.25}$ for M82 and NGC1068 from Rieke et al. (1980).

5) m refers to $\sigma_\lambda \propto \lambda^{-m}$.

The difference between the Planck spectra (full curves in Fig. 7) and the actually observed MIR spectra (dashed curves) in M82 and NGC253 indicates, that there must be some contribution from hot, circumstellar dust. But only in the Seyfert 2 galaxy NGC1068 are "warm" and "hot" dust clearly separated into two components, which have about equal luminosities. The 5-34 μm part of the spectrum can be explained by a spherical dust cloud of $A_V \simeq 10$ mag and radius ~ 300 pc, surrounding a central source of ~ 2 E11 $L_\odot$ which may be observed at wavelengths <2 μm. (Jones et al., 1977; Lebofsky et al., 1978). Emission from dust with temperatures $\lesssim 1500$ K appears to be the signature of Seyfert galaxies in the

IR continuum. However, the emission longward of 40 μm, which accounts for roughly one half of the total IR luminosity, cannot be connected to this central source but appears to be associated with an extended ($\lesssim 3$ kpc) dust cloud heated by star light. The (compared to M82 and NGC253) lower dust temperature in NGC1068 indicates, that this emission feature is due to a mixture of both dust heated by the general ISRF and OB stars, respectively. This situation hence is similar to the FIR emission of our Galaxy, where the components d2 and d3 are also observationally not separated.

iv) Normal galaxies

The analysis (Papers I, II) of the FIR/submm emission from our Galaxy predicts, that 1/10 of the luminosity is emitted from its central part (~ 0.5-1 kpc) and 9/10 from a disk of ~ 16 kpc diameter. About one half of the luminosity comes from dust grains associated with diffuse atomic hydrogen, which has a rather uniform space and surface density between galactocentric distances D_G of 4 and 16 kpc. Nevertheless, the surface brightness of the emission of this dust component would drop off rather rapidly with increasing D_G because of the corresponding decrease of the intensity of the ISRF. In Fig. 8 is shown, as a dashed curve, the expected relative variation with D_G of the surface brightness of this dust component, whose spectrum has been labelled d2 in Fig. 1. It has been assumed that $I_{FIR}(d2) \propto \int 4\pi J_\lambda (ISRF) d\lambda$, with $J_\lambda(ISRF, D_G)$ taken from paper II.

The other half of the galactic FIR emission comes from dust associated with ELD HII regions. The grain temperature of this dust component, labelled d3 in Fig. 1, is supposed to be constant and the surface brightness being proportional to the mean surface density of the production rate N_{Lyc} of Lyc photons of O stars ionizing the ELD HII regions. The expected relative variation of the surface brightness of this dust component d3 is shown in Fig. 8 as solid curve. It has been computed assuming $I_{FIR}(d3) \propto N_{Lyc}(D_G)$ with N_{Lyc} taken from Mezger (1978). The two curves are nearly identical, hence our Galaxy would be observed as a FIR/submm source of linear diameter ~ 12 kpc. It would not be possible to separate the emission from dust components d2 and d3, respectively on the basis of their different dependence on the galactocentric distance D_G. This predicted spatial variation of the galactic FIR emission is corroborated by observations, as shown by also plotting in Fig. 8 $\varepsilon/\varepsilon_{max}$, the volume emissivity derived by Boissé et al. from their FIR survey of the galactic plane (Fig. 2). It should be stressed, however, that the atomic gas in the galactic disk, with nearly constant surface density, extends out to galactocentric distances of ~ 16 kpc. Therefore curve d2 in Fig. 8, which shows the radial variation of the surface brightness of dust associated with atomic gas, reflects the variation of the intensity of the ISRF rather than that of the surface density of HI. Since up to 70% of all O stars in the Galaxy appear to be formed in spiral arms (Smith et al., 1978) our analysis further predicts, that spiral arms should be strong FIR features at wavelengths <100 μm. Until now, with two noticeable exceptions, all extragalactic sources observed in the FIR were objects with (at the angular resolu-

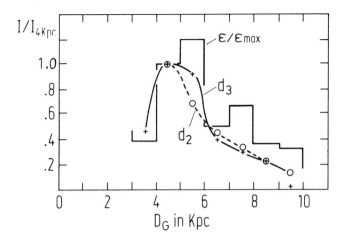

Fig. 8: The radial variation of the FIR surface brightness of the galactic disk, normalised to the surface brightness between D_G = 4-5 kpc. d2 and d3 refer to dust associated with diffuse atomic hydrogen and ionized hydrogen, respectively, $\varepsilon/\varepsilon_{max}$ is from Fig. 2. (For the dust components d2 and d3 see also Fig.1).

tion of $\sim$1 arc minute) unresolved luminous central sources of the type discussed in the previous sections III. ii) and iii). Since it is to be expected that the IRAS survey will turn out observations of low-luminosity, extended objects such as our Galaxy, I will briefly discuss MIR/FIR observations of two spiral galaxies, where emission features outside the central region have been resolved.

Telesco and Gatley (1981) mapped the nearly face-on seen barred spiral galaxy NGC1097 at λ10 µm. They resolved a nuclear source $\lesssim$560 pc, which contains $\sim$5% of the integrated 10 µm flux and a ring-like feature of $\sim$2 kpc diameter, which they identify with an optically observed, tightly wound spiral feature. They conclude that the extended MIR emission originates in extended vast complexes of HII regions which are associated with spiral features.

Smith (1982) made FIR maps and multicolor photometry of the face-on seen spiral galaxy M51 and its companion galaxy NGC5195. The spectrum of M51, longward of λ100 µm, corresponds to emission from dust with T $\sim$20(17) K, although the spectrum appears to deviate considerably from a Planck curve for λ<100 µm, revealing the presence of hot dust. The dust responsible for the emission of NGC5195 appears to be much warmer, 55-65 K. Emission in M51 occurs over an area 11x4 kpc. In many respects (see also Table 2) the FIR appearance of M51 ressembles that of our Galaxy. Intrinsic stellar luminosities appear to be the same, while the gas content of M51 as well as its FIR luminosity are about twice as high as in the Galaxy. As in our Galaxy, the HI disk stretches far beyond the FIR disk and has a central hole, which is due to the fact that most of the hydrogen exists in molecular form within the inner 4-5 kpc of M51. Unlike in our Galaxy, the distribution of H_2 (as outlined by the J=2$\rightarrow$1 transition of CO, Rickard et al., 1981) apparently does not show a ring structure, but such a structure may be smeared out by the beamwidth corresponding to a linear dimension of $\sim$3 kpc. This analogy suggests that the heating mechanism of dust in M51 is similar to that in the Galaxy and that the larger IR luminosity of M51 is primarily a result of its larger dust (and gas) content. With T_D = 17-20 K and the dust cross section and formulae given in Paper I the gas mass associated with the warm dust ranges from $(3-0.6)E10\ m_\odot$.

V. CONCLUSIONS

1) To date most observations of the dust continuum in external galaxies relate to pathological cases of galaxies with highly active nuclear regions. However, IRAS should detect most of the nearby regular spiral galaxies, whose FIR emission should be similar to that of our Galaxy or M51.

2) All FIR spectroscopy of external galaxies todate has been made from NASA's Kuiper Airborne Observatory. European IR astronomy will in my opinion be able to compete with IR astronomy carried out in the US only if European observers will have access to a similar astroplane to test new equipment, to train young astronomers, and to work continuously on the frontline of FIR astronomy.

3) The fact, for example, that the C^+ 157 μm line, expected to be the strongest line in our Galaxy, has not yet been detected in external galaxies is probably the result of the much lower detector sensitivity at wavelengths longward of 120 μm. Several groups in Europe and in the US are successfully operating heterodyne radiometers at wavelengths $\gtrsim 350$ μm, and there is no basic limitation for extending this technique to shorter wavelengths, with the potential of increasing detector sensitivities by orders of magnitude. The development of lightweight, reliable local oscillators surpasses, however, the financial and manpower capabilities of most European groups, and ESA certainly should take a strong lead in supporting this type of instrumental development.

4) Observations of compact, self-absorbed synchrotron sources will become the domain of the new generation of mm/submm telescopes mentioned above. Of special interest for space submm telescopes may be the carbon fiber reinforced epoxy material used in the construction of the MPIfR/U. of Arizona 10-m telescope, where test panels of size 1x1.2 m with an rms accuracy of ~ 4 μm have already been built.

ACKNOWLEDGEMENT

This review was written while I was a visiting scientist at the Steward Observatory of the U. of Arizona. I profited from discussions with Drs. P. Biermann, R. Genzel, M. Harwit, M. Lebofsky, J. Mathis, I. Pauliny-Toth, T. Phillips, G. Rieke, and J. Schmid-Burgk.

REFERENCES

Biermann, P. 1982, Proc. Torino Workshop on Astrophys. Jets
Boissé, P., Gispert, R., Coron, N., Wijnbergen, J., Serra, G., Ryter, C., Puget, J.L. 1981, Astron. Astrophys. <u>94</u>, 265
Eckart, A., Hill, P., Johnston, K.J., Pauliny-Toth, I.I.K., Spencer, J.H., Witzel, A. 1982, Astron. Astrophys. <u>108</u>, 157

Forrest, W.J., Houck, J.R. 1982, BAAS 14, 603
Gillett, F.C., Kleinmann, D.E., Wright, E.L., Capps, R. 1975, Astrophys. J. (Lett.) 198, L65
Glass, I.S. 1973, Mon. Not. Roy. Astr. Soc. 164, 155
Goldsmith, P.F., Langer, W.D. 1978, Astrophys. J. 222, 881
Güsten, R. 1981, Unpubl. thesis, U. of Bonn (and Güsten and Mezger, in prep.)
Hildebrand, R.H., Whitcomb, S.E., Wiston, R., Stiening, R.F., Harper, D.A., Moseley, S.H. 1978, Astrophys. J. 216, 698
Houck, J.R., Forrest, W.J., McCarthy, J.F. 1980, Astrophys. J. 242, L65
Jones, T.W., Leung, C.M., Gould, R.J., Stein, W.A. 1977, Astrophys. J. 212, 52
Kellermann, K.I., Pauliny-Toth, I.I.K. 1969, Astrophys. J. (Lett) 155, L71
Kellermann, K.I., Pauliny-Toth, I.I.K. 1981, Ann. Rev. Astron. Astrophys. 19, 373
Kreysa, E., Pauliny-Toth, I.I.K., Schultz, G.V., Sherwood, W.A., Witzel, A. 1980, Astrophys. J. 240, L17
Kühr, H., Pauliny-Toth, I.I.K., Witzel, A., Schmidt, J. 1981, Astron. J. 86, 854
Lebofsky, M.J., Rieke, G.H. 1979, Astrophys. J. 229, 111
Lebofsky, M.J., Rieke, G.H., Kemp, J.C. 1978, Astrophys. J. 222, 95
Mathis, J.S., Rumpl, W., Nordsiek, K.H. 1977, Astrophys. J. 217, 425 (MRN)
(MRN) 1978, ibid 219, L101
Mathis, J.S., Mezger, P.G., Panagia, N. 1982 (subm. to Astron. Astrophys.; Paper II)
Mezger, P.G., Smith, L.F. 1977, Proc. IAU Symp. 75 (T. de Jong, A. Maeder, eds., Reidel) p. 133
Mezger, P.G. 1978, Astron. Astrophys. 70, 565
Mezger, P.G., Mathis, J.S., Panagia, N. 1982, Astron. Astrophys. 105, 372 (Paper I)
Okuda, H. 1980, Proc. IAU Symp. 96 (C.G. Wynn-Williams and D.P. Cruikshank, eds., Reidel) p. 247
Phillips, T.G. and Keene, J. 1982, Proc. ESA Workshop "The Scientific Importance of Submm Observations", ESA SP-189, p. 45
Rickard, L.J., Palmer, P. 1981, Astron. Astrophys. 102, L13
Rieke, G.H., Lebofsky, M.J. 1979, Ann. Rev. Astron. Astrophys. 17, 477
Rieke, G.H. 1978, Astrophys. J. 226, 550
Rieke, G.H., Lebofsky, M.J., Thompson, R.I., Low, F.J., Tokunaga, A.T. 1980, Astrophys. J. 238, 24
Sherwood, W., Schultz, G.V., Kreysa, E. 1981, Nature 291, 301
Smith, L.F., Biermann, P., Mezger, P.G. 1978, Astron. Astrophys. 66, 65
Smith, J. 1982, Astrophys. J. 261, 463
Stacey, G., Smyers, S., Kurtz, N.T., Harwit, M. 1982 (preprint)
Telesco, C.M., Gatley, I. 1981, Astrophys. J. 247, L11.
Telesco, C.M., Harper, D.A. 1980, Astrophys. J. 235, 392
Watson, D., Dinerstein, H.L., Genzel, R., Lester, D.F., Storey, J.W.V., Townes, C.H., Werner, M.W. 1982 (preprint)
Werner, M.W., Salpeter, E.E. 1969, Mon. Not. Roy. Astr. Soc. 145, 249

MOLECULAR GAS DISTRIBUTION IN SPIRAL GALAXIES

N.Z. Scoville

Five College Radio Astronomy Observatory, U. of Massachusetts

I. INTRODUCTION

One of the major factors influencing both the morphology and evolution of spiral galaxies is the abundance and distribution of dense interstellar matter. Young stars formed from the gas clouds are responsible for the appearance and perhaps even the maintenance of the spiral pattern. Since these stars also generate a disproportionate share of the galactic luminosity, the long term evolution of galaxies is directly related to the exhausion of this matter (i.e., its absorption into low mass stars). The ability to probe all three gas components (molecular, atomic, and ionic) in addition to the luminosity from the young stars is a unique capability of the far infrared.

The importance of the molecular clouds has become evident only in the last few years with the advent of millimeter line observations--first in our own galaxy and most recently in nearby external galaxies. Studies in the Milky Way have shown both that the total abundance of H_2 in the inner region far exceeds that of HI and that most massive star formation (indicated by HII regions) is associated with molecular rather than atomic clouds. In the external galaxies, especially the higher luminosity mid to late type spirals, similar conclusions appear to hold; yet curiously the detailed distribution of clouds in these galaxies is often remarkably different from that in the Milky Way.

Though our present studies of molecular gas in external galaxies are still in a rather formative stage, I will attempt to review some of the initial results with emphasis toward the relevance of future far-infrared observations. (Since the best resolution now available for ground-based studies of the $J = 1 \rightarrow 0$ and $2 \rightarrow 1$ CO lines is 30-60", it is clear that a major advantage in the far-infrared will be higher angular resolution. To place the external galaxy data in perspective, I will first review the most recent results for our own galaxy reflecting on both the galactic distribution and the nature of the molecular clouds themselves.

II. GIANT MOLECULAR CLOUDS IN THE MILKY WAY

In view of its ubiquity the fundamental CO rotational transition (J = 1 → 0) at 115 GHz has become the primary probe of the galactic H_2 distribution. Even in clouds of low density ($n_{H_2} > 100$ cm^{-3}) and low kinetic temperature this transition is excited into emission significantly above the 2.7 K cosmic background.

The first CO surveys (Scoville and Solomon 1975, Gordon and Burton 1976) were restricted to the galactic equator (b = 0°) in the first quadrant of the galaxy. Later work has provided extensive latitude coverage (Δb ≥ 1°; Cohen and Thaddeus 1977; Sanders, Solomon, and Scoville 1983), ^{13}CO data (Solomon, Scoville, and Sanders 1979; Liszt, Xiang, and Burton, 1981), coverage in the outer galaxy (Cohen et al. 1981; Kutner and Mead 1981; Solomon, Stark, and Sanders 1983), and in the fourth galactic quadrant (Robinson, McCutcheon, and Whiteoak 1982).

a) Large-Scale Distribution in R

A common result of all the CO surveys in the inner galaxy in both northern and southern hemispheres is the bimodal form of the radial distribution--a central peak at R < 1 kpc and a secondary maximum in a ring at R = 4 to 8 kpc. In the breach between the two maxima the CO emission is < 20% of that seen in the ring peak at R = 5.5 kpc.

Figure 1 shows the surface density of molecular hydrogen derived from the most recent CO data in the northern hemisphere at $|b| < 2°$ and in the southern hemisphere at b = 0° (Sanders, Solomon, and Scoville 1983).[1] The surface densities of HI and the stellar disk (Caldwell and Ostriker 1981) are also shown. For all three species, the density was "integrated" perpendicular to the galactic disk; this figure therefore provides the best representation of the Milky Way for comparison with external galaxies.

In the region of our galaxy inside the solar circle at R = 10 kpc, Figure 1 indicates that the total integrated H_2 mass ($M_{H_2} \simeq 2 \times 10^9 M_\odot$) is about four times that of HI. However since the HI surface density is much flatter with radius, the HI and H_2 masses for the galaxy out to 20 kpc are very similar. Adding together both the HI and H_2 contributions, Sanders, Solomon, and Scoville (1983) find that the total ISM surface density is ~ 12% of the stellar disk surface density. Moreover, the total ISM curve follows rather closely the mass profile (solid curve) outside R = 5 kpc. The radial distribution of giant radio HII regions (not shown here) also shows a strong peak at R = 4-8 kpc similar to the H_2 distribution. This provides strong evidence that most of the O-B star formation occurs in molecular clouds. In fact all population zero tracers (CO, HII, γ-rays, and pulsars) except HI exhibit the 5 kpc peak.

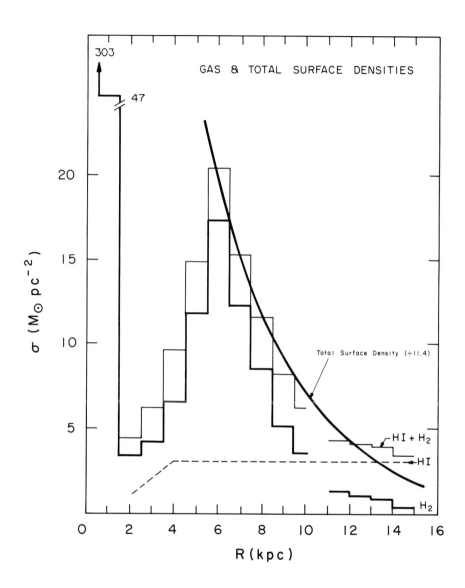

FIGURE 1: The surface density variation of H_2 in the Milky Way derived from the CO survey of Sanders, Solomon, and Scoville (1983) is compared with the HI surface density and stellar disk surface density from Caldwell and Ostriker (1981). The H_2 shows a strong peak in the galactic nucleus and a ring at R = 4 to 8 kpc neither of which are seen in HI. Note that the total ISM surface density (HI + H_2) follows the stellar disk at R ≃ 5 to 15 kpc.

b) Properties of the Molecular Clouds

The fact that most of the galactic CO emission originates from discrete clouds rather than diffuse, widespread structures can be easily seen from surveys with closely spaced observations using a small beam (eg. Solomon, Scoville, and Sanders 1979; Lizst, Xiang, and Burton 1981; Sanders, Scoville, and Solomon 1983). Most extensive is the last work which sampled 315 discrete molecular clouds with sizes in the range 10 to 80 pc. Virial analysis of the correlation between the measured CO linewidth and the apparent size suggested a mean density $n_{H_2} \simeq 200$ cm^{-3}, not strongly dependent on the cloud size.

Most remarkable are the very large sizes and masses of the clouds giving rise to the bulk of the CO emission. Figure 2 shows the mass function derived by Sanders, Scoville, and Solomon (1983); that is the fraction of the total H_2 mass contained in clouds in a specified logarithmic mass interval. For clouds in the size range sampled, they find a mean cloud mass $\sim 4 \times 10^5$ $M_\odot$ and over half the total mass is in clouds larger than 50 pc. Hence the name Giant Molecular Clouds! The total number of clouds with diameters 20 to 80 pc is estimated to be $\sim 5,000$ in the galaxy.

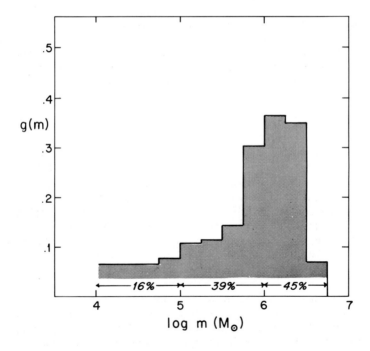

FIGURE 2: The fraction of total H_2 mass in clouds per unit logarithmic mass interval is shown based upon the sampling in the inner galaxy by Sanders, Scoville, and Solomon (1983). Most of the mass is contained in clouds of $M = 10^5 - 2 \times 10^6$ $M_\odot$, which are termed Giant Molecular Clouds.

III. MOLECULAR GAS IN EXTERNAL GALAXIES

Many of the earliest extragalactic CO studies concentrated on those galaxies with active nuclei or bright infrared emission (eg. Rickard et al. 1975, 1977). Galaxies such as M82, NGC 253, and NGC 1068 all showed strong CO emission in their nuclei and it was plausible to attribute at least some of the activity to a burst of star formation in the abundant molecular clouds (cf. Rieke et al. 1980, and Telesco and Harper 1980).

Given the peculiar radial distribution found for the molecular emission in the Milky Way--that is the molecular cloud ring--a central question in more recent extragalactic studies has been the extent to which our disk distribution might be typical of "normal" spiral galaxies. If it is not typical, then correlation of the ring morphology with other galactic properties might further our understanding of its origin--be it dynamic or evolutionary. A second question regarding the normal galaxies is the prevalence of H_2 over HI. Is H_2 > HI in the inner disk of all spiral galaxies or only in specific types?

a) CO in the Disks of External Galaxies

An extensive program to address these questions has been conducted by Judy Young and myself using the 14-m telescope at the Five College Radio Astronomy Observatory (University of Massachusetts). The resolution of this telescope is 50" in the $J = 1 \rightarrow 0$ CO line and in the last two years the sensitivity has reached the level where the disk emission in external galaxies can be mapped with $1/2$-2 hour integrations. A total of 40 out of 60 galaxies observed, have been detected and 20 have now been mapped at over 7 points.

Complete radial distributions out to ~ 15 kpc have been obtained for the late spirals NGC 6946, IC342, and M51 by Young and Scoville (1982a) and Scoville and Young (1983) and for M101 by Solomon et al. (1983). In Figure 3 the CO distributions are shown for comparison with the Milky Way. It is immediately evident the none of these galaxies show a ring like that in the Milky Way. Instead they exhibit a relatively monatonic falloff with radius similar to the behavior in the Milky Way at R > 5 kpc. This suggests that the significant feature of the Milky Way distribution is not the ring peak at 5.5 kpc but rather the breach or hole at 1-4 kpc. That is we should seek an origin for the paucity of gas in this zone rather than the excess of gas at 5.5 kpc.

Examining the types of galaxies shown in Figure 3, we believe it significant that all four external galaxies not showing the breach are probably of later type than the Milky Way. Thus the ring morphology could be characteristic of earlier type galaxies. To test this hypothesis major axis strips were recently observed in the two Sb galaxies NGC 7331 and NGC 2841 (Young and Scoville 1982b). The integrated intensities shown in Figure 4 clearly show molecular peaks at R ~ 5 kpc similar to our galaxy. In both NGC 7331 and NGC 2841 the peak occurs just outside the central galactic bulge, suggesting that the deficiency of gas inside R = 5 kpc is related to the extensive bulge. Perhaps the gas which "was" originally here has been exhausted early during formation of the bulge population. Neither of these galaxies shows a central peak like the Milky Way but this may be an effect of the limited spatial resolution.

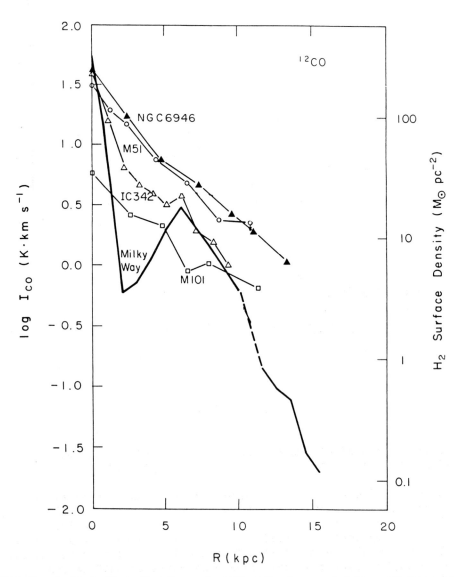

FIGURE 3: Radial distributions for CO emission and inferred H_2 surface density are shown for the Milky Way (Sanders, Solomon, and Scoville 1983), IC342 and NGC 6946 (Young and Scoville 1982a), M51 (Scoville and Young 1983), and M101 (Solomon et al. 1983). Note that only the Milk Way shows a gap at R ~ 3 kpc; all the others show a fairly monatonic falloff from the nucleus to R = 10-15 kpc.

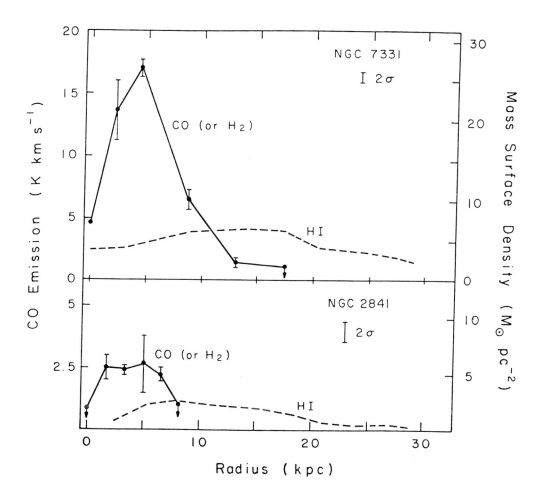

FIGURE 4: CO distributions observed along the major axis of two Sb galaxies NGC 7331 and NGC 2841 (Young and Scoville 1982b).

More significant than the comparisons with our own galaxy is the discovery that the CO in all four late spirals (IC342, NGC 6946, M51, and M101) closely follows the optical light profiles. In fact in all cases the CO dependence on radius is approximated well by an exponential falloff (with scale length R_0 = 4-5 kpc) which is similar to the exponential disk light distributions. In Figure 5 the CO surface density variations are shown for M51 and in Figure 6 the mean radial distribution in M51 is compared with other tracers such as HI, far-infrared and radio continuum emission, Hα, and the light profiles deduced by Schweizer (1976) for the disk and "arm" components of the stars. Since the far-infrared, the Hα, and to some extent the disk light are all probably indicators of the star formation rate, we interpret their similarity to the CO as an indication that the rate of star formation is directly proportional to the abundance of H_2 as measured by the CO.

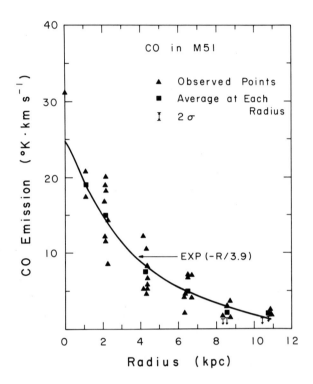

FIGURE 5: The CO intensities are plotted as a function radius in M51 (Scoville and Young 1983).

As found in many other galaxies with a high abundance of H_2, the HI distribution is relatively flat with radius and even shows a dip in the center where the molecules peak. Thus HI commonly shows an anticorrelation with most other population zero tracers in galaxies where $H_2 \gg HI$. An interesting question is what is the importance or the role of HI in such galaxies (cf. Seiden 1983). One might be inclined to dismiss the HI altogether if it were not for the fact that the "arm" light component in M51 shows a better correlation with the HI than H_2 (see Figure 6).

The discovery that the radial dependence of disk light and CO are similar in the four Sc-Scd galaxies suggests that one might look for such a correlation from one galaxy to another. For a sample of 8 Sc galaxies we have in fact found a nearly constant ratio of CO emission to blue luminosity in the central $R < 2.5$ kpc. The ratio remains approximately constant over a dynamic range of 100 in the luminosity if one considers only a single morphological class but there is some indication that the ratio may vary between morphological classes. The data for the Sc galaxies is given in Table 1. Here one may also see that the HI/L_B ratio varies by nearly a factor of 100 for the same galaxies with constant H_2/L_B.

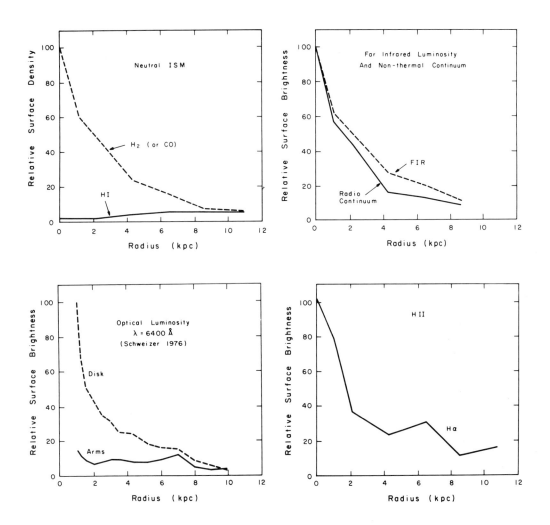

FIGURE 6: The mean radial distributions in M51 for H_2 (Scoville and Young 1983) are compared with the far-infrared (Smith 1982), HI (Weliachew and Gottesman 1973), non-thermal radio emission (Mathewson et al. 1972), and optical light (Schweizer 1976).

TABLE 1 -- Masses of H_2 and HI in Sc Galaxies

| Galaxy | Central R < 2.5 kpc[a] | | | $\frac{M(H_2)}{M(HI)}$ | $\frac{M(H_2)}{L_B}$ | $\frac{M(HI)}{L_B}$ |
	L_B (L_o)	$M(H_2)$[a] (M_o)	$M(HI)$[b] (M_o)		(M_o/L_o)	(M_o/L_o)
N5236	1.7×10^{10}	1.9×10^9	9.5×10^7	20	0.11	5.6×10^{-3}
N6946	1.1×10^{10}	2.4×10^9	1.2×10^8	20	0.22	1.1×10^{-2}
N5194	1.0×10^{10}	2.0×10^9	6.3×10^7	32	0.20	6.3×10^{-3}
N4321	7.4×10^9	1.5×10^9			0.20	
IC342	4.5×10^9	1.2×10^9	6.8×10^7	18	0.27	1.5×10^{-2}
N5457	2.8×10^9	5.6×10^8	9.1×10^7	6.2	0.20	3.2×10^{-2}
N2403	1.3×10^9	$<6.0 \times 10^7$	1.3×10^8	<0.46	<0.046	1.0×10^{-1}
N598	3.9×10^8	$<5.5 \times 10^7$	1.3×10^8	<0.42	<0.14	3.3×10^{-1}

[a]H_2 masses computed from from eqs. [3] and [4] in Young and Scoville (1982a).
[b]HI masses in central R < 2.5 kpc from aperture synethsis observations for seven galaxies (Gordon 1971; Rogstad and Shostak 1972; Weliachew and Gottesman 1973; Rogstad, Lockhart, and Wright 1974).

From the blue luminosities given in Table 1 it is possible to estimate the rate at which the ISM must be cycled into stars in order to maintain the observed energy generation rate. Under the very restrictive assumptions that all the blue luminosity is generated by O, B, and A stars and only those stars are formed, then the constant ratio $M_{H_2}/L_B \simeq 0.17\ M_o/L_o$ found for the Sc galaxies implies a cycle time for the interstellar gas into stars of $\sim 3 \times 10^9$ years! The fact that this time is already uncomfortably short compared to the galactic lifetime (without any allowance for low mass star formation) suggests that if anything we may be underestimating the H_2 masses derived from the CO line emission.

b) Galactic Nuclei

In view of the suggestions that the activity in galactic nuclei is linked to a burst of massive star formation, it is of great interest to determine both the abundance and the distribution of molecular clouds in these regions. If the star formation rate is proportional to the mass of star forming matter and the energy generation is proportional to the mass of young stars, then the nuclear activity should correlate with the mass of H_2.

To investigate whether such a correlation occurs, Rickard et al. (1977) and Rickard and Harvey (1983, cf. Morris and Rickard 1982) have compared the nuclear 10 μm, 100 μm and radio continuum fluxes (all of which are signposts of nuclear activity) with the observed CO emission in about a dozen galactic nuclei. For this selected sample, they do indeed find the continuum fluxes varying roughly as the square of the CO line emission.

An obvious deficiency with such gross correlation analysis is the fact that the angular resolution used in the CO measurements is 60" but the nuclear activity is known to occur on scales < 10" in many cases. In our own galactic center the far-infrared (Hoffmann, Frederick, and Emery 1971) is well correlated with a complex of molecular clouds at R ~ 300 pc (Scoville, Solomon, and Jefferts 1974), but our galactic nucleus would hardly classify as a very active one. In an attempt to overcome the inadequate telescope resolution available for the 2.6 mm line, we have recently undertaken a study of the CO emission in the Seyfert galaxy NGC 1068 where the detailed spatial distribution on the scale of ~ 5" was derived by "matching" the CO line profiles to the optical velocity field (Scoville, Young, and Lucy 1983). In this case our analysis indicates that most of the CO emission probably arises in the bright inner optical arms at radius ~ 12" or $R \simeq 1-1.5$ kpc. This is also similar to the spatial extent of the far-infrared emission with luminosity $L \simeq 1 \times 10^{11} = L_o$ (Telesco and Harper 1980). The total mass of H_2 here is estimated to be ~ 2×10^9 M_o and the time scale for cycling into O, B, and A stars can be no longer than 3×10^8 years if the far-infrared energy comes from young stars. There is no evidence of a large quantity of CO associated with the central non-thermal source; however, it is also not clear that our analysis is sensitive to this component in as much as the velocity of neutral gas is uncertain for the very center.

IV. CONCLUSION--THE NEED FOR HIGHER RESOLUTION

Briefly summarizing of the results from CO line studies in our own galaxy and several external galaxies we can safely assert that:

a) <u>Radial Distribution</u>--In the Milky Way and some Sb galaxies the molecular gas distribution exhibits a strong ring peak in the disk at $R \simeq 5$ kpc; in several later type galaxies only a smooth falloff (approximately exponential with R) is seen. In most cases studied so far the optical light from the disk is linearly correlated with the CO emission, suggesting that the star formation rate varies as the first power of the mean H_2 density.

b) <u>Cloud Properties</u>--From sampling the CO emission in the Milky Way disk it is evident that most of the gas is contained in Giant Molecular Clouds (GMC) with mean mass ~ 4×10^5 M_o and size ~ 30 pc. A spectrum of cloud sizes is seen with many more small clouds but these contain a small fraction of the gas.

c) <u>Galactic Nuclei</u>--At least half the galaxies observed (including our own), show a strong peak in the CO emission at the central R < 1 kpc.

The origin of this gas (even whether it is primeval or not) is not clear, but in several instances the high abundance of H_2 may be linked to the nuclear activity.

On the other hand there are also several very important (and embarrassingly basic) questions to which we have no clear answer:

d) Spiral Structure--What is the relationship of the molecular clouds to the spiral arms? Do the clouds exist only in arms or in interarm regions also (cf. Scoville and Hersh 1979 and Kwan 1979). If the later is the case, why does massive OB star formation occur preferentially along the arms?

e) H_2/HI Ratio--Why do some galaxies or some regions in galaxies have a greater efficiency for formation of molecular clouds out of their gas supply?

f) Nuclear Gas--Why do some galactic nuclei show an enormous mass of H_2 ($M_{H_2} > 2 \times 10^9 \, M_\odot$ in NGC 1068) and others with apparently similar morphological class show less H_2 by a factor of ~ 100. Is this gas residual in the nucleus or has it been recently accreted? How does the molecular gas spatially correlate with nuclear activity?

g) Molecular Cloud Characteristics in External Galaxies--At present we simply assume that the molecules in the external galaxies are in GMC's like those in our galaxy.

The major obstacle to answering all these questions is the insufficient angular resolution of existing telescopes. Resolution of $< 10"$ is needed to resolve the arms (d) and the nuclear structure (f). To sample individual clouds or even complexes requires resolution $< 5"$ except in M31. With the advent of millimeter interferometer systems and large submillimeter/IR telescopes we can look forward to great progress on these very fundamental questions.

This research is partially supported by a grant from the National Science Foundation AST 82-12252. This is contribution #543 of the Five College Radio Astronomy Observatory. It is a pleasure to acknowledge Sally Rule for preparation of this manuscript.

NOTES

[1] To convert the CO emission integrals to equivalent H_2 density a constant of proportionality was adopted based upon analysis for individual clouds of ^{13}CO, extinction, and the Virial Theorem (i.e. $N_{H_2} = 3.6 \times 10^{20} \int T_{CO} dv \, cm^{-2}$).

REFERENCES

Caldwell, J.A.R. and Ostriker, J.P. 1981, Ap.J., 251, 61.
Cohen, R.S., Cong, H.I., Dame, T.M., and Thaddeus, P. 1980, Ap.J.(Letters), 239, L53.
Cohen, R.S. and Thaddeus, P. 1977, Ap.J.(Letters), 217, L155.
Gordon, K.J. 1971, Ap.J., 169, 235.
Gordon, M.A. and Burton, W.B. 1976, Ap.J., 208, 346.
Hoffmann, W.F., Frederick, C.L., and Emery, R.J. 1971, Ap.J.(Letters), 164, L23.
Kutner, M. and Meade, K. 1981, Ap.J.(Letters), 249, L15.
Kwan, J. 1979, Ap.J., 229, 567.
Liszt, H.S. and Burton, W.B. 1981, Ap.J., 243, 778.
Liszt, H.S., Xiang, D., and Burton, W.B. 1981, Ap.J., 249, 532.
Mathewson, D.S., Vander Kruit, P.C., Brouw, W.N. 1972, Astron.Astro., 17, 468.
Morris, M. and Rickard, L.J. 1982, Ann. Rev. of Astronomy and Astrophysics, 20, 517.
Rickard, L.J. and Harvey, P. 1983 (in preparation).
Rickard, L.J., Palmer, P., Morris, M., Turner, B.E., and Zuckerman, B.Z. 1977, Ap.J., 213, 673.
Rickard, L.J., Palmer, P., Morris, M., Zuckerman, B., and Turner, B.E. 1975, Ap.J.(Letters), 199, L75.
Rieke, G.H., Lebofsky, M.J., Thompson, R.I., Low, F.J., and Tokunaga, A.T. 1980, Ap.J., 238, 24.
Robinson, B.M., McCutcheon, W.H., and Whiteoak, J.B. 1982, Int.J.Infrared Millimeter Waves, 3, 63.
Rogstad, D.H., Lockhart, I.A., and Wright, M.C.H. 1974, Ap.J., 193, 309.
Rogstad, D.H. and Shostak, G.S. 1972, Ap.J., 176, 315 (RS).
Sanders, D.B., Scoville, N.Z., and Solomon, P.M. 1983, Ap.J. (submitted).
Sanders, D.B., Solomon, P.M., and Scoville, N.Z. 1983, Ap.J. (submitted).
Schweizer, F. 1976, Ap.J.Suppl., 31, 313.
Scoville, N.Z. and Hersh, K. 1979, Ap.J., 229, 578.
Scoville, N.Z. and Solomon, P.M. 1975, Ap.J.(Letters), 199, L105.
Scoville, N.Z., Solomon, P.M., and Jefferts, K.B. 1974, Ap.J.(Letters), 187, L63.
Scoville, N.Z. and Young, J.S. 1983, Ap.J., 265, 148.
Scoville, N.Z., Young, J., and Lucy, L.B. 1983, Ap.J. (June 15, 1983).
Seiden, P. 1983, Ap.J. (in press).
Smith, J. 1982, Ap.J., 261, 463.
Solomon, P.M., Barrett, J.M., Sanders, D.B., and deZafra, R. 1983, Ap.J.(Letters), 266 (in press).
Solomon, P.M., Scoville, N.Z., and Sanders, D.B. 1979, Ap.J.(Letters), 232, L89.
Solomon, P.M., Stark, A.A., and Sanders, D.B. 1983, Ap.J.(Letters) (in press).
Telesco, C.M. and Harper, D.A. 1980, Ap.J., 235, 392.
Weliachew, L. and Gottesman, S.T. 1973, Astron.Astro., 24, 59.
Young, J.S. and Scoville, N.Z. 1982a, Ap.J., 258, 467.
Young, J.S. and Scoville, N.Z. 1982b, Ap.J.(Letters), 260, L41.

CONTRIBUTED PAPERS

The following papers were presented as posters at the symposium. The full text of these may be found in "Galactic and Extragalactic Spectroscopy", 1982, ESA SP-192, Eds. M.F. Kessler, J.P. Phillips and T.D. Guyenne.

1. RECOMBINATION & FINE STRUCTURE LINES

LINE EXCESS OBJECTS: A NEW INFRARED SPECTROSCOPIC CLASS
        R.I. Thompson.

INFRARED LINE AND RADIO CONTINUUM EMISSION OF VERY COMPACT MOLECULAR CLOUD SOURCES
        M. Felli & M. Simon.

INFRARED SPECTROSCOPY OF P-CYGNI'S STELLAR WIND
        M. Felli, E. Oliva, A Natta, R. Stanga & R. Beckwith.

CVF SPECTRALPHOTOMETRY OF THE HII REGION S 106
        C. Eiroa & H. Hefele.

ON THE NATURE OF THE EXCITING SOURCE IN G333.6-0.2
        M. Landini, A. Natta, E. Oliva, P. Salinari & A.F.M. Moorwood.

FROM INFRARED LINE FLUXES TO PYSICAL PROPERTIES OF HII REGIONS
        G. Stasinska & J.P. Baluteau.

2. MOLECULES & INTERSTELLAR CHEMISTRY

A MODEL OF HOT OH MASERS COOLED BY INFRARED LINE EMISSION AND THE FORMATION OF CH^+
        T.W. Hartquist & A. Dalgarno.

THE INFLUENCE OF METAL ION ASSOCIATION REACTIONS ON INTERSTELLAR CHEMISTRY
        T.J. Millar.

THE IMPORTANCE OF IMPROVED COLLISION CROSS-SECTIONS IN DERIVING INTERSTELLAR CLOUD PARAMETERS FROM ROTATIONAL LINE SPECTRA
        J.E. Beckman.

H_2 KINEMATICS IN THE PROTOPLANETARY NEBULA CRL 618
        S.E. Persson, P.J. McGregor, D.K. Duncan, H. Lanning, T.R. Geballe & C.J. Lonsdale.

THE STRENGTH OF HD INFRARED EMISSION IN ORION
        T.W. Hartquist, C.M. Mountain, M.J. Selby &
        E. Roueff.

CVF SPECTROSCOPY OF THE 2.3 μm CO BAND IN ACTIVE GALAXIES
        A.F.M. Moorwood.

4.6 μm ABSORPTION FEATURES DUE TO SOLID PHASE CO TOWARD COMPACT INFRARED SOURCES
        P.J. McGregor, S.E. Persson, J.H. Lacy, F. Baas,
        L. Allamandola, C.J. Lonsdale & T.R. Geballe.

3. OBSERVATIONS OF DUST

3-μm SPECTROSCOPY OF DUST
        D.C.B. Whittet, J.K. Davies, M.F. Bode, A. Evans &
        A.J. Longmore.

1.2 μm TO 13 μm − OBSERVATIONS OF CEP-A
        R. Lenzen & H. Hefele.

GL 2088 − A PECULIAR GALACTIC INFRARED SOURCE
        P. Persi & M. Ferrari-Toniolo.

HIGH-RESOLUTION INFRARED SPECTROSCOPY OF W51-IRS2
        J.P. Phillips. G.J. White, K.J. Richardson &
        P.M. Williams.

FAR-INFRARED OBSERVATIONS OF THE DIFFUSE COMPONENT OF THE GALACTIC DUST EMISSION
        H.L. Nordh, G. Olofsson & J.R. Houck.

EXTINCTION BY DUST IN DENSE INHOMOGENEOUS CLOUDS
        A. Natta & N. Panagia.

4. ATMOSPHERIC TRANSMISSION IN THE NEAR-INFRARED

EXTINCTION CORRECTIONS IN GROUND-BASED INFRARED SPECTROMETRY
        C.M. Mountain, M.J. Selby & C. Sanchez Magro.

LATE PAPERS

INFRARED, OPTICAL & UV-SPECTROSCOPY OF LOW REDSHIFT QSO's
        W. Kollatschny & K.J. Fricke.

THE RELATIONSHIP BETWEEN THE COLD AND DENSE REGIONS IN ρ OPH: THE DCO^+ LINES
        R.B. Loren & A. Wootten.

SUBJECT INDEX

Absorption Bands (Features)	13, 17, ??, 43-47, 66, 379-381
Absorption Efficiency	15, 29
Abundance Gradients	160, 282-283, 291, 303, 305
Abundances, Elemental	281-284, 327-328, 337
Abundances, Ionic	152, 296-303, 336, 342, 390
Abundances, Molecular	61, 70, 71, 74, 79, 80
Activation Energy	71, 77
Active Galaxies	374, 386-400, 406, 432-438
Active Galaxies, Classification of	386-387
Alfven Velocity	107, 110, 111
Ammonia Lines	181, 183, 187-188, 204, 211, 213-214
Atmospheric Absorption	85, 195, 222, 224,-226, 230, 289, 290, 310-311, 314, 320, 331
Autoionisation	92, 94
Beam Size Effects	180, 246, 298, 336-337
BL Lac Galaxies	387, 404-406
BL Lac Galaxies, Energy Distribution	405
Blazars	400-408
Boltzman Factor (Distribution)	70, 91
Carbon Chain Molecules	66, 84
Carbon Chemistry	84, 160, 161
Carbon Isotopes	69, 75, 273, 328
Carbon Radio Recombination Line Region	152, 200, 203
Carbon Rich Features	49, 51
Characteristic Frequency	8, 9
Characteristic Group	14
Charge Transfer	95-98
Chemical Bond	8
Chemistry, Laboratory Studies	66, 71, 77, 78, 79, 80, 81
Chemistry, Observational Constraints	83-86
Circular Variable Filters	309, 315, 343
Circumstellar Shells	43, 45, 47, 48, 49, 53, 312, 313, 314, 332-336
,Giant Stars	332-335
,Novae	344
,Symbiotic Stars	335-336
Clouds, Interstellar	145-164, 195-216
, Energy Budget	160, 161
Clumping (Density Inhomogeneity)	150, 151, 286, 298, 299, 304, 338, 351, 356-358

CO, Absorption Feature	46
CO, Abundance	128
CO, IR Rotational Transitions	104, 117, 126, 154-158, 196, 204, 207-209, 211, 212-213
CO, Ratio of Rotational Transition	118, 125, 126, 127, 243
CO, Rotation-Vibration Bands	9, 10, 170, 186, 190, 191, 269-276, 312-313, 316, 323, 374, 376, 377, 391, 393, 394
CO, Solid	10, 11, 20, 21, 47, 276
CO, Spiral Galaxies	447-453
CO, Sub-mm. Rotational Transitions	157, 168, 183, 196, 227-228, 234-246, 251, 252, 253, 255, 262, 433, 439, 443, 444, 447
CO Surveys	444, 445
C/O Ratio	333, 335, 340, 342
Collision Strengths	92, 337
Collisional (De)-Excitation	76, 84, 89, 91-92, 105, 116, 146-150, 172, 178, 198, 199, 241, 293, 295, 332, 338, 389, 408
Collisional Pumping	170-173, 182
Colour Excess	41
Colour Temperature	375
Column Density, Gas	255
, Ionic	151, 201, 203, 296-297, 302
, Molecular	170, 180, 271
Compact HII Regions	133-142, 258
, Dynamics	281, 286
, Excess Ionisation	137
, Excitation	281, 284-286
Complex Molecules, Formation	79-81
Concentration Effects	10
Condensed Phase, Spectral Properties	5
Continuum Emission, Mechanisms	402-403
Cool Phases of ISM	145
Cosmic Rays	70, 72, 74, 104, 106
Cosmological Investigations	408 - 414
Critical Density	147, 150, 151, 154, 174, 175, 241, 294, 295
Cross Sections	26, 163, 178, 429, 430, 432
CS Transitions	246-248
CII Fine Structure Emission	152, 154, 197-204
CII Regions	146
Detailed Balance	89, 92, 170
Detector Arrays	309, 372
Diffuse Galactic Emission	424, 428

SUBJECT INDEX

Diffuse ISM	146
Diffusion	11
Dissociation Region	198, 200, 202, 203
Dunham Coefficients	270
Dust Absorption, Differential	301
Dust, Excitation Mechanisms	25-27, 30-31, 339-340, 373, 379
, FIR Continuum	423, 424-427, 429-434, 436-440
, FIR Emissivity	42, 43
, Interstellar, Non-volatile component	18
, Spectroscopic Properties	5-33, 375
Effelsberg 100 m telescope	181, 191
Einstein Coefficients	163, 170, 171, 178, 197, 198, 235, 294,
Electron Number Density	135, 136, 138, 139, 140, 151, 293, 298, 299, 301, 303-305
, Maps	304
Electron Temperature	135, 136, 140
Elliptical Galaxies, Evolution	408
Emission Features	24-32, 42, 47, 51, 66, 158, 339, 364, 373, 395, 396, 407, 379-381
Emission Line Region, Physics of	387-388
Emission Processes, Atomic and Ionic	89-99
, Basic Atomic Theory	89-90
Emission Rate	155
Energy Level Diagram, CO	169, 190, 227
, CII	199
, CIII	95
, H_2	168, 189, 395
, OI	199
, SiO	227
Escape Probability	172, 206, 207
Evolved Objects	177, 331-345
Extinction	37-41, 104, 128, 276-277, 281, 295, 296, 317-318, 319, 352
Extinction Correction	284, 285
Extinction Efficiency	15, 16
Extinction Law	37-43, 377, 378
Extragalactic Objects, Classes of	370
Fabry-Pérot Interferometer	197, 215, 216, 272, 289, 310
Filter Wheel spectrometer	272
Fine Structure Lines	103, 116, 117, 128, 146, 148, 151, 152, 154, 197-204, 260, 281-286, 289-305, 331, 336, 342, 352, 356-357, 362, 363-372,

	382, 383, 387, 401, 423, 424, 426, 427, 433, 434
, Maps	286, 291, 304
, Wavelengths	163, 198, 294, 305
Fine Structure Transitions, Excitation of	98-99, 155
Flows, Bipolar	261-266
Fluorescence	26, 28, 31
Fourier Transform Spectrometers -Michelson Interferometers	189, 271, 272, 289, 305, 310, 316, 319
Free-Free Emission	426, 432, 433
, from Outflows	138
FU Ori Stars	259-261
Galactic Centre	352-366
, Density Distribution	351-353, 356
, Ionised Gas	356-358
, Mass Loss Bubble	362
, Temperature Distribution	353-354, 355, 356
, Total Luminosity	354-356
Galaxies, FIR Dust Emission from External	424, 427, 440
, IR Colours	376
Galaxy, FIR Dust Emission from Our	424, 426, 427, 438
Gas Phase Chemistry (Reaction)	61, 69-81, 85-86
Gaunt Factor	135
Giant Molecular Clouds	251-252, 383, 433, 444, 446, 453
Grains, Composition	65-66
, Core-mantle	6, 23, 24, 31, 61
, Shape	15
, Surfaces	59, 60, 61, 63, 65, 69, 70, 85-86
Graphite	333, 340, 341, 342, 429, 430, 431
Grating Spectrometer	215, 272, 309, 343, 372, 390, 396
Helium Lines	282, 284
Herbig-Haro Objects	251, 253, 260-261, 262, 263, 265, 266
Heterodyne Detection	216, 221-223, 423, 440
Hydrocarbons	18, 19, 23, 46, 70
Hydrogen Bond	10, 11, 12, 17
Hydrogen Recombination Line Ratios	133-137, 141, 377, 389, 392, 408
Hydrogen Recombination Lines	41, 99, 257, 273, 276, 385, 388-389, 390, 392-393, 401
, Emission Coefficients	134
, Intensities	134, 141
, IR Optical Depth	142, 286

SUBJECT INDEX

, Line Shapes	141
, Outflows	138-142
, Wavelengths	134
HD	206, 209, 210
HI Distribution	444
HII Regions	146, 182, 424-428, 432-434, 438-439
, FIR Spectroscopy	289-305
, Ionisation State	285, 293
, NIR Spectroscopy	281-286
Ice, Ammonia	44, 46
, Interstellar	6
, , Spectra	12, 13, 17, 22
, , Photolysed	18
, Laboratory	6, 14, 21, 23, 28
, Molecular	5, 6, 10, 13, 23, 29, 32
, Water	44, 45, 46, 53
Initial Mass Function	384
Interactions, Molecular	11
Interstellar Radiation Field	424, 425, 427, 431, 432, 438
Ion-Molecule Reactions	61, 69, 71-74, 77, 86
Ionisation Equilibrium of ISM	96
Ionisation Fronts	103, 146, 339, 340
Ionisation Structure	152, 281, 298, 301, 303
IR Colours	390, 410, 412
IR Hubble Diagram	411, 412
Isotope Fractionation	70, 71, 75, 78, 79
Jets	252, 261-266
Kuiper Airborne Observatory	189, 195, 197, 215, 216, 225-235, 285, 289, 291, 440
Late Type Stars	309-329, 391
Lear Jet	291
Line Broadening	9, 10
Line Emission, Regions	
Producing FIR	145-146, 153
,Production of FIR	146-151, 197-200, 293
Line Ratios, FIR	200, 203
Line Shape	182, 324-325
Line Shifting	9, 11
Local Oscillators	221-222
Long Wavelength Wing	12, 15, 17, 45

Mach Numbers	105, 107, 110
Magnetic Field	103, 104, 106-109
Magnetic Precursor	107, 110, 111, 114
Masers, Interstellar	99, 178, 251, 253, 261
Mass Loss, Eruptive	259-261, 326-327
, Steady	257-259, 326-327
Mauna Kea	222, 224-225, 230, 291
Microwave Background	414
Mira Variables	314, 331
Mixers	221
Molecular Clouds,	
Absorption Features	43, 44, 45
Chemistry	70, 73, 78, 79, 84
Contribution of Sub-mm.	
Molecular Lines	233-251
Density Structure	240-248
Dynamical Studies	252
Excitation Conditions	177-191, 196, 241, 269-272
Properties	446
Temperature Structure	234-240
Molecular Gas in Spiral	
Galaxies	443-454
Molecular Hydrogen	206, 338-339, 351, 360-362, 393, 394, 395
, Abundance	118, 244-245, 247
, Distribution	444
, Formation	59, 69, 86, 117
, Line Ratios	127, 276-277
, Rotation-Vibrational	
Transitions	104, 126, 186, 189, 269-272, 276-278
, Rotational Transition	125
, Shocked Emission	272, 277
Molecular Line Formation	
Mechanisms	178-186
Molecular Lines in IR and	
Sub-mm	112, 127, 128, 167-169, 181-186, 204-215
Molecular Rings	378, 439, 447
Molecules, Composition of	85
, Excitation of	170-174
, Formation of	59, 105
M82 Model	377, 384
N Galaxies	386
NeII-Clouds	356-357
N/O Ratio	291, 293, 302-303
Non-LTE Departure Coefficient	135
Normal Galaxies	374-386, 438-440
Novae	344-345

SUBJECT INDEX

OH, IR Lines	204, 208, 210, 211, 213
, Rotational Excitation	99
, Stretch	11, 12, 32
Optical Depth	170, 172, 180
, IR	96, 140, 146-150, 154, 197, 198, 354, 390
, Sub-mm	225, 236, 238, 241, 242, 255
Outflows	70, 103, 104, 335
, Collimation	265-266
, Free-Free Emission	138
, High Velocity	183, 186, 246, 251-267
, H Recombination Lines	138-142
, Mass Loss Rate	140
, Physical Characteristics	139-140
, Recombination Rate in	138
, from Young Stars	257-261
Oxygen Chemistry	158-160
Oxygen Isotopes	273
OI Emission	118, 122, 123, 127, 156, 197-204
OIII Emission, Interpretation Procedure	151
OIII Emission Rate	148
P-Branch	8, 9, 10
Photodissociation	59, 60, 61, 70
Photoionisation	92
Photon Trapping	171, 173, 174, 355, 388
Physical Conditions, Deduction of from Molecular Observations	170-174, 205-208
Planetary Nebulae	146, 336-341
, Dust Emission	339-341
, Emission Lines	336-338
, H_2 emission	338-339
, Precursors	319
Polarisation	15, 397-398, 404
Q-Branch	8, 9, 10
QSO's	146, 386, 400-408
, Dust	407
, Energy Distribution	401
, IR Properties	401
R-Branch	8, 9, 10
Radiative (De)-Excitation	146-150, 168, 169, 170
Radiative Precursor	105
Radiative Pumping	170-173, 182

SUBJECT INDEX

Radiative Transitions	178, 295
Radio-Emitting Levels	145
Radio Galaxies	386
Rate Coefficient	71, 78, 79, 80
Recombination, Dielectronic	93-94
, Radiative	92-93, 94
Recombination Lines (see also H Recom. Lines)	99, 189, 331, 336, 342, 344, 351, 362, 363, 373
Redshift	408
Refractory Materials	29
Relative Mobility	63
Residue, Non-volatile	18, 19
Reynolds Number	108
Rotation-Vibrational Bands (see also CO and Molecular Hydrogen)	8, 9, 84, 269-272
, Spectroscopy	7
Rotational Structure Supression	9
Rotational Temperature	271, 273, 276
Rotational Transitions	178, 295
Scattering Efficiency	15
Scattering Effects	15
Selection Rule	168, 270, 272
Self - Absorption	135, 149, 186, 423, 428, 429
Sensitivity, Instrumental	204, 214
Seyfert Galaxies	146, 386, 387-400, 437
, Dust	396-400
, Energy Distribution	391, 396
, H Emission	389-393
, IR Stars	391
, Molecules	393-395
Shocks	66, 69, 70, 71, 75-77, 103-129, 146, 183, 186, 197, 237, 358-362, 393
, C-type	103, 104, 105, 109-113, 118, 120-128
, Emission from J-type	114-119
, J-type	103, 104, 109, 110, 111, 112, 113, 114-118, 124
, Physical Conditions	112, 118, 120-122
Silicate Features	40, 41, 45, 47-49, 332-333, 335-336, 340, 341, 344, 373, 381, 390, 396, 397, 429, 430, 432, 434
Solid Phase	8, 11
Solid State	9, 13, 14, 29
Sound Speed	103, 106
Space Facilities	216, 230, 305, 423, 440
Spectral Index	396, 400, 401, 406, 428

SUBJECT INDEX

Spectrometers	167, 309-310
Spiral Galaxies, Evolution	443-454
, Mass of H_2	452, 453
, Molecular Gas	443-454
Star Formation	104, 251, 252, 254, 358, 364, 375, 378, 383, 384, 385, 412, 423, 432, 433, 434, 437, 447, 448, 449, 450, 452, 453
, in Elliptical Galaxies	377
, Traced in Dust	372
, Triggered by Gravitational Interaction	384
Starbursts	383-385, 437
Standards, IR	312-313
Statistical Equilibrium Equation	90
Stellar Atmospheres	323-325
Stellar Evolution	251, 254-257
Stellar Magnetic Fields	325
Stellar Photosphere	311, 324
, Molecules	312-314, 316-317, 320-322, 323
Stellar Pulsation	325-326
Stellar Winds, Chemistry	69
Sticking Probability	59, 61, 62
Stromgren Sphere	137
Supernovae, Energy Distribution	414
Surface Chemistry	85-86
Surface Mobility	59, 63, 64
Synchrotron Radiation	400, 401, 404, 405, 407, 408, 426, 428, 434, 435, 436
T Tauri Stars	257-259, 263, 319
Thermal Balance of ISM	99
Tracers for Different Densities	178, 240-242, 246
Tracers for Different Temperatures	174-175, 178, 234
10 µm Survey of Late-type Galaxies	379
Unidentified Bands	49-52, 364
University College Balloon Gondola	289, 291
Variability	398, 403, 404
Vibration Frequency	7, 8

Vibrations, Molecular,
Frequency   5, 7, 8, 21, 25, 26
VLA   263, 281, 286, 358, 359, 386, 404

Water, Amorphous   11
Wolf-Rayet Stars   341-344, 362

Young Stellar Objects   104, 177, 187, 257

OBJECT INDEX

Object	Page	Object	Page
λ And	325	χ Cyg	323, 326
R And	320, 321, 323	RW Cyg	313
AO 945-30	389	V645 Cyg	253, 254, 255
ζ Aqr	319	V1057 Cyg	239, 260
AS 205	257, 258	V1500 Cyg	344
AS 353	263	V1515 Cyg	259
B 335	430	30 Dor	377
BN Object	14, 17, 18, 104, 126, 142, 190, 191, 192, 237, 243, 271, 273, 274, 394	DR 21	181, 182, 188, 202, 240, 248, 254
		ELIAS 1	28
α BOO	323	G 12.8-0.2	238, 292
ρ BOO	316	G 29.9-0.0	283, 286, 292
RX BOO	314, 318	G 45.1+0.1	283, 292
		G 75.84+0.4	50, 292, 302, 303,
3C 84	404, 406	G 298.2-0.3	283, 284,
3C 120	380, 396	G 333.6-0.2	41, 283, 285, 286, 292
3C 273	380, 396, 397, 399, 403, 404, 406, 407, 408	GALAXY	424-428, 431, 437, 443-446, 448
3C 279	404		
3C 345	404, 406, 423	GALACTIC CENTRE	18, 19, 43, 104, 289, 351-368, 433
Cas A	96		
ρ Cas	327		
Cep A	254, 256	β Gem	316
μ Cep	313, 333, 334, 345	μ Gem	316
		GGD 12-15	254
O Cet	320, 321	GUM NEBULA	263
VY CMa	313	HD 44179	25, 50, 51
RS Cnc	313, 317	HD 50896	343
TY CrA	52		
V CrB	331	HD 97048	27, 28, 52
CRL 437	254	HDE 330036	335
CRL 490	136, 137, 138, 139, 140, 141, 254, 256	HARO 1-36	335
		He 2-10	387
		α Her	313
CRL 618	339	HH 1-2	260, 263
CRL 961	254, 256	HH 7-11	254, 256, 263, 264, 266
CRL 2591	254		
CRL 3053	50	HH 12	263
Y CVn	315, 317	HH 24-26	254, 256

469

OBJECT INDEX

HH 32	254	M51	386, 423, 437, 439, 447, 448, 449, 450, 451
HH 46	263		
HH 47	263		
U Hya	317		
IC 342	386, 447, 448, 449, 452	M82	202, 205, 377, 379, 381, 384, 385, 386, 423, 432, 433, 434, 436, 438
IC 359	28		
IC 418	48, 50, 51, 337		
IC 443	104	M101	379, 383, 447, 448, 449,
IC 2003	337		
IC 2165	337	MAFFEI 2	386
IC 4329A	380, 398	MCG-5-23-16	389
IC 4406	339	Mon R2	254, 255, 256, 292, 300,
IC 5063	401, 406		
IRC+10 216	319, 327		
IZw 187	403	MSH 15-52	104
K 3-50	14, 136, 137	MWC 297	276
KLEINMANN-LOW NEBULA	104, 126, 154, 157, 158, 182, 184, 185, 195, 196, 207, 209, 211, 212, 214, 215, 228, 229, 237, 243	N81	382, 434
		NGC 253	377, 379, 381, 384, 386, 423, 432, 436, 437, 438, 447
		NGC 598	452
		NGC 604	374, 382
		NGC 1052	401
L 183	184	NGC 1068	104, 372, 391, 392, 394, 396, 423, 434, 436, 438, 447, 454
L 1551	253, 254, 260, 262		
R Leo	318		
Lk Hα 101	254, 276	NGC 1275	391, 392, 406, 423
Lk Hα 198	254		
Lk Hα 208	254	NGC 1316	413, 414
Lk Hα 215	254	NGC 1365	389
Lk Hα 325	254	NGC 1614	381
R LYR	318	NGC 1808	380, 381, 386
M1	428	NGC 2023	27, 236, 254
M8	254, 292	NGC 2024	154, 202, 203, 248, 254, 292, 301
M17	154, 200, 202, 203, 248, 291, 292, 298, 302, 303, 305, 352		
		NGC 2068	27, 236
		NGC 2071	104, 128, 246, 254, 255, 256, 266
M17 (IRS 1)	138, 141		
M31	377, 454	NGC 2264	248, 254
M33	374, 382	NGC 2316	254
M42	133, 154, 160, 195, 196, 202, 203, 204, 283, 291, 292, 298, 301, 302, 305, 394	NGC 2392	337, 340
		NGC 2403	452
		NGC 2440	337
		NGC 2841	449
		NGC 2903	386
		NGC 2992	389

OBJECT INDEX

NGC 3132	339	NOVA CYGNI	
NGC 3310	379	1975	52
NGC 3690	384	NOVA VULPECULAE	
NGC 4151	371, 372, 391, 382, 395, 397, 398	1976	53
		OH 0739-14	17, 45, 46
		OJ 287	404, 406
NGC 4303	381	OMC-1	195, 236, 237, 244, 245, 246, 247, 273, 274, 276, 277
NGC 4321	452		
NGC 4419	381		
NGC 4438	381		
NGC 4536	381, 413, 414	ρ OPH	46, 254
NGC 4519	381	ζ OPH	204
NGC 5128	379, 383	ORION A	70, 79, 191, 248, 252, 253, 254, 256, 261, 277, 292, 303
NGC 5194	452		
NGC 5195	437		
NGC 5236	379, 452		
NGC 5457	452	ORION (IRc2)	184, 185, 190, 223
NGC 5461	381		
NGC 5506	388, 396	ORION NEBULA (SEE M42)	
NGC 6302	337, 338		
NGC 6334	136, 137, 254, 292	α_2 Ori	313, 320
		Θ^2 A Ori	237
NGC 6357	292	FU Ori	259, 260
NGC 6445	337	PARS 21	254
NGC 6537	202	ζ Per	65
NGC 6543	337, 338, 341	o Per	65
NGC 6720	338	S Per	313
NGC 6826	337	PG 1351 + 64	402
NGC 6946	386, 411, 447, 448, 449, 452	PKS 0735 + 178	405, 406
		TX PSC	321
NGC 7023	27	OJ 287	404, 406
NGC 7027	24, 25, 26, 28, 31, 49, 50, 51, 104, 202, 276, 337, 338, 339, 341, 393	QSO 0235 + 164	405
		QSO 0420-388	404, 423, 435
		QSO 1156+295	405, 406
		QSO 1413 + 135	405, 406, 407
		QSO 1921-293	405
NGC 7213	393	QSO 2251 + 15	405
NGC 7331	378, 447	S 87	254
NGC 7354	337	S 88B	254, 292
NGC 7469	395, 396	S 106	141, 184, 188, 189, 254, 283, 292
NGC 7538	202, 254, 301, 302, 304, 305		
		S 140	184, 248, 254, 256, 265
NGC 7538(IRS 9)	20		
NGC 7538(S158 A + G)	292	S 156	283, 292
		S 187	254
NGC 7552	386	S 235 B	254
NGC 7582	390, 396	S 252 A3	254
NGC 7662	337, 338	S 255 (IRS 1)	254
NOVA AQUILA 1982	344, 345	S 888	292

SERPENS MOL. CLOUD	254
α Sco	313
WX Ser	335
Sgr A	200, 202, 292
Sgr B2	181, 182, 210
VX Sgr	313
SMC	376
SN 1980 K	413, 414
SUN	322
α Tau	316, 320
T Tau	256, 319
Y Tau	334
DG Tau	254
RY Tau	254
119 Tau	313
RR Tel	387
TMC 1	79, 184, 188, 189
TRAPEZIUM	333, 334
UOA 27	276
γ Vel	343, 343
o Vir	328
SW Vir	316, 318
W3 A	283, 292
W3 (IRS 1)	41, 292, 302
W3 (OH)	97, 139, 140, 228, 229, 254
W33A	14, 20, 21, 22, 23, 240
W33B	240
W33	239, 283
W43	292, 303
W49	203, 210, 292
WSI	292

ASTROPHYSICS AND SPACE SCIENCE LIBRARY

Edited by

J. E. Blamont, R. L. F. Boyd, L. Goldberg, C. de Jager, Z. Kopal, G. H. Ludwig, R. Lüst,
B. M. McCormac, H. E. Newell, L. I. Sedov, Z. Švestka

1. C. de Jager (ed.), *The Solar Spectrum, Proceedings of the Symposium held at the University of Utrecht, 26–31 August, 1963.* 1965, XIV + 417 pp.
2. J. Orthner and H. Maseland (eds.), *Introduction to Solar Terrestrial Relations, Proceedings of the Summer School in Space Physics held in Alpbach, Austria, July 15–August 10, 1963 and Organized by the European Preparatory Commission for Space Research.* 1965, IX + 506 pp.
3. C. C. Chang and S. S. Huang (eds.), *Proceedings of the Plasma Space Science Symposium, held at the Catholic University of America, Washington, D.C., June 11–14, 1963.* 1965, IX + 377 pp.
4. Zdeněk Kopal, *An Introduction to the Study of the Moon.* 1966, XII + 464 pp.
5. B. M. McCormac (ed.), *Radiation Trapped in the Earth's Magnetic Field. Proceedings of the Advanced Study Institute, held at the Chr. Michelsen Institute, Bergen, Norway, August 16–September 3, 1965.* 1966, XII + 901 pp.
6. A. B. Underhill, *The Early Type Stars.* 1966, XII + 282 pp.
7. Jean Kovalevsky, *Introduction to Celestial Mechanics.* 1967, VIII + 427 pp.
8. Zdeněk Kopal and Constantine L. Goudas (eds.), *Measure of the Moon. Proceedings of the 2nd International Conference on Selenodesy and Lunar Topography, held in the University of Manchester, England, May 30–June 4, 1966.* 1967, XVIII + 479 pp.
9. J. G. Emming (ed.), *Electromagnetic Radiation in Space. Proceedings of the 3rd ESRO Summer School in Space Physics, held in Alpbach, Austria, from 19 July to 13 August, 1965.* 1968, VIII + 307 pp.
10. R. L. Carovillano, John F. McClay, and Henry R. Radoski (eds.), *Physics of the Magnetosphere, Based upon the Proceedings of the Conference held at Boston College, June 19–28, 1967.* 1968, X + 686 pp.
11. Syun-Ichi Akasofu, *Polar and Magnetospheric Substorms.* 1968, XVIII + 280 pp.
12. Peter M. Millman (ed.), *Meteorite Research. Proceedings of a Symposium on Meteorite Research, held in Vienna, Austria, 7–13 August, 1968.* 1969, XV + 941 pp.
13. Margherita Hack (ed.), *Mass Loss from Stars. Proceedings of the 2nd Trieste Colloquium on Astrophysics, 12–17 September, 1968.* 1969, XII + 345 pp.
14. N. D'Angelo (ed.), *Low-Frequency Waves and Irregularities in the Ionosphere. Proceedings of the 2nd ESRIN-ESLAB Symposium, held in Frascati, Italy, 23–27 September, 1968.* 1969, VII + 218 pp.
15. G. A. Partel (ed.), *Space Engineering. Proceedings of the 2nd International Conference on Space Engineering, held at the Fondazione Giorgio Cini, Isola di San Giorgio, Venice, Italy, May 7–10, 1969.* 1970, XI + 728 pp.
16. S. Fred Singer (ed.), *Manned Laboratories in Space. Second International Orbital Laboratory Symposium.* 1969, XIII + 133 pp.
17. B. M. McCormac (ed.), *Particles and Fields in the Magnetosphere. Symposium Organized by the Summer Advanced Study Institute, held at the University of California, Santa Barbara, Calif., August 4–15, 1969.* 1970, XI + 450 pp.
18. Jean-Claude Pecker, *Experimental Astronomy.* 1970, X + 105 pp.
19. V. Manno and D. E. Page (eds.), *Intercorrelated Satellite Observations related to Solar Events. Proceedings of the 3rd ESLAB/ESRIN Symposium held in Noordwijk, The Netherlands, September 16–19, 1969.* 1970, XVI + 627 pp.
20. L. Mansinha, D. E. Smylie, and A. E. Beck, *Earthquake Displacement Fields and the Rotation of the Earth, A NATO Advanced Study Institute Conference Organized by the Department of Geophysics, University of Western Ontario, London, Canada, June 22–28, 1969.* 1970, XI + 308 pp.
21. Jean-Claude Pecker, *Space Observatories.* 1970, XI + 120 pp.
22. L. N. Mavridis (ed.), *Structure and Evolution of the Galaxy. Proceedings of the NATO Advanced Study Institute, held in Athens, September 8–19, 1969.* 1971, VII + 312 pp.

23. A. Muller (ed.), *The Magellanic Clouds. A European Southern Observatory Presentation: Principal Prospects, Current Observational and Theoretical Approaches, and Prospects for Future Research*, Based on the Symposium on the Magellanic Clouds, held in Santiago de Chile, March 1969, on the Occasion of the Dedication of the European Southern Observatory. 1971, XII + 189 pp.
24. B. M. McCormac (ed.), *The Radiating Atmosphere*. Proceedings of a Symposium Organized by the Summer Advanced Study Institute, held at Queen's University, Kingston, Ontario, August 3–14, 1970. 1971, XI + 455 pp.
25. G. Fiocco (ed.), *Mesospheric Models and Related Experiments*. Proceedings of the 4th ESRIN-ESLAB Symposium, held at Frascati, Italy, July 6–10, 1970. 1971, VIII + 298 pp.
26. I. Atanasijević, *Selected Exercises in Galactic Astronomy*. 1971, XII + 144 pp.
27. C. J. Macris (ed.), *Physics of the Solar Corona*. Proceedings of the NATO Advanced Study Institute on Physics of the Solar Corona, held at Cavouri-Vouliagmeni, Athens, Greece, 6–17 September 1970. 1971, XII + 345 pp.
28. F. Delobeau, *The Environment of the Earth*. 1971, IX + 113 pp.
29. E. R. Dyer (general ed.), *Solar-Terrestrial Physics/1970*. Proceedings of the International Symposium on Solar-Terrestrial Physics, held in Leningrad, U.S.S.R., 12–19 May 1970. 1972, VIII + 938 pp.
30. V. Manno and J. Ring (eds.), *Infrared Detection Techniques for Space Research*. Proceedings of the 5th ESLAB-ESRIN Symposium, held in Noordwijk, The Netherlands, June 8–11, 1971. 1972, XII + 344 pp.
31. M. Lecar (ed.), *Gravitational N-Body Problem*. Proceedings of IAU Colloquium No. 10, held in Cambridge, England, August 12–15, 1970. 1972, XI + 441 pp.
32. B. M. McCormac (ed.), *Earth's Magnetospheric Processes*. Proceedings of a Symposium Organized by the Summer Advanced Study Institute and Ninth ESRO Summer School, held in Cortina, Italy, August 30–September 10, 1971. 1972, VIII + 417 pp.
33. Antonin Rükl, *Maps of Lunar Hemispheres*. 1972, V + 24 pp.
34. V. Kourganoff, *Introduction to the Physics of Stellar Interiors*. 1973, XI + 115 pp.
35. B. M. McCormac (ed.), *Physics and Chemistry of Upper Atmospheres*. Proceedings of a Symposium Organized by the Summer Advanced Study Institute, held at the University of Orléans, France, July 31–August 11, 1972. 1973, VIII + 389 pp.
36. J. D. Fernie (ed.), *Variable Stars in Globular Clusters and in Related Systems*. Proceedings of the IAU Colloquium No. 21, held at the University of Toronto, Toronto, Canada, August 29–31, 1972. 1973, IX + 234 pp.
37. R. J. L. Grard (ed.), *Photon and Particle Interaction with Surfaces in Space*. Proceedings of the 6th ESLAB Symposium, held at Noordwijk, The Netherlands, 26–29 September, 1972. 1973, XV + 577 pp.
38. Werner Israel (ed.), *Relativity, Astrophysics and Cosmology*. Proceedings of the Summer School, held 14–26 August 1972, at the BANFF Centre, BANFF, Alberta, Canada. 1973, IX + 323 pp.
39. B. D. Tapley and V. Szebehely (eds.), *Recent Advances in Dynamical Astronomy*. Proceedings of the NATO Advanced Study Institute in Dynamical Astronomy, held in Cortina d'Ampezzo, Italy, August 9–12, 1972. 1973, XIII + 468 pp.
40. A. G. W. Cameron (ed.), *Cosmochemistry*. Proceedings of the Symposium on Cosmochemistry, held at the Smithsonian Astrophysical Observatory, Cambridge, Mass., August 14–16, 1972. 1973, X + 173 pp.
41. M. Golay, *Introduction to Astronomical Photometry*. 1974, IX + 364 pp.
42. D. E. Page (ed.), *Correlated Interplanetary and Magnetospheric Observations*. Proceedings of the 7th ESLAB Symposium, held at Saulgau, W. Germany, 22–25 May, 1973. 1974, XIV + 662 pp.
43. Riccardo Giacconi and Herbert Gursky (eds.), *X-Ray Astronomy*. 1974, X + 450 pp.
44. B. M. McCormac (ed.), *Magnetospheric Physics*. Proceedings of the Advanced Summer Institute, held in Sheffield, U.K., August 1973. 1974, VII + 399 pp.
45. C. B. Cosmovici (ed.), *Supernovae and Supernova Remnants*. Proceedings of the International Conference on Supernovae, held in Lecce, Italy, May 7–11, 1973. 1974, XVII + 387 pp.
46. A. P. Mitra, *Ionospheric Effects of Solar Flares*. 1974, XI + 294 pp.
47. S.-I. Akasofu, *Physics of Magnetospheric Substorms*. 1977, XVIII + 599 pp.

48. H. Gursky and R. Ruffini (eds.), *Neutron Stars, Black Holes and Binary X-Ray Sources*. 1975, XII + 441 pp.
49. Z. Švestka and P. Simon (eds.), *Catalog of Solar Particle Events 1955–1969. Prepared under the Auspices of Working Group 2 of the Inter-Union Commission on Solar-Terrestrial Physics*. 1975, IX + 428 pp.
50. Zdeněk Kopal and Robert W. Carder, *Mapping of the Moon*. 1974, VIII + 237 pp.
51. B. M. McCormac (ed.), *Atmospheres of Earth and the Planets. Proceedings of the Summer Advanced Study Institute, held at the University of Liège, Belgium, July 29–August 8, 1974*. 1975, VII + 454 pp.
52. V. Formisano (ed.), *The Magnetospheres of the Earth and Jupiter. Proceedings of the Neil Brice Memorial Symposium, held in Frascati, May 28–June 1, 1974*. 1975, XI + 485 pp.
53. R. Grant Athay, *The Solar Chromosphere and Corona: Quiet Sun*. 1976, XI + 504 pp.
54. C. de Jager and H. Nieuwenhuijzen (eds.), *Image Processing Techniques in Astronomy. Proceedings of a Conference, held in Utrecht on March 25–27, 1975*. 1976, XI + 418 pp.
55. N. C. Wickramasinghe and D. J. Morgan (eds.), *Solid State Astrophysics. Proceedings of a Symposium, held at the University College, Cardiff, Wales, 9–12 July, 1974*. 1976, XII + 314 pp.
56. John Meaburn, *Detection and Spectrometry of Faint Light*. 1976, IX + 270 pp.
57. K. Knott and B. Battrick (eds.), *The Scientific Satellite Programme during the International Magnetospheric Study. Proceedings of the 10th ESLAB Symposium, held at Vienna, Austria, 10–13 June 1975*. 1976, XV + 464 pp.
58. B. M. McCormac (ed.), *Magnetospheric Particles and Fields. Proceedings of the Summer Advanced Study School, held in Graz, Austria, August 4–15, 1975*. 1976, VII + 331 pp.
59. B. S. P. Shen and M. Merker (eds.), *Spallation Nuclear Reactions and Their Applications*. 1976, VIII + 235 pp.
60. Walter S. Fitch (ed.), *Multiple Periodic Variable Stars. Proceedings of the International Astronomical Union Colloquium No. 29, held at Budapest, Hungary, 1–5 September 1976*. 1976, XIV + 348 pp.
61. J. J. Burger, A. Pedersen, and B. Battrick (eds.), *Atmospheric Physics from Spacelab. Proceedings of the 11th ESLAB Symposium, Organized by the Space Science Department of the European Space Agency, held at Frascati, Italy, 11–14 May 1976*. 1976, XX + 409 pp.
62. J. Derral Mulholland (ed.), *Scientific Applications of Lunar Laser Ranging. Proceedings of a Symposium held in Austin, Tex., U.S.A., 8–10 June, 1976*. 1977, XVII + 302 pp.
63. Giovanni G. Fazio (ed.), *Infrared and Submillimeter Astronomy. Proceedings of a Symposium held in Philadelphia, Penn., U.S.A., 8–10 June, 1976*. 1977, X + 226 pp.
64. C. Jaschek and G. A. Wilkins (eds.), *Compilation, Critical Evaluation and Distribution of Stellar Data. Proceedings of the International Astronomical Union Colloquium No. 35, held at Strasbourg, France, 19–21 August, 1976*. 1977, XIV + 316 pp.
65. M. Friedjung (ed.), *Novae and Related Stars. Proceedings of an International Conference held by the Institut d'Astrophysique, Paris, France, 7–9 September, 1976*. 1977, XIV + 228 pp.
66. David N. Schramm (ed.), *Supernovae. Proceedings of a Special IAU-Session on Supernovae held in Grenoble, France, 1 September, 1976*. 1977, X + 192 pp.
67. Jean Audouze (ed.), *CNO Isotopes in Astrophysics. Proceedings of a Special IAU Session held in Grenoble, France, 30 August, 1976*. 1977, XIII + 195 pp.
68. Z. Kopal, *Dynamics of Close Binary Systems*. XIII + 510 pp.
69. A. Bruzek and C. J. Durrant (eds.), *Illustrated Glossary for Solar and Solar-Terrestrial Physics*. 1977, XVIII + 204 pp.
70. H. van Woerden (ed.), *Topics in Interstellar Matter*. 1977, VIII + 295 pp.
71. M. A. Shea, D. F. Smart, and T. S. Wu (eds.), *Study of Travelling Interplanetary Phenomena*. 1977, XII + 439 pp.
72. V. Szebehely (ed.), *Dynamics of Planets and Satellites and Theories of Their Motion. Proceedings of IAU Colloquium No. 41, held in Cambridge, England, 17–19 August 1976*. 1978, XII + 375 pp.
73. James R. Wertz (ed.), *Spacecraft Attitude Determination and Control*. 1978, XVI + 858 pp.

74. Peter J. Palmadesso and K. Papadopoulos (eds.), *Wave Instabilities in Space Plasmas. Proceedings of a Symposium Organized Within the XIX URSI General Assembly held in Helsinki, Finland, July 31–August 8, 1978.* 1979, VII + 309 pp.
75. Bengt E. Westerlund (ed.), *Stars and Star Systems. Proceedings of the Fourth European Regional Meeting in Astronomy held in Uppsala, Sweden, 7–12 August, 1978.* 1979, XVIII + 264 pp.
76. Cornelis van Schooneveld (ed.), *Image Formation from Coherence Functions in Astronomy. Proceedings of IAU Colloquium No. 49 on the Formation of Images from Spatial Coherence Functions in Astronomy, held at Groningen, The Netherlands, 10–12 August 1978.* 1979, XII + 338 pp.
77. Zdeněk Kopal, *Language of the Stars. A Discourse on the Theory of the Light Changes of Eclipsing Variables.* 1979, VIII + 280 pp.
78. S.-I. Akasofu (ed.), *Dynamics of the Magnetosphere. Proceedings of the A.G.U. Chapman Conference 'Magnetospheric Substorms and Related Plasma Processes' held at Los Alamos Scientific Laboratory, N.M., U.S.A., October 9–13, 1978.* 1980, XII + 658 pp.
79. Paul S. Wesson, *Gravity, Particles, and Astrophysics. A Review of Modern Theories of Gravity and G-variability, and their Relation to Elementary Particle Physics and Astrophysics.* 1980, VIII + 188 pp.
80. Peter A. Shaver (ed.), *Radio Recombination Lines. Proceedings of a Workshop held in Ottawa, Ontario, Canada, August 24–25, 1979.* 1980, X + 284 pp.
81. Pier Luigi Bernacca and Remo Ruffini (eds.), *Astrophysics from Spacelab.* 1980, XI + 664 pp.
82. Hannes Alfvén, *Cosmic Plasma,* 1981, X + 160 pp.
83. Michael D. Papagiannis (ed.), *Strategies for the Search for Life in the Universe,* 1980, XVI + 254 pp.
84. H. Kikuchi (ed.), *Relation between Laboratory and Space Plasmas,* 1981, XII + 386 pp.
85. Peter van der Kamp, *Stellar Paths,* 1981, XXII + 155 pp.
86. E. M. Gaposchkin and B. Kołaczek (eds.), *Reference Coordinate Systems for Earth Dynamics,* 1981, XIV + 396 pp.
87. R. Giacconi (ed.), *X-Ray Astronomy with the Einstein Satellite. Proceedings of the High Energy Astrophysics Division of the American Astronomical Society Meeting on X-Ray Astronomy held at the Harvard-Smithsonian Center for Astrophysics, Cambridge, Mass., U.S.A., January 28–30, 1980.* 1981, VII + 330 pp.
88. Icko Iben Jr. and Alvio Renzini (eds.), *Physical Processes in Red Giants. Proceedings of the Second Workshop, helt at the Ettore Majorana Centre for Scientific Culture, Advanced School of Agronomy, in Erice, Sicily, Italy, September 3–13, 1980.* 1981, XV + 488 pp.
89. C. Chiosi and R. Stalio (eds.), *Effect of Mass Loss on Stellar Evolution. IAU Colloquium No. 59 held in Miramare, Trieste, Italy, September 15–19, 1980.* 1981, XXII + 532 pp.
90. C. Goudis, *The Orion Complex: A Case Study of Interstellar Matter.* 1982, XIV + 306 pp.
91. F. D. Kahn (ed.), *Investigating the Universe. Papers Presented to Zdenek Kopal on the Occasion of his retirement, September 1981.* 1981, X + 458 pp.
92. C. M. Humphries (ed.), *Instrumentation for Astronomy with Large Optical Telescopes, Proceedings of IAU Colloquium No. 67.* 1981, XVII + 321 pp.
93. R. S. Roger and P. E. Dewdney (eds.), *Regions of Recent Star Formation, Proceedings of the Symposium on "Neutral Clouds Near HII Regions – Dynamics and Photochemistry", held in Penticton, B.C., June 24–26, 1981.* 1982, XVI + 496 pp.
94. O. Calame (ed.), *High-Precision Earth Rotation and Earth-Moon Dynamics. Lunar Distances and Related Observations.* 1982, XX + 354 pp.
95. M. Friedjung and R. Viotti (eds.), *The Nature of Symbiotic Stars,* 1982, XX + 310 pp.
96. W. Fricke and G. Teleki (eds.), *Sun and Planetary System,* 1982, XIV + 538 pp.
97. C. Jaschek and W. Heintz (eds.), *Automated Data Retrieval in Astronomy,* 1982, XX + 324 pp.
98. Z. Kopal and J. Rahe (eds.), *Binary and Multiple Stars as Tracers of Stellar Evolution,* 1982, XXX + 503 pp.
99. A. W. Wolfendale (ed.), *Progress in Cosmology,* 1982, VIII + 360 pp.
100. W. L. H. Shuter (ed.), *Kinematics, Dynamics and Structure of the Milky Way,* 1983, XII + 392 pp.